T0156005

Springer Series in Statistics

Series Editors:
Peter Bühlmann, Peter Diggle, Ursula Gather, Scott Zeger

Springer Series in Statistics (SSS) is a series of monographs of general interest that discuss statistical theory and applications.

The series editors are currently Peter Bühlmann, Peter Diggle, Ursula Gather, and Scott Zeger. Peter Bickel, Ingram Olkin, and Stephen Fienberg were editors of the series for many years.

More information about this series at http://www.springer.com/series/692

Nicolas Chopin • Omiros Papaspiliopoulos

An Introduction to Sequential Monte Carlo

Nicolas Chopin
ENSAE, Institut Polytechnique de Paris
Palaiseau Cedex, France

Omiros Papaspiliopoulos
ICREA and Department of Economics
and Business
Universitat Pompeu Fabra
Barcelona, Spain

ISSN 0172-7397 ISSN 2197-568X (electronic)
Springer Series in Statistics
ISBN 978-3-030-47847-6 ISBN 978-3-030-47845-2 (eBook)
https://doi.org/10.1007/978-3-030-47845-2

Mathematics Subject Classification: 62, 60

© Springer Nature Switzerland AG 2020
This work is subject to copyright. All rights are reserved by the Publisher, whether the whole or part of the material is concerned, specifically the rights of translation, reprinting, reuse of illustrations, recitation, broadcasting, reproduction on microfilms or in any other physical way, and transmission or information storage and retrieval, electronic adaptation, computer software, or by similar or dissimilar methodology now known or hereafter developed.
The use of general descriptive names, registered names, trademarks, service marks, etc. in this publication does not imply, even in the absence of a specific statement, that such names are exempt from the relevant protective laws and regulations and therefore free for general use.
The publisher, the authors, and the editors are safe to assume that the advice and information in this book are believed to be true and accurate at the date of publication. Neither the publisher nor the authors or the editors give a warranty, expressed or implied, with respect to the material contained herein or for any errors or omissions that may have been made. The publisher remains neutral with regard to jurisdictional claims in published maps and institutional affiliations.

This Springer imprint is published by the registered company Springer Nature Switzerland AG.
The registered company address is: Gewerbestrasse 11, 6330 Cham, Switzerland

Contents

List of Symbols

$\{X_t\}$	Hidden stochastic process (in a state-space model)	11, 35
$\{Y_t\}$	Observed stochastic process (in a state-space model)	11, 35
$\mathcal{X}$	State space of a state-space model	11, 35
$\mathcal{Y}$	State space of the observation process	11
$:=$	Equal by definition	11
$p_t^\theta(x_t \vert x_{t-1})$	State transition density	12
$f_t^\theta(y_t \vert x_t)$	Observation/emission density	12
$\mathbb{R}$	Set of real numbers	14
$\mathcal{N}_d$	d-dimensional Gaussian distribution	15
d_y	Dimension of observation Y_t (in case $\mathcal{Y} \subset \mathbb{R}^{d_y}$)	16, 22
$\mathcal{P}$	Poisson distribution	16
d_x	Dimension of state X_t (in case $\mathcal{X} \subset \mathbb{R}^{d_x}$)	22
$\{(X_t, Y_t)\}$	Jointly Markov stochastic process	35, 42
$\mathcal{B}(\mathcal{X})$	A σ-algebra on the state space of the state process	35
$\gg$	Dominates (measures)	37
$\mathbb{Q}(\varphi)$	Expectation of ϕ under $\mathbb{Q}$, i.e., $\int_{\mathcal{X}} \varphi(x)\mathbb{Q}(\mathrm{d}x)$	38, 88
$\propto$	Proportional to (left-hand side must integrate or sum to one)	39, 76
$\delta_x(\mathrm{d}y)$	Dirac delta measure	39
$P_t(x_{t-1}, \mathrm{d}x_t)$	Transition kernel of the state process $\{X_t\}$	42
$f_t(y_t \vert x_t)$	Observation/emission density	43
$p_t(y_t \vert y_{0:t-1})$	Likelihood factor	44
L_t	Normalising constant of a Feynman-Kac model	52
$\mathbb{Q}_t$	Law of a Feynman-Kac model	52
l_t	Ratios of successive normalising constants of a Feynman-Kac model, $l_t = L_t/L_{t-1}$	52
$M_t(x_{t-1}, \mathrm{d}x_t)$	Transition kernel in the definition of a Feynman-Kac model	52

$G_t(x_{t-1}, x_t)$	Potential function in the definition of a Feynman-Kac model	52
$m_t(x_t\|x_{t-1})$	Transition density in the definition of a Feynman-Kac model	52
$H_{t:T}(x_t)$	Cost-to-go function in the Feynman-Kac model	58
$Q_{t\|T}(x_{t-1}, \mathrm{d}x_t)$	Transition kernel of the Markov process whose law is given by a Feynman-Kac model	58
$\mathcal{O}$	Big O notation	69
$\mathcal{O}_P$	Big O in Probability notation	82, 178
$\Rightarrow$	Convergence in distribution	82, 167
$\mathcal{U}$	Uniform distribution	83
W^n	Auto-normalised importance sampling weight, $W^n = w(X^n)/\sum_{k=1}^n w(X^k)$	89
w^n	Shorthand notation for $w(X^n)$	89
$\mathbb{Q}^N$	Particle approximation of $\mathbb{Q}$	89
$D_{\mathrm{KL}}(\cdot\|\|\cdot)$	Kullback-Leibler divergence	94
$\mathcal{M}$	Multinomial distribution	106
$\mathcal{E}$	Exponential distribution	113
frac	Fractional part: $\mathrm{frac}(x) = x - \lfloor x \rfloor$	115
$\mathcal{B}in$	Binomial distribution	115
Φ	$\mathcal{N}(0, 1)$ cumulative distribution function	141
$\xrightarrow{a.s.}$	Almost sure convergence	167
$\xrightarrow{P}$	Convergence in probability	167
$\mathcal{C}_b(\mathcal{X})$	Set of measurable, bounded functions $\varphi : \mathcal{X} \to \mathbb{R}$	167
$\\|\varphi\\|_\infty$	Supremum norm for functions	167
$\\| \cdot \\|_{\mathrm{TV}}$	Total variation norm	179
d_θ	Dimension of parameter θ; $\Theta \subset \mathbb{R}^{d_\theta}$	251
∇	Gradient	253
Cor	Correlation (between two random variables)	280
$\mathcal{B}eta$	Beta distribution	295
$\mathcal{IG}$	Inverse-Gamma distribution	314
$\mathcal{G}$	Gamma distribution	317

List of Figures

List of Algorithms

Chapter 1
Preface

Particle filters are about 25 years old. Initially confined to the so-called "filtering problem" (the sequential analysis of state-space models), they are now routinely applied to a large variety of sequential and non-sequential tasks and have evolved to the broader Sequential Monte Carlo (SMC) framework. This greater applicability comes at the price of a greater technicality. To make matters worse, literature on particle filters spans several scientific fields, mainly engineering (in particular signal processing), but also statistics, machine learning, probability theory, operations research, physics and econometrics. Of course, each field uses slightly different notations and terms to describe the same algorithms. As a result, tracking this literature has become a challenge for non-experts.

It thus seems timely to write an introduction to particle filters for a large audience, which would cover their potential uses (and misuses), the underlying theory, and even their computer implementation.

This book is our attempt at writing such an introduction. It relies on several deliberate choices on our parts, which we hope the reader will appreciate:

- We describe the different SMC algorithms found in the literature as special cases of one generic algorithm. What really varies from one case to the next is the underlying Feynman-Kac model.
- We do not take the view that Feynman-Kac models are scary beasts that only Probabilists should look into the eyes. Properly tamed, they are simply the proper way to describe the distributions a particle filter approximates. We spend some time in this book to explore the properties of these models. The big advantage of doing so is that many practical algorithms may be derived directly from such properties.
- In the same spirit, and although we tried to keep mathematical sophistication to a low level, we did not try to avoid at all prices measure-theoretic concepts such as Radon-Nikodym derivatives and Markov kernels. Several important practical problems feature Markov processes $\{X_t\}$ such that X_t given X_{t-1} *does not* admit a density, and trying to explain how the corresponding algorithms work (or do not

© Springer Nature Switzerland AG 2020

N. Chopin, O. Papaspiliopoulos, *An Introduction to Sequential Monte Carlo*, Springer Series in Statistics, https://doi.org/10.1007/978-3-030-47845-2_1

work!) in this case becomes very confusing if everything is expressed in terms of "densities". For readers less familiar with these concepts, we provide a gentle introduction to these notions.

Speaking of mathematical prerequisites, this book is, we hope, accessible to any reader who had a previous exposure to non-trivial Probability; some familiarity with probability spaces, σ-algebras, and related concepts should help but is not essential.

1.1 Relation to Other Books

As said above, our objective is to provide a general introduction to all the facets of SMC: the underlying theory, the actual algorithms, their complexities, their limitations, how to implement them in practice, and the wide range of their applications.

We hope that this book may also serve as an introduction to books that go deeper into the theory, such as the books of Del Moral (2004) and Cappé et al. (2005).

Regarding the former, we recommend in particular Chaps. 2–5 and Chap. 8, which are particularly relevant for the theoretical study of SMC algorithms. Regarding the latter, we recommend in particular Part II, as it covers a topic which we only allude to in this book: the asymptotic theory of parametric estimation for state-space models.

1.2 Structure of the Book

Most chapters have the following plan: a short abstract, followed by the core of the chapter than contains the main methods and theory, a section on numerical experiments, a set of exercises (for self-study, or for covering more advanced points), a bibliography, and a "Python corner" that discusses the implementation of the considered methods in Python; see the end of this chapter for more information on the Python corners.

The first half of the book covers the prerequisites for describing particle algorithms at a certain level of generality:

- Chapters 2 and 3 give a general, non-technical overview of the applications of particle filters: Chap. 2 focuses on the sequential analysis of state-space models, while Chap. 3 describes non-sequential problems that may be tackled with these algorithms.
- Chapters 4 and 5 describe the mathematical tools and concepts that underlie particle algorithms; e.g. Markov processes, Feynman-Kac models, the related forward-backward recursions, and so on. These two chapters may require more efforts from the readers than the rest, but the upside is that many ideas and

algorithms found in subsequent chapters become obvious when seen as an application of such tools.

- Chapters 6 and 7 cover respectively the forward-backward algorithm (for finite state-space models), and the Kalman filter (for linear-Gaussian state-space models). These two algorithms correspond to the two most common cases where the Feynman-Kac distributions may be computed exactly (without resorting to particle filtering). These two chapters may be skipped at first reading.
- Chapters 8 and 9 give prerequisites in Monte Carlo sampling, particularly importance sampling (Chap. 8) and resampling (Chap. 9).

The core of the book consists of Chaps. 10 (on particle filtering, and its application to state-space models), 12 (on particle smoothing), and 13 (quasi-Monte Carlo particle filter). The rest of the book covers various advanced topics that the readers may be read selectively based on their interests:

- Chapter 11 gives an introduction to the asymptotic theory of particle filters.
- Chapters 14 and 16 cover respectively maximum likelihood and Bayesian estimation of state-space models. Since the latter relies on MCMC (Markov chain Monte Carlo) algorithms, Chap. 15 contains a brief introduction on MCMC.
- Chapter 17 covers SMC samplers, that is, SMC algorithms that may be applied to a general sequence of distributions.
- Chapter 18 describes SMC^2, an advanced SMC sampler for the sequential estimation of state-space models.

1.3 Note to Instructors

This book grew out of courses on particle filtering taught at different places, at different levels, and for different durations. Slides may be requested from the authors. An example of such a course (in a very short, two-hour format, at PhD level) may be viewed at: https://library.cirm-math.fr/Record.htm?idlist=4&record=19285595124910037779. The more common format for the first author is a 18-h course taught at a graduate (M2) level at the ENSAE.

We make the following recommendations to potential instructors. First, it is important to spend at least 1 h on the various applications covered in Chaps. 2 and 3. This part is very accessible, and should give the students a good idea of what the course is about and what SMC methods are capable of.

The next step is to cover Chaps. 4 and 5, on Markov processes and Feynman-Kac models respectively. Students may struggle with the notations, so it is important to explain them very carefully, using e.g. simple illustrative examples (where the density of the Markov kernel exists or does not exist, and so on). You may want to skip the second part of Chap. 7 (on backward decompositions) in order to spend more time on the simpler forward recursion.

We typically skip Kalman filtering (Chap. 7) and cover briefly the finite case (Chap. 6), mostly as a way to illustrate the ideas of the previous chapter.

Then, one should cover importance sampling and resampling (Chaps. 8 and 9). This can go quite fast if students have already some background on Monte Carlo.

Once all these notions have been properly introduced, covering particle filtering (Chap. 10) is surprisingly straightforward, and may be done in about 1 h.

At this stage, it is very helpful to discuss with the students how the different particle algorithms may be implemented in practice. We typically spend at least one hour going through the material covered in the Python corners of up to Chap. 10. This part really helps students to get a more concrete understanding of the underlying concepts.

The rest of the course should depend on your preferences and time constraints. Going through all the different approaches for smoothing (Chap. 12) is very tedious, so you may want to skip this chapter entirely, or cover only one algorithm, e.g. FFBS. On the other hand, parameter estimation for state-space models (Chaps. 14 and 16) is a topic which is simple to review, and which students enjoy quite a lot. For the latter chapter, it is nice to focus on PMMH as students find this algorithm particularly intuitive.

Certain years, we manage to cram one more chapter, and go for SMC samplers, as this chapter may be the basis for several interesting projects.

Speaking of projects, we typically provide the students with a recent research paper, ask them to implement one or several SMC samplers (e.g. the one described in the paper plus another one they have to figure out given some indications), and discuss their results. They are allowed (and even encouraged) to use `particles`, the Python package we developed to go with this book (see below).

1.4 Acknowledgements

As said above, this book is partly based on material developed over the years while teaching particle filters in various places (Université Paris Dauphine, ENSAE Paris, University of Copenhagen, a summer school at Warwick University, and another one at CIRM in Marseille). We are very grateful to our colleagues who kindly invited us to teach this material (Susanne Ditlevsen, Joaquín Miguez, Murray Pollock and Christian P. Robert) and to students who (enthusiastically, or sometimes stoically) attended the courses, and helped us to improve the presentation.

We are also most grateful to all our colleagues who kindly accepted to read and provide feedback on a preliminary version of the book (in alphabetical order): Adrien Corenflos, Hai-Dang Dau, Arnaud Doucet, Gabriel Ducrocq, Pierre Jacob, and Jonty Rougier. Special thanks to Sergios Agapiou, Mathieu Gerber, Umberto Picchini, and Daniel Sanz-Alonso for their detailed comments.

The first author would like to dedicate this book to his family, in particular his father who passed recently, his mother, his wife and his two particles Alice and Antoine. The second would like to acknowledge the inspiration from another tamed FK beast, who has also actively participated in some of the training activities mentioned above, the Female Kanine (sic) Xiska.

The book received its final revision and editing during the SARS-Cov-2 lockdown: $ουδεν\ κακον\ αμιγες\ καλου$!

1.5 Python Corner

While writing this book, we developed a Python package, called `particles`, which implements in a modular way a large class of particle algorithms. This package is available on GitHub (at https://github.com/nchopin/particles); see below for installation instructions.

Each chapter of this book ends up with a Python corner like this one, where we discuss how to implement in Python the methods covered in that chapter, in particular by considering excerpts from `particles`.

The rest of this Python corner is a very brief introduction to Python and its scientific libraries. It is not a proper tutorial on Python and its numerical libraries, such as e.g. the Scipy lectures (http://www.scipy-lectures.org), the on-line courses at the software carpentry (https://software-carpentry.org/), or the book of McKinney (2012) (which is also a good introduction to some of the scientific libraries discussed in the next section). Rather, it is meant to explain a few quirks of the Python language that, sometimes, confuse beginners. Fortunately, Python has been designed to be highly readable, so the following should be enough to make beginners able to make most of the Python corners.

Other advantages of Python is that it is free and open source (unlike Matlab), and that it is becoming the most popular language in machine learning and in Engineering (to the detriment of Matlab). In Statistics, R remains more popular, in particular because it features such a large number of statistical libraries. Note however that it is easy to call any R library from within Python, using the rpy2 library.

1.5.1 Syntax

The following piece of Python code defines a function for computing factorials:

```
def factorial(n, verbose=False):
    "Compute factorial of n (the product of 1, 2, ... to n)."

    p = 1
    for k in range(2, n + 1):
        p = p * k
        if verbose:
            print("factorial of %i is %i" % (k, p))
    return p

print(factorial(5))
```

These few lines reveal some noteworthy aspects of Python:

- Python uses whitespace indentation to delimit blocks (rather than e.g. curly braces). This is a great way to enforce readability. Common practice is to use four spaces per indentation level. It is also possible to use one tab to define an indentation level. Be aware of however that the number of spaces used to represent one tab is arbitrary. As a result, Python is not able to process text files that mix tabs and spaces. The best way to avoid 'tab headaches' is to make the following changes in the settings of your favourite text editor: (a) tab width is four spaces; (b) automatically translate tabs into spaces. That way, when you press tab once (resp. backspace), you automatically get to the next (resp. previous) level, without either entering tabs in your document. (These settings are actually sensible for *any* language: clear indentation always makes code more readable, but this cannot be achieved using tabs, given again their ambiguous meaning.)
- Python is dynamically typed: the type of the arguments of `factorial` are not specified.
- Named arguments are optional: if `verbose` is not specified, it is automatically set to `False`; to set to true, the proper syntax is `x = factorial(10, verbose=True)`.
- The string below the `def` statement defines the *documentation* of function `factorial`. Once the code above is executed, `factorial` becomes an object, and `help(factorial)` returns its documentation.
- Python uses 0-based indexing (like C). For instance, `range(n)` iterates over integers $1, \ldots, n-1$. In the code block above, `range(2, n+1)` is the half-open interval $2, \ldots, n$.

1.5.2 *Scientific Computation in Python*

Here are the main scientific libraries in Python:

- `NumPy` implements 'ndarrays' (multidimensional numerical arrays), and functions that operates on them. It also includes routines for linear algebra, Fourier analysis and random number generation. Most scientific libraries are built upon `NumPy`.
- `SciPy` is a comprehensive package for scientific computing, with modules for linear algebra, optimization, integration, special functions, signal and image processing, statistics, ODE solvers, among other things.
- `Matplotlib` is a 2D plotting library with a syntax similar to Matlab.
- `seaborn` complements `Matplotlib` with statistical plots, in a spirit somehow similar to the R library `ggplot`.
- `pandas` implements data frames (à la R) and various tools for data analysis.
- `rpy2` makes it possible to call R functions and libraries within Python.

- Numba is an experimental library that speeds up the execution of Python; see Performance below.
- scikit-learn is a popular machine learning library.

All these libraries and many others are available in any good scientific Python distribution; see below under 'Installation'. The only strict dependencies for particles are NumPy and SciPy; Matplotlib and seaborn are optional dependencies, as they are required only to run certain examples.

The code blocks provided in the Python corners assume that the following 'import' have been performed:

```
import numpy as np
```

The first line makes all NumPy functions and objects available with the syntax np.func. (If we had typed import numpy, then they would be available as numpy.func.) For instance:

```
x = np.arange(100)
y = np.exp(x[:10])
```

creates two 1D arrays, x containing 0, 1, ..., 99, and y containing $e^0, \ldots, e^9$; recall that Python uses 0-based indexing, hence x[:10] is an array containing x[0] to x[9]. The second line specifically calls the NumPy version of exp, which accepts ndarrays as inputs and returns ndarrays as outputs.

It is of course a bit annoying (especially if you are coming from Matlab or R) to type np. in front of all standard functions, e.g. log, sin, and so on. One may be tempted to use instead an implicit import:

```
from numpy import *

x = arange(100)
y = exp(x[:10])
```

so that all NumPy objects and functions are available into the base *namespace*. This is fine for interactive work, but in general it is better to be explicit about where does any object comes from; importing an awful lot of functions and objects into the namespace is typically called *namespace pollution* and it is not very *Pythonic*! It leads to code that it is hard to follow. In particular, implicitly importing two libraries with overlapping functionalities (such as NumPy and SciPy) is a good way to write obscure and buggy code.

1.5.3 Random Number Generation

NumPy has a module called random, which contains procedures to sample from many probability distributions. To generate an array containing 100 draws from the $\mathcal{N}(2, 3^2)$ distribution:

```
from numpy import random

x = random.normal(loc=2., scale=3., size=100)
```

SciPy has a module called stats, which represents probability distributions as *objects*, from which may sample from, compute the probability density, and so on.

```
from scipy import stats

law = stats.norm(loc=2., scale=3.)
x = law.rvs(size=100)
y = law.logpdf(x)  # log of probability density function at x
z = law.cdf(x) # cumulative distribution function at x
```

The second approach is clearly more versatile. particles implements probability distributions in a similar way; see the documentation of module distributions for more details.

1.5.4 Performance

Let's return to our factorial example; here is a more concise and more efficient version based on NumPy arrays:

```
def factorial(n):
    return np.prod(np.arange(2, n+1))
```

Conventional wisdom is that loops are very slow in interpreted languages (such as Python, Matlab or R), and that they should be 'vectorized', i.e. replaced by array operations. This is true to some extent, and it certainly applies to our factorial example. But it is more accurate to say that loops *with a slim body* are slow. Loops with a body that does expensive operations are typically fine, and should be left alone. Another point to take into account is that 'vectorizing' a piece of code often makes it less readable, sometimes very much less so. For instance, try to figure out what result you get if you type factorial(-3), by looking at both versions of factorial.

We will discuss this point more in detail when practical examples come up in the following chapters. For now, we mention that loops may also be sped-up using Numba, which makes it possible to compile on the fly selected parts of a program; see Chap. 9 for an illustrative example.

1.5.5 Installing Python

A simple way to install in one go Python and its main scientific libraries is to install the Anaconda distribution from Continuum Analytics, which is available on all major platforms; see https://www.continuum.io. Two versions are proposed: 2.7 and 3.7 at the time of writing. We recommend the latter. Python 2 will soon be deprecated, and most libraries are now compatible with Python 3. `particles` is currently compatible with both, but we may drop support for Python 2 in the future.

The Anaconda distribution comes with all the libraries mentioned above (and many more), and also contains the following useful tools:

- `ipython`: a replacement for the standard Python shell, which is very popular among scientists. It also has a notebook mode, which makes it possible to mix live code, equations and figures in a browser window. Very handy for teaching. See also `jupyter`.
- `spyder`: a development environment in the spirit of Rstudio or the Matlab interface.
- `pip`: a utility to install packages available in the huge repository pypi (the Python package index, see https://pypi.python.org).

1.5.6 Installing `particles`

You may download `particles` from https://github.com/nchopin/particles, (manually or by cloning the repository, if you are familiar with the version control system `git`).

The package is organised as follows: the scripts that were used to perform the numerical experiments discussed in this book (and to generate the plots) are in folder `book`; the various datasets used in these experiments are in `datasets`; the documentation is in `docs`, although it is more convenient to consult it directly online, at https://particles-sequential-monte-carlo-in-python.readthedocs.io/en/latest/.

The package itself is made of modules which may be found in folder `particles`. To use these modules, you must first install the package, using for instance `pip`; i.e. on the command line (assuming you are already inside folder `particles`):

```
pip install --user .
```

Once the package is installed, you should be able to import any module from `particles`; e.g.

```
from particles import resampling as rs
```

See page https://github.com/nchopin/particles, for more details on the installation.

1.5.7 Other Software

`particles` is meant to be easy to use and extend. For raw performance, we particularly recommend LibBi (http://libbi.org, see also its successor at https://

birch-lang.org/) which takes as an input a description of the considered model, and implements in C++ the corresponding particle algorithms, using multi-core CPUs or even GPUs.

Other noteworthy pieces of software are `SMCTC`, a C++ template class library, see Johansen (2009); `pomp`, a R package that focuses on parameter estimation of state-space models, see https://kingaa.github.io/pomp/; and `SequentialMonteCarlo.jl`, a Julia package developed by Antony Lee, see https://github.com/awllee.

Bibliography

Cappé, O., Moulines, E., & Rydén, T. (2005). *Inference in hidden Markov models. Springer series in statistics*. New York: Springer.

Del Moral, P. (2004). *Feynman-Kac formulae. Genealogical and interacting particle systems with applications. Probability and its applications*. New York: Springer.

Johansen, A. M. (2009). SMCTC: Sequential Monte Carlo in C++. *Journal of Statistical Software, 30*(6), 1–41.

McKinney, W. (2012). *Python for data analysis*. Sebastopol, CA: O'Reilly Media.

Chapter 2
Introduction to State-Space Models

Summary The sequential analysis of state-space models remains to this day the main application of Sequential Monte Carlo. The intent of this Chapter is to define informally state-space models, and discuss several typical examples of such models from different areas of Science.

However, we warn readers beforehand that we will need the mathematical machinery developed in the following chapters to define in sufficient generality state-space models, to develop recursions for filters and smoothers, and design a variety of simulation algorithms, including particle filters.

2.1 A First Definition

Since state-space models are used in so many different fields of science, there is some variability in the literature regarding how formally and how generally they are defined. The purpose of this section is to introduce informally their main features, while deferring technicalities to the following chapters.

A state-space model is a time series model that consists of two discrete-time processes $\{X_t\}:= (X_t)_{t\geq 0}$, $\{Y_t\}:= (Y_t)_{t\geq 0}$, taking values respectively in spaces $\mathcal{X}$ and $\mathcal{Y}$; in this book ":=" will mean "equal by definition". These spaces can be multi-dimensional Euclidean spaces, discrete spaces but often also less standard. A simplified specification of the model is done by means of a parameter vector $\theta \in \Theta$, and a set of densities that define the joint density of the processes via a factorisation:

$$p_0^\theta(x_0) \prod_{t=0}^{T} f_t^\theta(y_t|x_t) \prod_{t=1}^{T} p_t^\theta(x_t|x_{t-1}) .$$

This describes a generative probabilistic model, where X_0 is drawn according to the initial density $p_0^\theta(x_0)$, and then each X_t is drawn conditionally on the previously drawn $X_{t-1} = x_{t-1}$ according to the density $p_t^\theta(x_t|x_{t-1})$, and each Y_t conditionally on the most recent $X_t = x_t$, according to the density $f_t^\theta(y_t|x_t)$.

© Springer Nature Switzerland AG 2020

N. Chopin, O. Papaspiliopoulos, *An Introduction to Sequential Monte Carlo*, Springer Series in Statistics, https://doi.org/10.1007/978-3-030-47845-2_2

Notation/Terminology Upper case letters will denote random variables and lower case their realisations. The semi-colon notation will be used extensively in this book for collections of random variables, e.g. $Y_{0:t} = (Y_0, \ldots, Y_t)$ or their realisations, e.g. $y_{0:t} = (y_0, \ldots, y_t)$. The functions $p_t^\theta(x_t|x_{t-1})$ and $f_t^\theta(y_t|x_t)$ are densities in x_t and y_t respectively, and belong to parametric families indexed by two sets of parameters, (θ, x_{t-1}) and (θ, x_t) respectively. When the model parameters are assumed known we will drop θ from the notation and simply write $p_t(x_t|x_{t-1})$, $f_t(y_t|x_t)$ and so on and so forth.

An assumption that it is unsatisfactory in this definition is that the conditional distribution of X_t given $X_{t-1} = x_{t-1}$ has a "density". Section 2.4 discusses practical examples where this assumption does not hold. It is fairly straightforward, at the cost of a small increase in mathematical sophistication, to move away from this assumption. We do so in Chap. 4 and in particular in Sect. 4.5, where we provide a more general definition of state-space models. A major step towards the mathematical and computational treatment of such models is the recognition of the properties of both $\{X_t\}$ and $\{(X_t, Y_t)\}$ as Markov processes. This is also done in Sect. 4.5.

2.2 A Second Definition

Another common way to define a state-space model is by writing equations that relate the variables X_t and Y_t to sequences of independent "shocks" or "noise terms":

$$
\begin{aligned}
X_0 &= K_0(U_0, \theta) \\
X_t &= K_t(X_{t-1}, U_t, \theta), \quad t \geq 1 \\
Y_t &= H_t(X_t, V_t, \theta), \quad t \geq 0
\end{aligned}
$$

where K_0, K_t, H_t, are deterministic functions, and $\{U_t\}$, $\{V_t\}$ are sequences of i.i.d. random variables. We shall see in Sect. 2.4 that state-space models arising in Engineering are often formulated in this way. On the other hand, this "functional" formulation of state-space models is also restrictive, and is quite cumbersome for certain state-space models that arise in other fields; see e.g. Sect. 2.4.2.

2.3 Sequential Analysis of State-Space Models

The phrase 'state-space model' refers not only to a certain collection of processes, but also to a certain type of 'inferential scenario': process $\{Y_t\}$ is observed, at least at certain times, process $\{X_t\}$ is not, and the objective is to recover the X_t's given the Y_t's; or more formally to derive the distribution of certain X_t's conditional on certain components of $\{Y_t\}$. This justifies the informal definition of a state-space model as a 'Markov chain observed with noise'. In the Statistics literature, the X_t's are often called latent variables.

In particular, the notion of sequential analysis refers to the following scenario: one would like to draw inference on the process X_t sequentially in time, that is, at each time t, given the data $y_{0:t}$ (the realisation of $Y_{0:t}$) collected up to date t. Traditionally, this is done without taking into account parameter uncertainty; i.e. θ is known and fixed. Specifically, the following tasks are distinguished:

- **filtering**: deriving the distribution of X_t conditional on $Y_{0:t} = y_{0:t}$, for $t = 0, 1, \ldots$
- **state prediction**: deriving the distribution of X_{t+1} (more generally $X_{t+1:t+h}$, $h \geq 1$) conditional on $Y_{0:t} = y_{0:t}$, for $t = 0, 1, \ldots$
- **fixed-lag smoothing**: deriving the distribution of $X_{t-h:t}$, $h \geq 1$, conditional on $Y_{0:t} = y_{0:t}$, for $t = h, h + 1, \ldots$
- **(complete) smoothing**: deriving the distribution of $X_{0:t}$ given $Y_{0:t} = y_{0:t}$, for $t = 0, 1, \ldots$

We are not very clear for the moment on what we mean by deriving these distributions: do we wish to compute expectations with respect to them, to simulate from them, or compute their probability distribution? This point is left for later. On top of state inference, sequential analysis may also include the following objectives relative to the observed process itself:

- **data prediction**: deriving the law of $Y_{t+1:t+h}$, conditional on $Y_{0:t} = y_{0:t}$, for $h \geq 1$ and $t = 0, 1, \ldots$
- **computation of the likelihood factor** $p_t^\theta(y_t|y_{0:t-1})$, that is, the probability density of Y_t at point y_t, conditional on $Y_{0:t-1} = y_{0:t-1}$ and for a given parameter θ.

With respect to the latter task, note that the full likelihood (probability density function) of a dataset $y_{0:T}$ is obtained from the T first likelihood factors through a simple chain rule decomposition

$$p_T^\theta(y_{0:T}) = \prod_{t=0}^{T} p_t^\theta(y_t|y_{0:t-1}),$$

with the convention that $p_t^\theta(y_t|y_{0:t-1}) = p^\theta(y_0)$ for $t = 0$. Hence, these likelihood factors are of direct interest for parameter inference.

Why are all these quantities difficult to compute? Because they are integrals of large dimension. For instance, the full likelihood may be written as

$$p_T^\theta(y_{0:T}) = \int_{\mathcal{X}^{T+1}} p_0^\theta(x_0) \prod_{t=0}^{T} f_t^\theta(y_t|x_t) \prod_{t=1}^{T} p_t^\theta(x_t|x_{t-1}) \mathrm{d}x_{0:T} \,.$$

assuming as in Sect. 2.1 that $X_t|X_{t-1}$ and $Y_t|X_t$ admit densities $p_t^\theta(x_t|x_{t-1})$ and $f_t^\theta(y_t|x_t)$, respectively. In the same way, the filtering distribution at time t may be written as a ratio of two integrals with support $\mathcal{X}^t$ for the numerator, and $\mathcal{X}^{t+1}$ for the denominator; see Exercise 2.1.

Of course, trying to compute directly such integrals of growing dimension, recursively in time, is bound to be very ineffective. Fortunately, it is possible to relate all the above quantities through recursive operations that involve integrals of fixed dimension, as explained in the coming chapters.

2.4 Some Examples of State-Space Models

2.4.1 *Signal Processing: Tracking, Positioning, Navigation*

State-space modelling plays a central role in target tracking and related tasks in signal processing (positioning, navigation). There, X_t represents the position of a moving object (a car, a ship, a plane, ...), and Y_t some measurement that provides imperfect information on X_t. The distinction between tracking, positioning and navigation relates to whether the observer tracks its own position (e.g. GPS positioning in a car), or that of a different object (tracking of a hostile plane), and may be ignored for now.

For instance, assume that $\mathcal{X} = \mathbb{R}^2$; throughout the book $\mathbb{R}$ denotes the set of real numbers, and $\mathbb{R}^d$ the d-dimensional Euclidean space. Assume that X_t is the 2D position of a ship, whose movement is modelled by a random walk:

$$X_t = X_{t-1} + U_t, \quad U_t \sim \mathcal{N}_2(0, \sigma_X^2 I_2), \tag{2.1}$$

and Y_t is a (noisy) angular measurement obtained by a radar or a similar device (sonar, ultrasounds, ...), i.e.

$$Y_t = \operatorname{atan}\left(\frac{X_t(2)}{X_t(1)}\right) + V_t, \quad V_t \sim \mathcal{N}(0, \sigma_Y^2).$$

In this example, $\theta = (\sigma_X^2, \sigma_Y^2)$. Figure 2.1 illustrates this bearings-only tracking model.

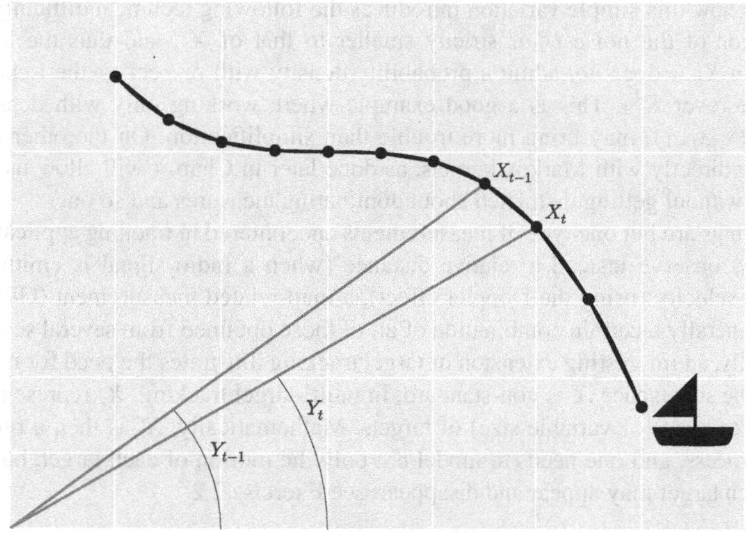

Fig. 2.1 Pictorial description of the bearing-only tracking problem

Notation/Terminology For a vector X_t in $\mathbb{R}^d$, $X_t(k)$ denotes its k-th component. I_d denotes the d-dimensional identity matrix, 0_d denotes the d-dimensional vector of 0's, etc. $\mathcal{N}_d(\cdot,\cdot)$ denotes a d-dimensional Gaussian distribution and $\mathcal{N}(\cdot,\cdot)$ a univariate Gaussian.

Motion model (2.1) is too basic for most problems; in particular a random walk may be too erratic to represent the motion of e.g. a ship. In certain applications, such as in-car GPS positioning, one measures exactly the velocity v_t, leading to motion model (again in 2D):

$$X_t = X_{t-1} + v_t + U_t, \quad U_t \sim \mathcal{N}_2(0, \sigma_X^2 I_2). \tag{2.2}$$

In applications where the velocity is not observed (typically tracking), one may assume that it is the velocity and not the position that follows a random walk; so that the object moves according to an integrated random walk (which generates more regular trajectories than a random walk). To that effect, take $\mathcal{X} = \mathbb{R}^4$, X_t to be the 4-dimensional vector comprised of the position (two first components) and the velocity (two last components) of the target, and, in matrix notations

$$X_t = \begin{pmatrix} I_2 & I_2 \\ 0_2 & I_2 \end{pmatrix} X_{t-1} + \begin{pmatrix} 0_2 & 0_2 \\ 0_2 & U_t \end{pmatrix}, \quad U_t \sim \mathcal{N}_2(0, \sigma_X^2 I_2). \tag{2.3}$$

Note how this simple variation introduces the following technical difficulty: the dimension of the noise U_t is strictly smaller to that of X_t, and thus the law of X_t given X_{t-1} does *not* admit a probability density with respect to the Lebesgue measure (over $\mathbb{R}^4$). This is a good example where working only with densities, i.e., $p_t^\theta(x_t|x_{t-1})$, may bring more trouble than simplification. On the other hand, working directly with Markov kernels, as done later in Chap. 4 will allow us to be precise without getting distracted about dominating measures and so on.

Bearings are but one type of measurements encountered in tracking applications; one may observe instead a relative distance (when a radio signal is emitted), a relative velocity (using the Doppler effect), a map-related measurement (GPS), or more generally a certain combination of all of these obtained from several sensors.

Finally, an interesting extension of target tracking illustrates the need for models where the state space $\mathcal{X}$ is non-standard. In multi-target tracking, X_t represents the motion of a set (of variable size) of targets. Mathematically, X_t is then a random point process, and one needs to model not only the motion of each target, but also how each target may appear and disappear; see Exercise 2.2.

2.4.2 *Time Series of Counts in Neuroscience, Astrostatistics, and Genetics*

An important class of problems in neuroscience is neural decoding, that is, drawing inference about the interaction of a subject with her environment, based on her neural activity. Neural decoding is an essential component of designing brain-machine interface devices, such as neural motor prostheses.

Some of the models used in neural decoding are remarkably similar to tracking models. Consider for instance the following model, which may be used to infer the motion of the subject's hand based on measurements from electrodes implanted in the subject's brain.

Generalising motion model (2.3) to three dimensions, one may take $\mathcal{X} = \mathbb{R}^6$, X_t to consist of the position and velocity of the subject's hand, and assume

$$X_t = \begin{pmatrix} I_3 & I_3 \\ 0_3 & I_3 \end{pmatrix} X_{t-1} + \begin{pmatrix} 0_3 & 0_3 \\ 0_3 & U_t \end{pmatrix}, \quad U_t \sim \mathcal{N}_3(0, \sigma^2 I_3).$$

The observation Y_t is a vector of spike counts $Y_t(k)$, for each neuron k, which are independent and Poisson-distributed:

$$Y_t(k)|X_t \sim \mathcal{P}(\lambda_k(X_t)), \quad \log \lambda_k(X_t) = \alpha_k + \beta_k X_t,$$

for $k = 1, \ldots, d_y$, where d_y denotes the dimension of the observation vector at each time t, and $\mathcal{P}$ the Poisson distribution. In this model $\theta = (\sigma^2, \alpha_{1:d_y}, \beta_{1:d_y})$.

One interesting difference with the previous section is that it becomes much less natural to write Y_t as a deterministic function of X_t and some noise term, $Y_t = H_t(X_t, V_t, \theta)$. Working directly with the law of Y_t given X_t proves to be more convenient.

Similar models have been used in astrostatistics to model the varying intensity of photon emissions in time; i.e. $Y_t|X_t$ is as above, but with $d_y = 1$, and $X_t|X_{t-1}$ is also an auto-regressive process, but of dimension one.

More complex state-space models for counts have been constructed in recent years to deal with sequencing data in genetics; there Y_t is the number of 'reads', which is a noisy measurement of the transcription level X_t at position t in the genome; in the most basic of those models, $Y_t|X_t$ is Poisson with rate ϕX_t (although a negative-binomial distribution is often used instead, to allow for over-dispersion).

2.4.3 Stochastic Volatility Models in Finance

In Finance, one is often interested in modelling the volatility of log-returns, $Y_t = \log(p_t/p_{t-1})$, where p_t is the price of a certain asset at time t. A standard approach is ARCH (Auto-regressive conditional heteroscedastic) or GARCH (Generalised ARCH) modelling, where the variance of Y_t is a deterministic function of past data $Y_{0:t-1}$. A competing class of models takes the volatility to be a stochastic process, as this seems a more natural assumption, and also something more compatible with the data. A basic stochastic volatility model is:

$$Y_t|X_t = x_t \sim \mathcal{N}(0, \exp(x_t))$$

where $\{X_t\}$ is an auto-regressive process:

$$X_t - \mu = \rho(X_{t-1} - \mu) + U_t, \quad U_t \sim \mathcal{N}(0, \sigma^2)$$

and $\theta = (\mu, \rho, \sigma^2)$. Often one imposes that $|\rho| < 1$ to make $\{X_t\}$ stationary; in fact one may expect $\rho \approx 1$ as financial data often exhibit volatility clustering (i.e. volatility remains high or low for long periods of time).

Many variants have been considered in the literature. $Y_t|X_t$ may be attributed heavier than Gaussian tails (i.e. Student). One may account for a 'leverage effect' by assuming that the Gaussian noises of $Y_t|X_t$ and $X_t|X_{t-1}$ are correlated; $Y_t = \exp(X_t/2)V_t$, and $(U_t, V_t)^T$ is a centred bivariate Gaussian vector such that $\mathrm{Corr}(U_t, V_t) = \phi$, with $\phi \leq 0$. Formally, this is the same model as before, but with a slight modification for $Y_t|X_t$; see Exercise 2.4. One may introduce skewness by taking $Y_t = \alpha X_t + \exp(X_t/2)V_t$. One may assume the volatility to be an auto-regressive process of order $k > 1$. In that case, we can retain the basic structure of state-space models by increasing the dimension of $\{X_t\}$; e.g. $\mathcal{X} = \mathbb{R}^2$,

$$X_t - \begin{pmatrix} \mu \\ \mu \end{pmatrix} = \begin{pmatrix} \rho_1 & \rho_2 \\ 1 & 0 \end{pmatrix} \left(X_{t-1} - \begin{pmatrix} \mu \\ \mu \end{pmatrix} \right) + \begin{pmatrix} U_t \\ 0 \end{pmatrix}, \quad U_t \sim \mathcal{N}(0, \sigma^2)$$

and then $X_t(1)$ is an auto-regressive process of order 2, $X_t(1) - \mu = \rho_1(X_{t-1}(1) - \mu) + \rho_2(X_{t-2}(1) - \mu) + U_t$, and one may take the variance of $Y_t|X_t$ to be $\exp(X_t(1))$. Note however that, as in the tracking example, $X_t|X_{t-1}$ does no longer admit a probability density with respect to a common (e.g. Lebesgue) measure, as the first component is a deterministic function of X_{t-1}.

A lot of research is now devoted to multivariate extensions of the basic model above, in order to model the joint evolution of log-returns of several assets.

2.4.4 Hierarchical State-Space Models for Panel Data

State-space modelling also extends to panel data (also known as longitudinal data), that is bi-dimensional data y_{tk} that vary both with time t and with 'unit' k (e.g. an individual).

Say we wish to forecast the outcome of the next US presidential election based on state-level survey data (assuming two candidates, one Democrat, one Republican): Y_{tk} is the number of respondents willing to vote for the democratic candidate in a survey conducted at time t in state k:

$$Y_{tk} \sim \mathcal{B}in(\pi_{tk}, n_{tk}),$$

n_{tk} is the number of respondents in that survey, and π_{tk} is the proportion of voters of state k willing to vote Democrat, which is modelled as:

$$\pi_{tk} = \text{logit}^{-1}(\delta_t + X_{tk})$$

where $\text{logit}^{-1}(x) = 1/(1 + e^{-x})$, δ_t is a national-level effect, and X_{tk} a state-level effect, which are both random walks:

$$\delta_t = \delta_{t-1} + \sigma_\delta U_t, \quad X_{tk} = X_{(t-1)k} + \sigma_X V_{tk}.$$

The goal is to predict the Y_{Tk}'s, where T is the time of the election. One may view the whole model as a single state-space model, with state $X_t = (\delta_t, X_{t1}, \ldots, X_{tK})$, and observation $Y_t = (Y_{t1}, \ldots, Y_{tK})$. Alternatively, it may be computationally more convenient to decompose the model into K conditionally independent state-space models, with observed process $\{Y_{tk}\}$, Markov process $\{X_{tk}\}$, coupled via the common national-level process $\{\delta_t\}$. In the spirit of the other examples in this chapter this is a simplified hierarchical state-space model since a realistic predictive model for electoral results would typically include covariates both at national and state level and potentially further random effects.

In a completely different context, a more involved hierarchical state-space model can be used for the synthesis of different sources of data for evaluating the effects of investments done at early age in one's life. Each individual k is associated with two coupled latent stochastic processes that capture the evolution of their cognitive and manual skills, $\{X_{tk}(i)\}$ for $i = 1, 2$, where $i = 1$ is cognitive and $i = 2$ is manual. These processes evolve according to the so-called constant elasticity of substitution (CES) dynamics with multiplicative noise and depend on control processes that measure the amount of investment done in each period, for each person and for the development of each skill, $I_{tk}(i)$:

$$X_t(i) = \left(\gamma_{i,1} X_{t-1}(1)^{\rho_i} + \gamma_{i,2} X_{t-1}(2)^{\rho_i} + \gamma_{i,3} I_{(t-1)k}(i)\right)^{1/\rho_i} \exp\{U_t(i)\}$$

for $U_t \sim \mathcal{N}(0, \sigma_{i,x}^2)$ and $\gamma_{i,1} + \gamma_{i,2} + \gamma_{i,3} = 1$. The skills are measured indirectly via skill-specific tests. In the simplest specification, we can assume that at each period we have a test result (or the average of a series of tests)

$$Y_{tk}(i) \sim \mathcal{N}(\mu_t(i) + \log X_{tk}(i), \sigma_{i,y}^2)$$

where $\{\mu_t\}$ is a common latent process that captures the overall stochastic trend in test performance over time. Again, this is a prototype of a coupled skills model to understand how the investments I_{tk} affect the development of skills. More realistic specifications might involve hierarchical prior on the hyper-parameters, further random effects and multiple test measurements in each period each of which has different correlation with the measured skills, a structure akin to a factor model.

2.4.5 Nonlinear Dynamical Systems in Ecology, Epidemiology, and Other Fields

Most of the models we have seen so far assume that $X_t|X_{t-1}$ is linear and Gaussian, and that $Y_t|X_t$ is nonlinear. The converse is also common, in particular in fields of science interested in modelling natural phenomena with dynamic systems. There, typically $Y_t = X_t + V_t$, where V_t is a Gaussian noise, while $\{X_t\}$ admits some complex non-linear dynamics.

For instance, in population ecology, the theta-logistic model assumes that:

$$X_t = X_{t-1} + \tau_0 - \tau_1 \exp(\tau_2 X_{t-1}) + U_t \tag{2.4}$$

where X_t is the logarithm of the population size (of e.g. bats in a cave), U_t is a noise term, and the τ_i's are all positive; $\theta = (\tau_0, \tau_1, \tau_2)$. Without noise, and with $\tau_1 = \tau_2 = 0$, the population increases exponentially at rate τ_0. The additional term $-\tau_1 \exp(\tau_2 X_t)$ is here to curb this growth when X_t reaches a certain threshold, so as to model the "carrying capacity" of the environment. If $\tau_1 > 0$, the population size admits a stable equilibrium at $(\tau_0/\tau_1)^{1/\tau_2}$. For $\tau_2 = 1$, (2.4) is known as the Ricker model. For certain values of the parameters, $\{X_t\}$ shows highly non-linear or even chaotic behaviour; see Fig. 2.2.

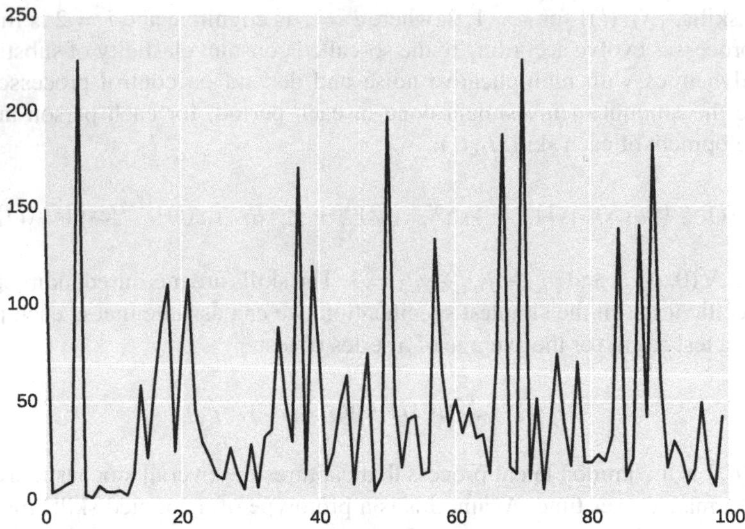

Fig. 2.2 Population size $\exp(X_t)$ as a function of t, where $\{X_t\}$ is simulated according to the Ricker model, with $\tau_0 = 1$, $\tau_1 = 0.02$, $\tau_2 = 1$

Another popular model for population dynamics is the Lotka–Volterra predator-prey model, where $\mathcal{X} = (\mathbb{Z}^+)^2$, $X_t(1)$ is the number of preys, $X_t(2)$ is the number of predators, and, working in continuous-time:

$$\begin{aligned} X_t(1) &\xrightarrow{\tau_1} 2X_t(1) \\ X_t(1) + X_t(2) &\xrightarrow{\tau_2} 2X_t(2), \quad t \in \mathbb{R}^+ \\ X_t(2) &\xrightarrow{\tau_3} 0 \end{aligned}$$

where the chemical notations above mean: the probability that, during period $[t, t + dt]$, one prey is replaced by two preys is $\tau_1 dt$, and so on. Process $\{X_t\}$ is now defined for $t \in \mathbb{R}^+$, but observations are still collected at discrete times; e.g. $Y_t = X_t(2) + V_t$ if only the number of predators is observed (with noise).

Lotka-Volterra is a good example of a Markov process with intractable dynamics; that is a Markov process we can simulate from (using Gillespie algorithm) but whose Markov transition is intractable; i.e. the probability density of $X_t|X_{t-1}$ (for t restricted to observation times, i.e. $t \in \mathbb{Z}^+$) does not admit a tractable expression.

The Lotka-Volterra predator-prey model is a simple instance of the more general class of stochastic kinetic models, which have uses not only in population ecology but also in chemistry (so as to model chemical reactions), and systems biology.

epidemiology is another field interested in population dynamics; for a certain infectious disease (e.g. measles), one considers populations of, say, susceptible, infected, and recovered cases, and models how these populations interact (in a

Markovian way). Typically one observes the evolution of only one such population (e.g. data are death records), possibly with noise. Thus, one ends up again with a state-space model with an intractable Markov transition for $X_t|X_{t-1}$.

2.4.6 State-Space Models with an Intractable or Degenerate Observation Process

We have seen several examples of state-space models with intractable dynamics, i.e. $X_t|X_{t-1}$ may be simulated from, but does not admit a tractable density. There also exist practical examples of models such that $Y_t|X_t$ is intractable in the same way. This precludes particle filtering (as it will become clear in Chap. 10).

Dean et al. (2014) provide an elegant (but approximate) solution to this problem, which is related to the current research on likelihood-free inference. Let $X_t' = (X_t, Y_t)$, and let $Y_t' = Y_t + V_t$ with $V_t \sim \mathcal{N}(0, \sigma^2)$. The artificial state-space model $\{(X_t', Y_t')\}$ can now be used in lieu of $\{(X_t, Y_t)\}$; for instance, the law of $X_t'|Y_{0:t}'$ (or rather of its first component, X_t) provides an approximation of the true filtering distribution $X_t|Y_{0:t}$. The approximation error is determined by σ^2: the smaller σ^2, the smaller the approximation error (but also the poorer the performance of the corresponding particle algorithm, unfortunately; this point will be discussed in Sect. 10.3).

A similar solution may be adopted for state-space models such that $Y_t|X_t$ has very small or even zero variance; e.g., in the Lotka-Volterra example of the previous section, $Y_t = X_t(2)$ exactly. Then again we may introduce an approximate state-space model with larger noise for $Y_t|X_t$ in order to be able to carry out sequential inference at least approximately.

An intriguing proposition in this context is to perform inference *exactly* by adding noise to the data; i.e. replace data Y_t by $Y_t' = Y_t + V_t$, with $V_t \sim \mathcal{N}(0, \sigma^2)$ (the same noise distribution as in the artificial observation process $Y_t'|X_t'$). Then we trade off an approximation error for an increase in variance (for the law of the signal given the data).

2.4.7 Linear Gaussian State-Space Models

An important class of state-space models is the set of linear Gaussian models, that is, models such that both $X_t|X_{t-1}$ and $Y_t|X_t$ are Gaussian distributions, with fixed variance, and an expectation that is a linear function of the conditioning variables, i.e. X_{t-1} and X_t respectively:

$$X_t = AX_{t-1} + U_t \tag{2.5}$$

$$Y_t = BX_t + V_t \tag{2.6}$$

where $\mathcal{X} = \mathbb{R}^{d_x}$, $\mathcal{Y} = \mathbb{R}^{d_y}$, A and B are matrices of size $d_x \times d_x$, and $d_y \times d_x$, and U_t, V_t are independent Gaussian vectors. (One may also assume that U_t and V_t are correlated, see Exercise 2.3.)

Sequential inference may be performed exactly for such models, using an algorithm known as the Kalman filter; see Chap. 7. Thus, linear Gaussian models will not play a central role in this book, as they do not require any type of Monte Carlo approximation. (On the other hand, models only partially linear and Gaussian may mix Kalman and particle filtering, see Sect. 10.4).

However, it is worth mentioning that linear Gaussian state-space models (and Kalman filtering) have proven extremely popular in many areas, in particular in navigation and control; in fact the first publicly known application of the Kalman filter is the navigation of the Apollo space capsule by NASA in the early 1960s.

Another very popular class of time-series models is that of ARMA (autoregressive moving average models), where the observed process $\{Y_t\}$ verifies:

$$Y_t - \sum_{i=1}^{p} \phi_i Y_{t-i} = U_t - \sum_{i=1}^{q} \gamma_i U_{t-i}, \quad U_t \sim \mathcal{N}(0, \sigma^2)$$

and the U_t's are i.i.d. $\mathcal{N}(0, \sigma^2)$ random variables. As a matter of fact, ARMA models may be cast as linear Gaussian state-space models; see Exercise 2.5. This reformulation has several practical benefits. For instance, the most commonly used method to evaluate the likelihood of an ARMA model (given parameters $\phi_{1:p}$, $\gamma_{1:q}$ and σ^2) is the Kalman filter.

2.4.8 Hidden Markov Models, Change-Point Models, and Mixture of Dirichlet Processes

State-space models with a finite state space $\mathcal{X}$, e.g. $\mathcal{X} = \{1, \ldots, K\}$, are often called hidden Markov models. They represent the second important class of state-space models for which sequential inference may be performed exactly; see Chap. 6.

Historically, hidden Markov models have been used as early as the seventies in speech processing and related branches of signal processing. There, X_t is typically a word (hence K is the size of the vocabulary, and may be very large), and Y_t an acoustic measurement. The task is to recover the uttered sentence from the data. That consecutive words follow a Markov process is of course a convenient simplification.

HMM have also found many uses in domains dealing with time series showing some form of time heterogeneity. In economics for instance (where HMM have been rediscovered independently, and called Markov-switching models), X_t typically represents the state of the Economy (e.g. $K = 2$, state 1 is recession, state 2 is expansion), and $Y_t|X_t = k$ follows some parametric model, with a parameter that depends on k; say, for $X_t = k$,

$$Y_t - \mu_k = \phi_k(Y_{t-1} - \mu_k) + V_t, \quad V_t \sim \mathcal{N}(0, \sigma_k^2).$$

As for linear Gaussian models, there is no point implementing a particle filter for (basic) hidden Markov models, as filtering distributions may be computed exactly. But particle filters remain useful for certain variants of these models. For instance, when one integrates out the parameters of a hidden Markov model (e.g. the $(\phi_k, \mu_k, \sigma_k)$'s above), one obtains a new model, where Y_t depends on the whole trajectory $X_{0:t}$. In that case, one may implement a particle filter that carries forward these complete trajectories; see Sect. 10.4 and the bibliography. This approach may be applied in particular to learning sequentially MDP (mixture of Dirichlet processes) models. These Bayesian non-parametric models, which are popular in particular in machine learning, amount to an infinite mixture, where the mixture index X_t evolves according to a Polya urn process:

$$\mathbb{P}(X_t = k | X_{0:t-1} = x_{0:t-1}) = \begin{cases} n_{t-1}^k/(t+\alpha) & k = 1, \cdots, m_{t-1} \\ \alpha/(t+\alpha) & k = m_{t-1}+1 \end{cases}$$

with $\alpha > 0$, n_{t-1}^k is the number of occurrences of state k in $x_{0:t-1}$, and m_{t-1} is the number of states that have appeared at least once in $x_{0:t-1}$. Given X_t, the datapoints follow a certain parametric distribution, e.g. $Y_t|X_t = k \sim \mathcal{N}(\mu_k, \sigma_k^2)$, where the (μ_k, σ_k^2) are i.i.d. a priori. Again, if we integrate out these parameters, we end up with a model which may be tackled with a particle filter that essentially propagates complete state trajectories.

Change-point models are another class of models closely related to hidden Markov models. Again, one assumes $\{Y_t\}$ follows some parametric model, but with the parameter changing over time. One way to model this is to take $\{X_t\}$ to be piece-wise constant; but then $\{X_t\}$ is not Markov. A better approach is to define X_t to be the date since last change. Then the state space evolves over time: $\mathcal{X}_t = \{0, \ldots, t\}$. For more details on the connection between change-point and hidden Markov models, see Exercise 6.2, and for how particle filters may be applied to change-point models, see the bibliography of Chap. 6.

Exercises

2.1 *Following the discussion of Sect. 2.3, show that the filtering density, that is the probability density of X_t conditional on $Y_{0:t} = y_{0:t}$ may be written as a ratio of two integrals, of support $\mathcal{X}^T$ and $\mathcal{X}^{T+1}$, respectively. (You may assume that both $X_t|X_{t-1}$ and $Y_t|X_t$ admit probability densities.)*

2.2 *Discuss how to construct a state-space model for multi-target tracking (Sect. 2.4.1), such that (a) X_t represents a finite set of targets, of random size, which may appear or disappear at any time; (b) Y_t consists of noisy observations of these targets, which in addition may be occulted at certain times, or on the contrary detected.*

2.3 *Consider the linear Gaussian model defined by* (2.5) *and* (2.6)*. Show that if we assume that the noise terms U_t and V_t are correlated, (i.e. (U_t, V_t) is a Gaussian vector such that U_t and V_t are not independent), the model may be reformulated as a standard state-space model.*

2.4 *Same exercise as before for the univariate stochastic volatility model described in Sect. 2.4.3: $X_t - \mu = \rho(X_{t-1} - \mu) + \sigma U_t$, $Y_t = \exp(X_t/2)V_t$, and assume (U_t, V_t) is a bivariate Gaussian vector such that* $\mathrm{Cov}(U_t, V_t) = \rho$.

2.5 *(From Sect. 2.4.7). Show that any Gaussian ARMA (Auto-Regressive Moving Average) model, that is any model for a process $\{Y_t\}$ such that*

$$Y_t - \sum_{i=1}^{p} \phi_i Y_{t-i} = U_t - \sum_{i=1}^{q} \gamma_i U_{t-i}, \quad U_t \sim \mathcal{N}(0, \sigma^2)$$

can be cast as a linear Gaussian state-space model.

Bibliographical Notes

Our discussion of the use of state-space models in different scientific areas is partly based on the following references: Caron et al. (2007) for GPS positioning; Vo et al. (2003) for multi-target tracking; Koyama et al. (2010) for neural decoding; Yu and Meng (2011) for astrostatistics; Kim et al. (1998) for stochastic volatility modelling; Linzer (2013) for state-level US election forecasting and Cunha et al. (2010) for estimating skill-formation using state-space models; Mirauta et al. (2014) on state-space modelling in genetics, and the underlying biology; Wood (2010) and Fasiolo et al. (2016) for chaotic systems and their use in ecology; Rasmussen et al. (2011) for compartmental models in epidemiology; the book of Wilkinson (2006) for stochastic kinetic models and systems biology; the standard text of Rabiner (1989) for finite state-space models and their use in speech processing; see also e.g. Frühwirth-Schnatter (2006) for other statistical applications of finite state-space models. For applications of the nonlinear state-space models we have discussed in this chapter in Economics see e.g. Flury and Shephard (2011). Regarding the application of particle filtering to mixtures of Dirichlet processes, see Fearnhead (2004) and Griffin (2017). The mathematical problem of filtering Bayesian non-parametric models is studied in Papaspiliopoulos et al. (2016) where even some exact filters are obtained.

We also recommend the following standard textbooks for some of the topics mentioned in this chapter: Francq and Zakoian (2011) for GARCH models, Brockwell and Davis (2006) for ARMA models, and Durbin and Koopman (2012) for linear Gaussian models and their uses outside of Engineering.

Bibliography

Brockwell, P. J., & Davis, R. A. (2006). *Time series: Theory and methods. Springer series in statistics*. New York: Springer. Reprint of the second (1991) edition.

Caron, F., Davy, M., Duflos, E., & Vanheeghe, P. (2007). Particle filtering for multisensor data fusion with switching observation models: Application to land vehicle positioning. *IEEE Transactions on Signal Processing, 55*(6, part 1), 2703–2719.

Cunha, F., Heckman, J., & Schennach, S. (2010). Estimating the technology of congitive and noncognitive skill formation. *Econometrica, 78*, 883–931.

Dean, T. A., Singh, S. S., Jasra, A., & Peters, G. W. (2014). Parameter estimation for hidden Markov models with intractable likelihoods. *Scandinavian Journal of Statistics, 41*(4), 970–987.

Durbin, J., & Koopman, S. J. (2012). *Time series analysis by state space methods. Oxford statistical science series* (Vol. 38, 2nd ed.). Oxford: Oxford University Press.

Fasiolo, M., Pya, N., & Wood, S. N. (2016). A comparison of inferential methods for highly nonlinear state space models in ecology and epidemiology. *Statistical Science, 31*(1), 96–118.

Fearnhead, P. (2004). Particle filters for mixture models with an unknown number of components. *Statistics and Computing, 14*(1), 11–21.

Flury, T., & Shephard, N. (2011). Bayesian inference based only on simulated likelihood: Particle filter analysis of dynamic economic models. *Econometric Theory, 27*, 933–956.

Francq, C., & Zakoian, J.-M. (2011). *GARCH models: Structure, statistical inference and financial applications*. Hoboken, NJ: Wiley.

Frühwirth-Schnatter, S. (2006). *Finite mixture and Markov switching models. Springer series in statistics*. New York: Springer.

Griffin, J. E. (2017). Sequential Monte Carlo methods for mixtures with normalized random measures with independent increments priors. *Statistics and Computing, 27*(1), 131–145.

Kim, S., Shephard, N., & Chib, S. (1998). Stochastic volatility: Likelihood inference and comparison with ARCH models. *The Review of Economic Studies, 65*(3), 361–393.

Koyama, S., Castellanos Pérez-Bolde, L., Shalizi, C. R., & Kass, R. E. (2010). Approximate methods for state-space models. *Journal of the American Statistical Association, 105*(489), 170–180. With supplementary material available on line.

Linzer, D. A. (2013). Dynamic Bayesian forecasting of presidential elections in the states. *Journal of the American Statistical Association, 108*(501), 124–134.

Mirauta, B., Nicolas, P., & Richard, H. (2014). Parseq: Reconstruction of microbial transcription landscape from RNA-Seq read counts using state-space models. *Bioinformatics, 30*, 1409–16.

Papaspiliopoulos, O., Ruggiero, M., & Spanò, D. (2016). Conjugacy properties of time-evolving Dirichlet and gamma random measures. *Electronic Journal of Statistics, 10*(2), 3452–3489.

Rabiner, L. R. (1989). A tutorial on hidden Markov models and selected applications in speech recognition. *Proceedings of the IEEE, 77*, 257–284.

Rasmussen, D. A., Ratmann, O., & Koelle, K. (2011). Inference for nonlinear epidemiological models using genealogies and time series. *PLOS Computational Biology, 7*(8), e1002136, 11.

Vo, B.-N., Singh, S., & Doucet, A. (2003). Sequential Monte Carlo implementation of the PHD filter for multi-target tracking. In *Proceedings of the International Conference on Information Fusion* (pp. 792–799).

Wilkinson, D. J. (2006). *Stochastic modelling for systems biology. Chapman & Hall/CRC mathematical and computational biology series*. Boca Raton, FL: Chapman & Hall/CRC.

Wood, S. N. (2010). Statistical inference for noisy nonlinear ecological dynamic systems. *Nature, 466*(7310), 1102–1104.

Yu, Y., & Meng, X.-L. (2011). To center or not to center: that is not the question—An ancillarity-sufficiency interweaving strategy (ASIS) for boosting MCMC efficiency. *Journal of Computational and Graphical Statistics, 20*(3), 531–570.

Chapter 3
Beyond State-Space Models

Summary We have stated in the previous chapter that the main application of Sequential Monte Carlo is the sequential analysis of state-space models. However, SMC may be used in other contexts. Notably, it may be applied to any computational task that may formulated as an artificial filtering problem. We briefly review in this chapter some of these extra applications of SMC and their connection to state-space models. In particular, we discuss rare-event simulation, Bayesian sequential (and non-sequential) inference, likelihood-free inference, and more generally the problem of simulating from, and computing the normalising constant of, a given probability distribution.

3.1 Rare-Event Simulation for Markov Processes

Rare-event simulation refers to two related problems: (a) computing the probability of an event, when this probability is very small (typically $< 10^{-8}$); and (b) simulating random variables conditionally on such an event. Rare-event simulation is (among others) an important sub-domain of operations research, and a typical application is the computation of the probability of failure of a complex network or system.

When the rare event is defined with respect to a Markov process, the connection to state-space modelling is straightforward. Consider a Markov chain $\{X_t\}$ taking values in $\mathcal{X}$, and the event

$$E_T = \{X_t \in \mathcal{A}_t,\ \forall t \in 0 : T\}$$

for certain sets $\mathcal{A}_t \subset \mathcal{X}$. That is, we would like either to compute the probability of E_T, or simulate the random variables $X_{0:T}$ conditional on E_T, even in situations such that $\mathbb{P}(E_T)$ is very small.

If we now introduce the variables $Y_t = \mathbb{1}\{X_t \in \mathcal{A}_t\}$, and the observations $y_t = 1$, then we obtain an (artificial) state-space model such that (a) its marginal likelihood $p_T(y_{0:T})$ is the probability of the event E_T; and (b) its smoothing distribution, $X_{0:T}|Y_{0:T} = y_{0:T}$, is the distribution of $X_{0:T}$ conditional on event E_T.

© Springer Nature Switzerland AG 2020

N. Chopin, O. Papaspiliopoulos, *An Introduction to Sequential Monte Carlo*, Springer Series in Statistics, https://doi.org/10.1007/978-3-030-47845-2_3

As a result, the particle algorithms presented in this book may be applied directly to this type of rare-event simulation problem. However, they can also be applied to rare-event simulation even when the events do not refer to an underlying Markov process; this is discussed in Sect. 3.5.

3.2 Bayesian Sequential Learning

Now consider a state-space model where the latent process $\{X_t\}$ would be constant: i.e. there exists a random variable Θ such that $\Theta = X_0 = X_1 = \ldots$ (almost surely). This particular model reduces to a Bayesian model, where the datapoints $Y_0, Y_1, \ldots$, are independent conditional on $\Theta = \theta$, and Θ is an unknown parameter, which, in the Bayesian view of Statistics, is treated as an unobserved random variable with (prior) distribution $\nu(\mathrm{d}\theta)$.

In addition, the filtering distribution, that is the law of $X_t|Y_{0:t}$ now turns into the posterior distribution of Θ given the data up to t. This suggests one may use SMC algorithms to perform sequential Bayesian inference, or more generally some form of on-line data processing.

In practice, extending SMC algorithms to sequential learning is less straightforward than it appears at first sight. In particular, having a hidden process that remains constant is detrimental to the good performance of particle algorithms, and one needs some way to make Θ "move" from time to time. We will return to this point in Chap. 17.

3.3 Simulating from a Given Distribution Through Tempering

In the previous Section, we suggested that SMC may be used to approximate, for a given Bayesian model, the sequence of posterior distributions of parameter Θ given data $y_{0:t}$, for $t = 0, 1, \ldots, T$. One may be interested only in the "final" posterior distribution (given the full dataset). Still, one may use SMC to move progressively from an 'easy' distribution (the prior) to a 'difficult' distribution (the posterior), through intermediate steps. By easy and difficult, we refer either to the required CPU effort, or the actual difficulty to come up with a sampling algorithm. In particular, this SMC approach makes it possible to deal with very large datasets, for which simulating directly from the full posterior becomes too expensive.

There are other useful ways to define such artificial sequences of probability distributions, that start at $\nu(\mathrm{d}\theta)$ and end at $\pi(\mathrm{d}\theta)$, where π is the target distribution. One such way is known as *tempering*. Assume that

$$\pi(\mathrm{d}\theta) = \frac{1}{L} \exp\{-V(\theta)\}\, \nu(\mathrm{d}\theta)$$

where V may be computed point-wise, but the normalising constant $L > 0$ may be unknown. (This is equivalent to assuming that $\nu \gg \pi$, and that the density $\pi(\mathrm{d}\theta)/\nu(\mathrm{d}\theta)$ may be computed up to a multiplicative constant.) Then define

$$\mathbb{P}_\lambda(\mathrm{d}\theta) = \frac{1}{L_\lambda} \exp\{-\lambda V(\theta)\}\nu(\mathrm{d}\theta), \quad \lambda \in [0, 1], \tag{3.1}$$

where L_λ is the normalising constant, $L_\lambda = \int_\Theta \exp\{-\lambda V(\theta)\}\nu(\mathrm{d}\theta)$. This sequence that smoothly interpolates between ν and π is called a geometric bridge, or a tempering path. We may use SMC to sample sequentially from $\mathbb{P}_{\lambda_0}, \mathbb{P}_{\lambda_1}, \ldots, \mathbb{P}_{\lambda_T}$ for a certain discrete grid $0 = \lambda_0 < \lambda_1 < \ldots < \lambda_T = 1$, as we will explain in detail in Chap. 17. Again, in most cases, we are interested only in the final distribution $\mathbb{P}_1 = \pi$.

A typical application of such an approach is the simulation of a distribution π that has well separated modes. Since simulating directly from such a distribution is difficult, one may use SMC instead to sample initially from some distribution ν, which covers the support of π, and move progressively towards π, through intermediate distributions $\mathbb{P}_{\lambda_t}$ that are progressively more and more multimodal. Figure 3.1 illustrates this idea in a toy example.

Tempering is related to the thermodynamic integration identity:

$$\log \frac{L_1}{L_0} = -\int_0^1 \mathbb{P}_\lambda(V)\, \mathrm{d}\lambda \tag{3.2}$$

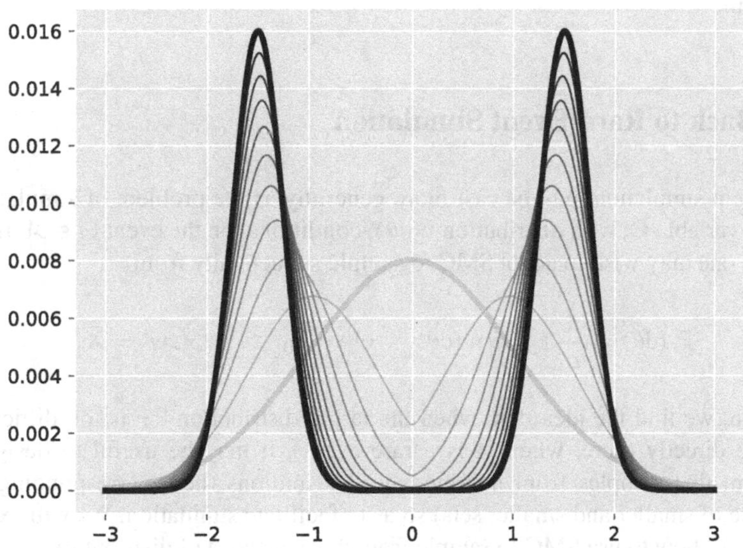

Fig. 3.1 Tempering sequence when $\nu(\mathrm{d}\theta)$ is $\mathcal{N}(0, 1)$ and $\pi(\mathrm{d}\theta)$ is a Gaussian mixture distribution. Each line corresponds to the density of $\mathbb{P}_\lambda$, see (3.1), for $\lambda = 0, 0.1, \ldots, 1$ (the darker the line, the larger λ)

where $\mathbb{P}_\lambda(V)$ denotes the expected value of $V(\Theta)$ under $\mathbb{P}_\lambda$, $\mathbb{P}_\lambda(V) := \int V(\theta)\mathbb{P}_\lambda(d\theta)$ (see Sect. 4.2 for a systematic treatment of these notations). The proof of the identity is left as Exercise 3.1. This identity may be used to approximate L_1: replace the integral above by a Riemann sum (again based on a discrete grid $0 = \lambda_0 < \ldots < \lambda_T = 1$)

$$\log \frac{L_1}{L_0} \approx -\sum_{t=1}^{T} (\lambda_t - \lambda_{t-1})\mathbb{P}_{\lambda_t}(V)$$

then, to compute each term of the above sum, use the same type of tempering SMC algorithm to recursively sample from $\mathbb{P}_{\lambda_t}$.

3.4 From Tempering to Optimisation

Tempering also provides a heuristic for non-convex optimisation. Say we want to maximise a certain function V, which may have several modes (maxima). Consider the same type of tempering sequence as in the previous section, except now $\lambda \in [0, \infty)$, instead of $[0, 1]$. Then, as $\lambda \to +\infty$, $\mathbb{P}_\lambda(d\theta)$ progressively concentrates around the global maxima of V. The corresponding algorithm will simulate variables from some initial distribution $\nu(d\theta)$ (the choice of which is arbitrary), and then evolve these variables towards the modal regions of function V. In practice, some rule of thumb is required to determine when to stop such an algorithm.

3.5 Back to Rare-Event Simulation

Rare-event simulation may be cast more generally as the problem of simulating a random variable Θ, with distribution $\nu(d\theta)$, conditional on the event $\Theta \in \mathcal{A}$. In that context, one may wish to adapt SMC to sample sequentially from

$$\mathbb{P}_t(d\theta) = \frac{1}{L_t}\mathbb{1}_{\mathcal{A}_t}(\theta)\nu(d\theta), \quad \Theta = \mathcal{A}_0 \supset \ldots \supset \mathcal{A}_T = \mathcal{A}.$$

Again, we find the idea that, when the target distribution $\mathbb{P}_T$ is too difficult to simulate directly (here, when $\mathcal{A}$ is a rare event), it may be useful to design an algorithm that samples from intermediate distributions (here corresponding to a sequence of smaller and smaller sets) so as to facilitate simulation. We will explain in Chap. 17 how to use SMC to sample from this sequence of distributions.

3.6 Likelihood-Inference, ABC Algorithms

Likelihood-free inference refers to the problem of performing inference on models defined through a black-box simulator; that is, one is able to sample data Y for any given parameter θ, but any other operation (including the calculation of the likelihood) is intractable. There has been a lot of interest in this topic in recent years, in particular in population genetics, phylogenics, and epidemiology. In population genetics for instance, one is interested in recovering the history of a population given genetic data. This history is represented by a genealogical tree, which is easy to sample given parameters such as mutation and growth rates, but the likelihood of the corresponding model is intractable.

A popular approach to likelihood inference is to implement an ABC (Approximate Bayesian Computation) algorithm, which samples in some way from distribution:

$$\pi_\epsilon(\mathrm{d}\theta, \mathrm{d}y) \propto \nu(\mathrm{d}\theta)\mathbb{M}^\theta(\mathrm{d}y)\mathbb{1}\{d(y, y^\star) \leq \epsilon\} \tag{3.3}$$

where $\nu(\mathrm{d}\theta)$ is the prior distribution, $\mathbb{M}^\theta(dy)$ represents the simulator (for a given parameter value θ), $y^\star$ is the actual data, and $d(y, y^\star)$ is a certain distance or pseudo-distance.

ABC is close in spirit to rare-event simulation. When ϵ is too small, sampling directly from (3.3) becomes too difficult. One may instead implement an SMC sampler that tracks a sequence of ABC distributions that corresponds to a decreasing sequence of ϵ's: $\mathbb{P}_t(\mathrm{d}\theta, \mathrm{d}y) = \pi_{\epsilon_t}(\mathrm{d}\theta, \mathrm{d}y)$, with $\epsilon_0 > \epsilon_1 > \ldots$. Again, see Chap. 17 for how to implement an SMC-ABC sampler in practice.

3.7 Probabilistic Graphical Models

Finally, as yet another illustration of the idea of applying SMC to a sequence of intermediate distributions, consider probabilistic graphical models. These are models where the dependencies between the considered variables are represented by a certain graph, such as Fig 3.2. The joint distribution of the variables (denoted by $X_1, \ldots, X_n$) admits a density that may be factorised as:

$$\pi(x_{1:n}) = \frac{1}{L}\prod_{c \in \mathcal{C}} \psi_c(x_c) \tag{3.4}$$

where $L > 0$ is a (typically unknown) normalising constant, $\mathcal{C}$ is the set of cliques (maximal subsets of nodes such that each pair of nodes is connected) and x_c means the set of variables such that the corresponding nodes are in clique $c \in \mathcal{C}$.

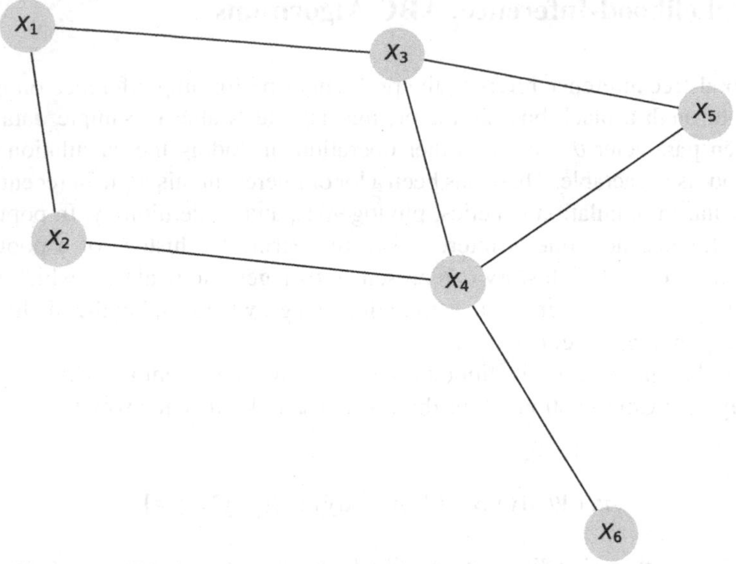

Fig. 3.2 Probabilistic (undirected) graphical model with six nodes; in this case the only clique with more than two nodes is (X_3, X_4, X_5)

One sequence of distribution that has been considered in the literature is:

$$\pi_t(x_{\mathcal{D}_t}) = \frac{1}{L_t} \prod_{c \in \mathcal{C}_t} \psi_c(x_c), \quad \mathcal{D}_t = \cup_{c \in \mathcal{C}_t} c,$$

where $\mathcal{C}_0 \subset \mathcal{C}_1 \ldots \subset \mathcal{C}_T = \mathcal{C}$ is an increasing sequence of subsets of $\mathcal{C}$. An SMC sampler applied to such a sequence makes it possible in particular to estimate L, the normalising constant of (3.4), which may be used in order to estimate hyper-parameters or to perform model choice.

Exercises

3.1 *Thermodynamic integration, path sampling and a strategy for tempering:*

1. *Show that the function* $\lambda \to \log L_\lambda$ *is convex.*
2. *Prove identity* (3.2).

Bibliographical Notes

The idea of adapting SMC to, respectively, tempering, and sequential Bayesian inference, appeared initially as annealed importance sampling in Neal (2001) and as IBIS in Chopin (2002). A more general perspective on non-sequential applications of SMC, with a discussion on geometric bridges and other sequences of target distributions, may be found in Del Moral et al. (2006). Good reviews on rare-event simulation and likelihood-free inference are respectively Kroese et al. (2011, Chap. 10) and Marin et al. (2012).

The thermodynamic integration identity is also called the path sampling identity, following Gelman and Meng (1998) who derived it independently it to compute marginal likelihoods in Statistics; see also Ogata (1989). Zhou et al. (2016) discuss its use in the SMC context.

SMC algorithms tailored to optimisation tasks appear in particular in Johansen et al. (2008) and Schäfer (2013). These algorithms bear close resemblance to certain members of the family of optimisation algorithms known as "genetic algorithms" (see e.g. the book of Goldberg 1989). In return, the SMC framework (and its asymptotic theory) seems to offer an interesting insight into the properties of these heuristic algorithms.

SMC algorithms for probabilistic graphical models appeared in particular in Hamze and de Freitas (2006), Carbonetto and Freitas (2007) and Naesseth et al. (2014).

Bibliography

Carbonetto, P., & Freitas, N. D. (2007). Conditional mean field. In B. Schölkopf, J. C. Platt, & T. Hoffman (Eds.), *Advances in neural information processing systems 19* (pp. 201–208). Cambridge, MA: MIT Press.

Chopin, N. (2002). A sequential particle filter method for static models. *Biometrika, 89*(3), 539–551.

Del Moral, P., Doucet, A., & Jasra, A. (2006). Sequential Monte Carlo samplers. *Journal of the Royal Statistical Society: Series B (Statistical Methodology), 68*(3), 411–436.

Gelman, A., & Meng, X.-L. (1998). Simulating normalizing constants: From importance sampling to bridge sampling to path sampling. *Statistical Science, 13*(2), 163–185.

Goldberg, D. E. (1989). *Genetic algorithms in search, optimization and machine learning* (1st ed.). Boston, MA: Addison-Wesley.

Hamze, F., & de Freitas, N. (2006). Hot coupling: A particle approach to inference and normalization on pairwise undirected graphs. In Y. Weiss, B. Schölkopf, & J. C. Platt (Eds.), *Advances in neural information processing systems 18* (pp. 491–498). Cambridge, MA: MIT Press.

Johansen, A. M., Doucet, A., & Davy, M. (2008). Particle methods for maximum likelihood estimation in latent variable models. *Statistics and Computing, 18*(1), 47–57.

Kroese, D. P., Taimre, T., & Botev, Z. I. (2011). *Handbook of Monte Carlo methods*. Hoboken, NJ: Wiley.

Marin, J.-M., Pudlo, P., Robert, C. P., & Ryder, R. J. (2012). Approximate Bayesian computational methods. *Statistics and Computing, 22*(6), 1167–1180.

Naesseth, C. A., Lindsten, F., & Schön, T. B. (2014). Sequential Monte Carlo for graphical models. In Z. Ghahramani, M. Welling, C. Cortes, N. D. Lawrence, & K. Q. Weinberger (Eds.), *Advances in neural information processing systems 27* (pp. 1862–1870). Red Hook, NY: Curran Associates, Inc.

Neal, R. M. (2001). Annealed importance sampling. *Statistics and Computing, 11*(2), 125–139.

Ogata, Y. (1989). A Monte Carlo method for high-dimensional integration. *Numerische Mathematik, 55*(2), 137–157.

Schäfer, C. (2013). Particle algorithms for optimization on binary spaces. *ACM Transactions on Modeling and Computer Simulation, 23*(1), Art. 8, 25.

Zhou, Y., Johansen, A. M., & Aston, J. A. D. (2016). Toward automatic model comparison: An adaptive sequential Monte Carlo approach. *Journal of Computational and Graphical Statistics, 25*(3), 701–726.

Chapter 4
Introduction to Markov Processes

Summary We introduce Markov processes using probability kernels. This allows us to define state-space models with wildly different state-spaces and dynamics in a common framework. We study two basic sets of properties of the probability distributions of such processes: the evolution of marginal distributions via recursions, and the structure of conditional distributions. In terms of the latter, we discuss the notion of conditional independence; when the Markov process consists of two components, $\{(X_t, Y_t)\}$, we study the distribution of $\{X_t\}$ conditional on $\{Y_t\}$; we call the process whose distribution is this conditional distribution a partially observed Markov process. We show that state-space models are instances of this framework.

4.1 Probability Kernels

We have already seen that interesting models might involve both continuous and discrete variables in the state process or the observations, or even random variables that take values in more general spaces, such as continuous-time stochastic process or point processes. This remark also applies to particle filters. These different settings cannot be treated neatly by the elementary definition of state-space models in terms of "densities", as in Sect. 2.1. More flexible tools are required.

Recall that the pair of a space of outcomes $\mathcal{X}$, and a σ-algebra $\mathcal{B}(\mathcal{X})$ that defines measurable subsets of $\mathcal{X}$, is called a measurable space.

Notation/Terminology $\mathcal{B}(\mathcal{X})$ will denote a σ-algebra (typically a Borel σ-algebra) of a set $\mathcal{X}$.

© Springer Nature Switzerland AG 2020
N. Chopin, O. Papaspiliopoulos, *An Introduction to Sequential Monte Carlo*,
Springer Series in Statistics, https://doi.org/10.1007/978-3-030-47845-2_4

Consider now two measurable spaces $(\mathcal{X}, \mathcal{B}(\mathcal{X}))$ and $(\mathcal{Y}, \mathcal{B}(\mathcal{Y}))$. These could be those involved in the definition of the state-space model, but here we think about them in abstract terms; indeed, $\mathcal{Y}$ might be equal to $\mathcal{X}$ in the following definition.

Definition 4.1 A probability kernel from $(\mathcal{X}, \mathcal{B}(\mathcal{X}))$ to $(\mathcal{Y}, \mathcal{B}(\mathcal{Y}))$, $P(x, \mathrm{d}y)$, is a function from $(\mathcal{X}, \mathcal{B}(\mathcal{Y}))$ to $[0, 1]$ such that

(a) for every x, $P(x, \cdot)$ is a probability measure on $(\mathcal{Y}, \mathcal{B}(\mathcal{Y}))$,
(b) for every $A \in \mathcal{B}(\mathcal{Y})$, $x \to P(x, A)$ is a measurable function in $\mathcal{X}$.

Probability kernels provide a nice mathematical tool to define conditional distributions. To see this, note that if $\mathbb{P}_0(\mathrm{d}x_0)$ is a probability measure on $(\mathcal{X}, \mathcal{B}(\mathcal{X}))$ and $P_1(x_0, \mathrm{d}x_1)$ is a probability kernel from $(\mathcal{X}, \mathcal{B}(\mathcal{X}))$ to $(\mathcal{X}, \mathcal{B}(\mathcal{X}))$, then

$$\mathbb{P}_1(\mathrm{d}x_{0:1}) = \mathbb{P}_0(\mathrm{d}x_0)P_1(x_0, \mathrm{d}x_1) \tag{4.1}$$

is a probability measure on the product space $(\mathcal{X}^2, \mathcal{B}(\mathcal{X})^2)$. The above notation is standard and what it really means is that, for any $A \in \mathcal{B}(\mathcal{X})^2$,

$$\mathbb{P}_1(A) = \int_A \mathbb{P}_0(\mathrm{d}x_0)P_1(x_0, \mathrm{d}x_1) = \int_{\mathcal{X}^2} \mathbb{1}[x_{0:1} \in A]\mathbb{P}_0(\mathrm{d}x_0)P_1(x_0, \mathrm{d}x_1)$$

where $\mathbb{1}[x_{0:1} \in A]$ is the indicator function that equals 1 if $x_{0:1} \in A$ and 0 otherwise. Note that, by construction, $\mathbb{P}_1(\mathrm{d}x_0) = \mathbb{P}_0(\mathrm{d}x_0)$.

We say that a vector of random variables $X_{0:1}$ is distributed according to $\mathbb{P}_1$ if

$$\mathbb{P}_1(X_0 \in A_0, X_1 \in A_1) = \int_{A_0 \times A_1} \mathbb{P}_0(\mathrm{d}x_0)P_1(x_0, \mathrm{d}x_1)$$

for any $A_0, A_1 \in \mathcal{B}(\mathcal{X})$. We will also be writing $\mathbb{P}_1(X_0 \in \mathrm{d}x_0, X_1 \in \mathrm{d}x_1)$, to refer to the distribution of $X_{0:1}$, so that we avoid using arbitrary subsets A_0, A_1, as above. If $X_{0:1}$ is distributed according to $\mathbb{P}_1$, then it turns out that the conditional distribution of X_1 given $X_0 = x$, is precisely $P_1(x, \cdot)$:

$$\mathbb{P}_1(X_1 \in \mathrm{d}x_1 | X_0 = x_0) = P_1(x_0, \mathrm{d}x_1).$$

(Formal proof of this claim requires a careful measure-theoretic argument that we skip here.) The construction also implies that if $\varphi : \mathcal{X} \to \mathbb{R}$ is bounded and measurable, then

$$\mathbb{E}_{\mathbb{P}_1}[\varphi(X_1)|X_0 = x_0] = \int_{\mathcal{X}} \varphi(x_1)P_1(x_0, \mathrm{d}x_1).$$

Notation/Terminology We will index the expectation operator $\mathbb{E}$ by the probability measure that is used to compute the corresponding probabilities, $\mathbb{P}_1$ in the above expression, i.e. $\mathbb{E}_{\mathbb{P}_1}$. This formalism will become really important in the following section when different probability measures will be considered for the same set of variables, using the notion of change of measure.

4.2 Change of Measure and a Basic Lemma

We recall another useful tool from Probability, which turns out to be central in this book.

Definition 4.2 Let $(\mathcal{X}, \mathcal{B}(\mathcal{X}))$ be a measurable space, and $\mathbb{M}$ and $\mathbb{Q}$ two probability measures defined on this space. We then say that $\mathbb{Q}$ is absolutely continuous with respect to $\mathbb{M}$, if for any $A \in \mathcal{B}(\mathcal{X})$ for which $\mathbb{M}(A) = 0$, $\mathbb{Q}(A) = 0$. In this case, we also say that $\mathbb{M}$ dominates $\mathbb{Q}$, and we write $\mathbb{M} \gg \mathbb{Q}$, or $\mathbb{Q} \ll \mathbb{M}$.

One may think of $\mathbb{M}$ and $\mathbb{Q}$ as two alternative models for a given phenomenon. An implication of the definition is that $\mathbb{M}(A) = 1$ implies that $\mathbb{Q}(A) = 1$, by applying the property to $\mathcal{X} - A$. Therefore, when an event is impossible or certain under $\mathbb{M}$ then it is so under $\mathbb{Q}$. This definition treats the two measures asymmetrically, since it might be possible that $\mathbb{Q}(A) = 0$ whereas $\mathbb{M}(A) > 0$. When given two such measures, $\mathbb{Q}(A)/\mathbb{M}(A)$ can be well-defined for all $A \in \mathcal{B}(\mathcal{X})$, since there will not be the case of dividing a positive number by 0. The Radon-Nikodym theorem provides a powerful extension of this property by stating that for two such measures there exists a measurable function $w(x) \geq 0$ such that

$$w(x) = \frac{\mathbb{Q}(\mathrm{d}x)}{\mathbb{M}(\mathrm{d}x)},$$

a notation that really means the following integral equation:

$$\mathbb{Q}(A) = \int_A w(x)\mathbb{M}(\mathrm{d}x) = \int_{\mathcal{X}} \mathbb{1}[x \in A]w(x)\mathbb{M}(\mathrm{d}x), \quad \forall A \in \mathcal{B}(\mathcal{X}). \tag{4.2}$$

The function w is known as the Radon-Nikodym derivative or as the likelihood ratio. The reverse to the Radon-Nikodym theorem is elementary: if $\mathbb{Q}$ and $\mathbb{M}$ are related as in (4.2) then $\mathbb{Q}$ is absolutely continuous with respect to $\mathbb{M}$; this follows by basic properties of expectation since if $\mathbb{M}(A) = 0$ then $\mathbb{1}[x \in A]w(x)$ will integrate to 0. A restatement of (4.2) is that $\int_{\mathcal{X}} \mathbb{1}[x \in A]\mathbb{Q}(\mathrm{d}x) = \int_{\mathcal{X}} \mathbb{1}[x \in A]w(x)\mathbb{M}(\mathrm{d}x)$ for all

$A \in \mathcal{B}(\mathcal{X})$. It should not come as a surprise that more generally,

$$\int_{\mathcal{X}} \varphi(x)\mathbb{Q}(\mathrm{d}x) = \int_{\mathcal{X}} \varphi(x)w(x)\mathbb{M}(\mathrm{d}x)\,, \tag{4.3}$$

for all bounded measurable functions φ.

Notation/Terminology $\mathbb{E}_{\mathbb{Q}}[\varphi(X)]$ is used to denote the expected value of a test function φ applied on a random variable X distributed according to $\mathbb{Q}$. However, we will also write $\mathbb{Q}(\varphi)$ for the same thing, i.e.,

$$\mathbb{Q}(\varphi) := \int_{\mathcal{X}} \varphi(x)\mathbb{Q}(\mathrm{d}x)\,,$$

which is a standard notation and avoids referring specifically to an underlying random variable.

The following simple result will prove helpful throughout the book.

Lemma 4.3 *Suppose that $\mathbb{Q}$ and $\mathbb{M}$ are probability measures on a space $\mathcal{X}$, and $w(x) \propto \mathbb{Q}(\mathrm{d}x)/\mathbb{M}(\mathrm{d}x)$. Then, for any test function φ,*

$$\mathbb{M}(\varphi w) = \mathbb{Q}(\varphi)\mathbb{M}(w)\,.$$

Proof This follows simply by noting that $\mathbb{Q}(\mathrm{d}x) = \{w(x)/\mathbb{M}(w)\}\,\mathbb{M}(\mathrm{d}x)$. Multiplying both sides by $\varphi(x)$, integrating and re-arranging yields the result. □

4.3 Backward Kernels

Section 4.1 decomposed the joint distribution $\mathbb{P}(\mathrm{d}x_{0:1})$ into the marginal at time 0 and the conditional given in terms of the kernel. However, we can decompose the distribution in a "backwards" manner instead:

$$\mathbb{P}_1(\mathrm{d}x_0)P_1(x_0, \mathrm{d}x_1) = \mathbb{P}_1(\mathrm{d}x_1)\overleftarrow{P}_0(x_1, \mathrm{d}x_0)\,,$$

where the equality is understood as one of two joint distributions, and $\overleftarrow{P}_0$ will be called the backward kernel. If $P_1(x_0, \mathrm{d}x_1)$ is dominated by $\mathbb{P}_1(\mathrm{d}x_1)$ for all values of x_0, we can rearrange and express the backward kernel as a change of measure from $\mathbb{P}_1(\mathrm{d}x_0)$,

$$\overleftarrow{P}_0(x_1, \mathrm{d}x_0) = \frac{P_1(x_0, \mathrm{d}x_1)}{\mathbb{P}_1(\mathrm{d}x_1)}\mathbb{P}_1(\mathrm{d}x_0)\,, \tag{4.4}$$

where the fraction denotes the density between the forward kernel and $\mathbb{P}_1$. The expression can also be viewed as an application of Bayes theorem. The assumed domination holds when $P_1(x_0, \mathrm{d}x_1) = p_1(x_1|x_0)\nu(\mathrm{d}x_1)$, in which case the backward kernel can be rewritten as

$$\overleftarrow{P}_0(x_1, \mathrm{d}x_0) \propto p_1(x_1|x_0)\mathbb{P}_1(\mathrm{d}x_0)\,, \tag{4.5}$$

where the normalising constant is $\int_{\mathcal{X}} p_1(x_1|x_0)\mathbb{P}_1(\mathrm{d}x_0)$, and $\propto$ means that the probability kernel on the left is equal up to a normalising constant (which can depend on x_1) to that on the right. However, there are non-contrived situations for which $P_1(x_0, \mathrm{d}x_1)$ is not dominated by $\mathbb{P}_1(\mathrm{d}x_1)$; consider the trivial kernel $P_1(x_0, \mathrm{d}x_1) = \delta_{x_0}(\mathrm{d}x_1)$. When $\mathbb{P}_0(\mathrm{d}x_0)$ is such that $\mathbb{P}_0(\{x\}) = 0$ for all $x \in \mathcal{X}$ the kernel is not dominated the $\mathbb{P}_1(\mathrm{d}x_1)$; see the bearings-only example of Sect. 2.4.1 for models with this type of dynamics.

Notation/Terminology Here and throughout the book $\delta_x(\mathrm{d}y)$ defined on a measure space $(\mathcal{X}, \mathcal{B}(\mathcal{X}))$, with $\mathcal{B}(\mathcal{X})$ a Borel σ-algebra, is the so-called Dirac delta measure, which is a probability measure that assigns probability 1 to the set $\{x\}$.

4.4 Markov Processes and Recursions

Consider now a sequence of such probability kernels from $(\mathcal{X}, \mathcal{B}(\mathcal{X}))$ to $(\mathcal{X}, \mathcal{B}(\mathcal{X}))$, $P_1, P_2, \ldots, P_T$, for some $T \in \mathbb{Z}^+$ and a probability measure $\mathbb{P}_0(\mathrm{d}x)$ on $(\mathcal{X}, \mathcal{B}(\mathcal{X}))$.

Definition 4.4 A sequence of random variables $X_{0:T}$ with joint distribution given by

$$\mathbb{P}_T(X_{0:T} \in \mathrm{d}x_{0:T}) = \mathbb{P}_0(\mathrm{d}x_0)\prod_{s=1}^{T} P_s(x_{s-1}, \mathrm{d}x_s)\,, \tag{4.6}$$

is called a (discrete-time) Markov process with state-space $\mathcal{X}$, initial distribution $\mathbb{P}_0$ and transition kernel at time t, P_t. Likewise, a probability measure decomposed into a product of an initial distribution and transition kernels as in (4.6) will be called a Markov measure.

In the same spirit as the result mentioned earlier for $T = 1$, it can be shown that (4.6) implies that for all $t \leq T$,

$$\mathbb{P}_T(X_t \in \mathrm{d}x_t|X_{0:t-1} = x_{0:t-1}) = \mathbb{P}_T(X_t \in \mathrm{d}x_t|X_{t-1} = x_{t-1}) = P_t(x_{t-1}, \mathrm{d}x_t)\,. \tag{4.7}$$

This equation contains two important pieces of information. First, that under $\mathbb{P}_T$, X_t is conditionally independent from $X_{0:t-2}$ given X_{t-1}. Conditional independence refers to the fact that $\mathbb{P}_T(X_t \in \mathrm{d}x_t | X_{0:t-1}) = \mathbb{P}_T(X_t \in \mathrm{d}x_t | X_{t-1})$, i.e., that $X_{0:t-2}$ drops from the conditioning set once we condition on X_{t-1}. In other words, the "past", $X_{0:t-2}$, is irrelevant in computing probabilities for the future value of the process, X_t, given the information about its present value, X_{t-1}. Second, it identifies the conditional distribution with the corresponding transition kernel P_t. Actually, we have more generally that

$$\mathbb{P}_T(X_t \in \mathrm{d}x_t | X_{0:s} = x_{0:s}) = P_{s+1:t}(x_s, \mathrm{d}x_t), \quad \forall t \leq T, s < t, \tag{4.8}$$

where $P_{s+1:t}$ denotes the convolution of the kernels $P_{s+1}, \ldots, P_t$,

$$P_{s+1:t}(x_s, A) = \int_{\mathcal{X}^{t-s-1}} P_{s+1}(x_s, \mathrm{d}x_{s+1}) P_{s+2}(x_{s+1}, \mathrm{d}x_{s+2}) \cdots P_t(x_{t-1}, A).$$

Again, this implies that X_t is conditionally independent from $X_{0:s-1}$ given X_s for any $s < t$. As with $T = 1$, (4.8) implies that for every bounded measurable φ,

$$\mathbb{E}_{\mathbb{P}_T}[\varphi(X_t) | X_{0:s} = x_{0:s}] = \int_{\mathcal{X}} \varphi(x_t) P_{s+1:t}(x_s, \mathrm{d}x_t).$$

We can define a sequence of probability measures $\mathbb{P}_t$, for $t \leq T$, by replacing T with t in (4.6). The following is a simple but important property.

Proposition 4.5 *Consider a sequence of probability measures, indexed by t, defined as:*

$$\mathbb{P}_t(X_{0:t} \in \mathrm{d}x_{0:t}) = \mathbb{P}_0(\mathrm{d}x_0) \prod_{s=1}^{t} P_s(x_{s-1}, \mathrm{d}x_s),$$

where P_s are probability kernels. Then, for any $t \leq T$,

$$\mathbb{P}_T(\mathrm{d}x_{0:t}) = \mathbb{P}_t(\mathrm{d}x_{0:t}).$$

This means that when we compute the marginal distribution of $X_{0:t}$ under $\mathbb{P}_T$, by averaging over $X_{t+1:T}$, we recover the $\mathbb{P}_t$ distribution. This property follows directly from (4.6) by integrating out $x_{t+1:T}$, and the definition of $\mathbb{P}_t$. It implies that $\mathbb{E}_{\mathbb{P}_t}[\varphi(X_{0:t})] = \mathbb{E}_{\mathbb{P}_T}[\varphi(X_{0:t})]$, for any function $\varphi : \mathcal{X}^{t+1} \to \mathbb{R}$ and any $t \leq T$.

Using iterated conditional expectation, the Markov property and the definition of the convoluted kernels in (4.8) we can obtain a useful expression for the marginal distributions of a Markov process:

$$\mathbb{P}_t(X_t \in \mathrm{d}x_t) = \mathbb{E}_{\mathbb{P}_t}[\mathbb{P}_t(X_t \in \mathrm{d}x_t | X_{0:s})] = \mathbb{E}_{\mathbb{P}_t}[P_{s+1:t}(X_s, \mathrm{d}x_t)]. \tag{4.9}$$

Given that the expectation only involves X_s, we can use Proposition 4.5 to yield what is known as the Chapman-Kolmogorov equation:

$$\mathbb{P}_t(X_t \in \mathrm{d}x_t) = \mathbb{E}_{\mathbb{P}_s}[P_{s+1:t}(X_s, \mathrm{d}x_t)], \quad \forall s \leq t-1 . \tag{4.10}$$

When the kernels $P_s(x_{s-1}, \mathrm{d}x_s)$ have densities $p_s(x_{s-1}, x_s)$ with respect to some measure $\nu(\mathrm{d}x)$, then $\mathbb{P}_t(\mathrm{d}x_t)$ is also dominated by ν with densities that satisfy the recursion

$$p_t(x_t) = \int_{\mathcal{X}} p_{t-1}(x_{t-1}) p_t(x_{t-1}, x_t) .$$

Fixing the final horizon to T, (4.6) provides one of the possible factorisations of the joint distribution of a Markov process. The following result provides its time-reversal.

Proposition 4.6 *We have:*

$$\mathbb{P}_T(X_{0:T} \in \mathrm{d}x_{0:T}) = \mathbb{P}_T(\mathrm{d}x_T) \prod_{t=1}^{T} \overleftarrow{P}_{T-t}(x_{T-t+1}, \mathrm{d}x_{T-t}) , \tag{4.11}$$

where $\overleftarrow{P}_{t-1}(x_t, \mathrm{d}x_{t-1})$ *is the so-called backward kernel at time* $t-1$, *defined via the equation of measures*

$$\mathbb{P}_T(\mathrm{d}x_t) \overleftarrow{P}_{t-1}(x_t, \mathrm{d}x_{t-1}) = \mathbb{P}_T(\mathrm{d}x_{t-1}) P_t(x_{t-1}, \mathrm{d}x_t) . \tag{4.12}$$

Proof Recall from Sect. 4.1 that $\mathbb{P}_0(\mathrm{d}x_0)P_1(x_0, \mathrm{d}x_1) = \mathbb{P}_1(\mathrm{d}x_1)\overleftarrow{P}(x_1, \mathrm{d}x_0)$. This property can be generalised to this multi-period framework to yield (4.12). Recognising that due to Proposition 4.5 $\mathbb{P}_0(\mathrm{d}x_0) = \mathbb{P}_T(\mathrm{d}x_0)$, we can replace the first two terms in (4.6) by $\mathbb{P}_T(\mathrm{d}x_1)\overleftarrow{P}_0(x_1, \mathrm{d}x_0)$, and then $\mathbb{P}_T(\mathrm{d}x_1)P_2(x_1, \mathrm{d}x_2)$ by $\mathbb{P}_T(\mathrm{d}x_2)\overleftarrow{P}_1(x_2, \mathrm{d}x_1)$, and so forth, to obtain the alternative backward decomposition of the Markov measure. □

Remark 4.1 Integrating both sides of (4.12) with respect to x_t we obtain the following backward recursion for the marginal distributions,

$$\mathbb{P}_T(\mathrm{d}x_{t-1}) = \int \mathbb{P}_T(\mathrm{d}x_t) \overleftarrow{P}_{t-1}(x_t, \mathrm{d}x_{t-1})$$

which is complementary to the forward recursion provided in (4.10). Using the marginalisation property in Proposition 4.5, we obtain

$$\mathbb{P}_t(\mathrm{d}x_t) \overleftarrow{P}_{t-1}(x_t, \mathrm{d}x_{t-1}) = \mathbb{P}_{t-1}(\mathrm{d}x_{t-1}) P_t(x_{t-1}, \mathrm{d}x_t) . \tag{4.13}$$

When $P_t(x_{t-1}, \mathrm{d}x_t)$ is dominated by $\mathbb{P}_t(\mathrm{d}x_t)$ for all x_{t-1} we can re-arrange to get

$$\overleftarrow{P}_{t-1}(x_t, \mathrm{d}x_{t-1}) = \frac{P_t(x_{t-1}, \mathrm{d}x_t)}{\mathbb{P}_t(\mathrm{d}x_t)} \mathbb{P}_{t-1}(\mathrm{d}x_{t-1})$$

where the first term is the corresponding density.

4.5 State-Space Models as Partially Observed Markov Processes

Consider now a stochastic process that involves two components, $\{(X_t, Y_t)\}$, where $X_t \in \mathcal{X}$, and $Y_t \in \mathcal{Y}$, and $\mathcal{X} \times \mathcal{Y}$ is equipped with the product σ-algebra $\mathcal{B}(\mathcal{X}) \times \mathcal{B}(\mathcal{Y})$. The distribution of the process is specified in terms of an initial distribution $\mathbb{P}_0(\mathrm{d}x_0)$, and two sequences of kernels, $P_t(x_{t-1}, \mathrm{d}x_t)$, for $t = 1, 2, \ldots$, where $P_t(x_{t-1}, \mathrm{d}x_t)$ is a probability kernel from $(\mathcal{X}, \mathcal{B}(\mathcal{X}))$ to $(\mathcal{X}, \mathcal{B}(\mathcal{X}))$, and $F_t(x_t, \mathrm{d}y_t)$ from $(\mathcal{X}, \mathcal{B}(\mathcal{X}))$ to $(\mathcal{Y}, \mathcal{B}(\mathcal{Y}))$. The joint distribution of $(X_{0:T}, Y_{0:T})$ is then given by

$$\begin{aligned}
\mathbb{P}_T(X_{0:T} \in \mathrm{d}x_{0:T}, Y_{0:T} \in \mathrm{d}y_{0:T}) &= \mathbb{P}_0(\mathrm{d}x_0) \prod_{t=1}^{T} P_t(x_{t-1}, \mathrm{d}x_t) \prod_{t=0}^{T} F_t(x_t, \mathrm{d}y_t) \\
&= \{\mathbb{P}_0(\mathrm{d}x_0) F_0(x_0, \mathrm{d}y_0)\} \prod_{t=1}^{T} \{P_t(x_{t-1}, \mathrm{d}x_t) F_t(x_t, \mathrm{d}y_t)\} \\
&= \mathbb{P}_T(\mathrm{d}x_{0:T}) \prod_{t=0}^{T} F_t(x_t, \mathrm{d}y_t) \,.
\end{aligned} \tag{4.14}$$

We have expressed the measure in different ways, each of which is helpful to understand some of its properties. A comparison of the second line above with (4.6) should immediately convince the reader that $\{(X_t, Y_t)\}$ is a Markov process with initial distribution $\mathbb{P}_0(\mathrm{d}x_0) F_0(x_0, \mathrm{d}y_0)$ and transition kernels $P_t(x_{t-1}, \mathrm{d}x_t) F_t(x_t, \mathrm{d}y_t)$. The last line in (4.14) recognises that $\{X_t\}$ is marginally distributed as a Markov process with transition kernels $P_t(x_{t-1}, \mathrm{d}x_t)$; a marginal distribution that we will denote by $\mathbb{P}_T(\mathrm{d}x_{0:T})$.

What we have specified in (4.14) is just a multivariate Markov process with more detailed information about how different components of the state variable interact with each other. The separation into X_t's and Y_t's is done because we intend to use this stochastic process as a time series model for data $\{Y_t\}$. Note that the state-space model described in Sect. 2.1 is a special case of the process defined, provided one has dominating measures $\mu(\mathrm{d}x)$ and $\nu(\mathrm{d}y)$ such that

$$\begin{aligned} \mathbb{P}_0(\mathrm{d}x_0) = p_0(x_0)\mu(\mathrm{d}x_0)\,, \; P_t(x_{t-1}, \mathrm{d}x_t) &= p_t(x_t|x_{t-1})\mu(\mathrm{d}x_t)\,, \\ F_t(x_t, \mathrm{d}y_t) &= f_t(y_t|x_t)\nu(\mathrm{d}y_t)\,. \end{aligned} \tag{4.15}$$

In Sect. 2.1 we loosely defined the state-space model in terms of "densities", here we are being more explicit about the dominating measures that these are defined with respect to. However, we have already seen that the assumption of a measure $\mu(\mathrm{d}x_t)$ that dominates the distributions $P_t(x_{t-1}, \mathrm{d}x_t)$ for all x_{t-1} is not met in several models of interest. Therefore, we will mostly work under the framework that does not make this assumption. On the other hand, a good balance between theoretical rigour and practical relevance is achieved by assuming that the observation densities do exist with respect to a suitable dominating measure, be it Lebesgue, counting or something else. Hence, throughout this book we will work under the assumption that

$$F_t(x_t, \mathrm{d}y_t) = f_t(y_t|x_t)\nu(\mathrm{d}y_t) \quad \forall t\,,$$

where $f_t(y_t|x_t)$ will be referred to as the observation or emission density.

Notation/Terminology We shall use the phrase "the components of the state-space model" to refer collectively to the quantities that define it, i.e., T, distribution $\mathbb{P}_0$, kernels $P_t(x_{t-1}, \mathrm{d}x_t)$ for $1 \le t \le T$, and densities $f_t(y_t|x_t)$ for $0 \le t \le T$.

In Chap. 2 we showed that the dynamics of the state-space model might depend on further parameters θ. In the broader framework of this section, we think of θ as a set of parameters that define the parametric form of $f_t(y_t|x_t)$, and as parameters involved in the construction of the kernels $P_t(x_{t-1}, \mathrm{d}x_t)$. When it is relevant (e.g. when we discuss the estimation of such parameters) we emphasise this dependence by writing $f_t^\theta(y_t|x_t)$ and $P_t^\theta(x_{t-1}, \mathrm{d}x_t)$ for the observation densities and the state transition kernels respectively.

Summarising the above, we rewrite the state-space model in the following way that will dominate the analysis in the rest of this book:

$$\mathbb{P}_T(X_{0:T} \in \mathrm{d}x_{0:T}, Y_{0:T} \in \mathrm{d}y_{0:T}) = \mathbb{P}_T(\mathrm{d}x_{0:T}) \prod_{t=0}^{T} f_t(y_t|x_t) \prod_{t=0}^{T} \nu(\mathrm{d}y_t)\,. \tag{4.16}$$

Additionally, a direct marginalisation of $X_{0:t}$ yields that the joint distribution of $Y_{0:t}$ has density with respect to the measure $\prod_{s=0}^{t} \nu(\mathrm{d}y_s)$:

$$\mathbb{P}_T(Y_{0:t} \in \mathrm{d}y_{0:t}) = \mathbb{E}_{\mathbb{P}_t}\left[\prod_{s=0}^{t} f_t(y_s|X_s)\right] \prod_{s=0}^{t} \nu(\mathrm{d}y_s) .$$

Notation/Terminology The marginal density of the observations will be denoted by $p_t(y_{0:t})$ or by $p_t^{\theta}(y_{0:t})$ when emphasising its dependence on parameters θ. According to the equation above, it is obtained as

$$p_T(y_{0:t}) = p_t(y_{0:t}) = \mathbb{E}_{\mathbb{P}_t}\left[\prod_{s=0}^{t} f_s(y_s|X_s)\right] . \tag{4.17}$$

This function is known as the likelihood, a name which makes more sense when we view $p_t^{\theta}(y_{0:t})$ as a function of θ for given observations $y_{0:t}$. Ratios of consecutive densities correspond to conditional densities, $p_t(y_t|y_{0:t-1})= p_t(y_{0:t})/p_{t-1}(y_{0:t-1})$; such ratios will be called likelihood factors, since

$$p_t(y_{0:t}) = p_0(y_0) \prod_{s=1}^{t} p_s(y_s|y_{0:s-1}) .$$

Likewise, we can define ratios over different horizons,

$$p_{t+k}(y_{t:t+k}|y_{0:t-1}) = p_{t+k}(y_{0:t+k})/p_{t-1}(y_{0:t-1}) , \quad k \geq 0 , \, t \geq 1 .$$

Note that the definition of the state-space model and (4.17), implies that

$$\mathbb{P}_t(X_{0:t} \in \mathrm{d}x_{0:t}|Y_{0:t} = y_{0:t}) = \frac{1}{p_t(y_{0:t})} \left\{\prod_{s=0}^{t} f_s(y_s|x_s)\right\} \mathbb{P}_t(\mathrm{d}x_{0:t}) . \tag{4.18}$$

To see this, notice that by multiplying both sides of (4.18) by $p_t(y_{0:t}) \prod_{s=0}^{t} \nu(\mathrm{d}y_s)$ we recover the joint measure of $(X_{0:t}, Y_{0:t})$ in (4.16) on both sides, decomposed in different ways. These conditional distributions of the state process given the observations for different times t, are a central theme in this book.

We will also be interested in conditional distributions of the states given past data:

$$\mathbb{P}_t(X_{0:t} \in \mathrm{d}x_{0:t}|Y_{0:k} = y_{0:k}) = \frac{1}{p_k(y_{0:k})} \left\{\prod_{s=0}^{k} f_s(y_s|x_s)\right\} \mathbb{P}_t(\mathrm{d}x_{0:t}) , \quad k < t ,$$

Statistical analysis with such a model involves dealing with the tasks outlined in Sect. 2.1, and which we recast in this new notation:

- **filtering**: deriving $\mathbb{P}_t(X_t \in dx_t | Y_{0:t} = y_{0:t})$;
- **state prediction**: deriving $\mathbb{P}_t(X_{t+1:t+h} \in dx_{t+1:t+h} | Y_{0:t} = y_{0:t})$, for $h \geq 1$;
- **fixed-lag smoothing**: deriving $\mathbb{P}_t(X_{t-l:t} \in dx_{t-l:t} | Y_{0:t} = y_{0:t})$ for some $l \geq 1$;
- **(complete) smoothing**: deriving $\mathbb{P}_t(X_{0:t} \in dx_{0:t} | Y_{0:t} = y_{0:t})$;
- **likelihood computation**: deriving $p_t(y_{0:t})$.

The solution to these problems requires understanding of the properties of the hidden process conditionally on the observations, when they jointly evolve according to (4.14). We will deal with such questions in Sect. 5.1, where we will develop a theoretical framework—the Feynman-Kac formalisation—for their study, and we will be able to provide some explicit answers.

However, a different kind of tool can also be used to provide valuable intuition for some of these questions. This is the theory of graphical models, elements of which are discussed next.

4.6 The Markov Property: Graphs

The model whose distribution is described in (4.14) can also be represented graphically, as follows. Every variable in the model, that is X_0, Y_0, X_1 and so forth, is represented as a node in a graph. Whenever two variables are related by a kernel in (4.14), an edge is drawn to connect them; hence, there will be an edge between X_1 and X_2, or X_1 and Y_1, or X_1 and X_0, but not between X_1 and Y_2. This procedure yields the graph in Figure 4.1.

Note that this representation is a skeleton of the model: it retains the information about how the variables interact to yield the factorisation of the joint distribution in (4.14), but it discards the details about this interaction, i.e., the functional forms of the kernels. A graph of the type shown in Fig. 4.1 is known as an undirected graphical model. The qualification refers to the fact that the edges that link the variables are directionless.

Graphical models, and the theory developed to work with them, are perfectly suited to explore conditional independence structures in a probabilistic model. The result that will be extremely useful to us is the following. First, we define the notion of a path between two nodes as a consecutive sequence of edges in the graph. For example, a path from Y_0 to X_2 involves the edges between Y_0 and X_0, X_0 and

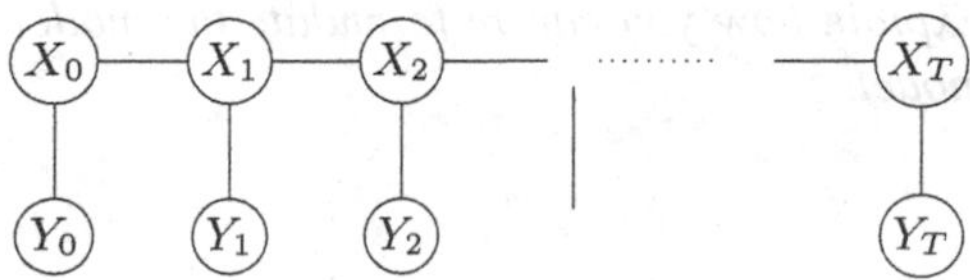

Fig. 4.1 Partially observed Markov process as an undirected graphical model

X_1 and X_1 and X_2. Consider now three sets of variables, A, B and C, e.g., $A = \{X_t, Y_t\}$, $B = \{X_{t-1}, Y_{t-1}\}$ and $C = \{X_{0:t-2}, Y_{0:t-2}\}$, for some $t > 1$ in Figure 4.1.

Notation/Terminology Throughout the book we use $\{\cdot\}$ as a shorthand for sequences, e.g. $\{X_t\}$ as a shorthand notation for the stochastic process $(X_t)_{t\geq 0}$. In this section alone, $\{\cdot\}$ is used to denote a set of variables, e.g. $\{X_t\}$ denotes the set that contains X_t only.

If every path from a node in A to a node in C passes through a node in B, then the variables in A are conditionally independent of the variables in C given those in B. In other words, according to this probabilistic model, the conditional distribution of variables in A given those in B and C, is the same as the conditional distribution of variables in A given those in B. One can check that this property holds for $A = \{X_t, Y_t\}$, $B = \{X_{t-1}, Y_{t-1}\}$ and $C = \{X_{0:t-2}, Y_{0:t-2}\}$, for any $t > 1$ in Figure 4.1, which is conditional independence property of Markov processes that we have already discussed. Actually, we can now have a more general definition of Markov processes, as those characterised by the conditional independence property that the "future" is independent of the "past" given the "present".

Graphical models can help us see structures that are not obvious in the definition in (4.14). For instance, examination of the graph shows that X_t itself is a Markov process, in the sense that X_t is conditionally independent from $X_{0:t-2}$ given X_{t-1} for each $t > 1$; we already noticed this property in the discussion following (4.14). Additionally, the process X_t conditionally on $Y_{0:T}$ is a Markov process; to see this apply the criterion given above for $A = \{X_t\}$, $B = \{X_{t-1}, Y_{0:T}\}$ and $C = \{X_{0:t-2}\}$ for any $t > 1$. On the other hand, the process Y_t is not necessarily Markov: note that there exists a path from $\{Y_t\}$ to $\{Y_{0:t-2}\}$ that does not pass through $\{Y_{t-1}\}$, but instead through X_t, X_{t-1} and X_{t-2}. We encourage the reader to return to this graph and this perspective several times while reading this book.

Exercises

4.1 *Compute kernel $P_{s+1:t}(x_s, \mathrm{d}x_t)$ when each kernel $P_s(x_{s-1}, \mathrm{d}x_s)$ corresponds to a $\mathcal{N}_1(\rho x_{s-1}, 1)$ distribution.*

4.2 *Consider two processes $\{X_t\}$ and $\{Y_t\}$, with the same properties as in a state-space model, except that Y_t conditional on $X_{0:t}$ depends on both X_t and X_{t-1}. Explain how you can re-formulate the model to cast it as a standard state-space model.*

Python Corner

We have seen that a state-space model is entirely defined by the following components: distribution $\mathbb{P}_0(\mathrm{d}x_0)$, kernels $P_t(x_{t-1}, \mathrm{d}x_t)$, and kernels $F_t(x_t, \mathrm{d}y_t)$.

In `particles`, state-space models may be defined as follows:

```
import numpy as np
from particles import state_space_models as ssm
from particles import distributions as dists

class StochVol(ssm.StateSpaceModel):
    default_params = {'mu': -1., 'rho':0.9, 'sigma':1.}

    def PX0(self):
        sig0 = self.sigma / np.sqrt(1. - self.rho**2)
        return dists.Normal(loc=self.mu, scale=sig0)

    def PX(self, t, xp):
        return dists.Normal(loc=self.mu + self.rho * (xp - self.mu),
                            scale=self.sigma)

    def PY(self, t, xp, x):
        return dists.Normal(scale=np.exp(0.5 * x))
```

One recognises the following stochastic volatility model:

$$X_0 \sim N\left(\mu, \frac{\sigma^2}{1-\rho^2}\right)$$

$$X_t|X_{t-1} = x_{t-1} \sim N(\mu + \rho(x_{t-1} - \mu), \sigma^2)$$

$$Y_t|X_t = x_t \sim N(0, e^{x_t})$$

The code above actually defines a *class*; that is, a generic way to define an object (referred to as `self` in the code) which contains 'attributes' (i.e. data) and 'methods' (functions that operate on these attributes). Here, the attributes of object `self` are parameter values (e.g. `self.mu` for parameter μ), and the methods are functions which define distributions. For example, method `PXt` returns the distribution of $X_t|X_{t-1}$ at time t, given X_{t-1} =xp.

To actually define a given model, one needs to *instantiate* (create an instance of) `StochVol`:

```
my_sv_model = StochVol(mu=0.8, rho=0.95, sigma=1.)
```

This gives a stochastic volatility model, with parameter values $\mu = 0.8$, $\rho = 0.95$, and $\sigma = 1$. From that, one may access its attributes directly:

```
my_sv_model.mu
> 0.8
```

or even its functions:

```
x0 = my_sv_model.PX0().rvs(size=30)  # generate 30 draws from PX0
```

More interestingly, we may now simulate data from our model:

```
x, y = my_sv_model.simulate(100)  # simulate (X_t, Y_t) from the model

from matplotlib import pyplot as plt
plt.plot(y)  # plot the simulated data
```

Where does the `simulate` method come from? It is *inherited* from parent class `StateSpaceModel`; the fact that `StochVol` descends from `StateSpaceModel` is specified in the first line of the definition of class `StochVol` which is itself defined in the `state_space_models` module of `particles`. This base class defines several generic methods, such as `simulate`.

People already familiar with Python classes may wonder why `StochVol` does not have an `__init__` method (in Python, the `__init__` method is always called when the corresponding class is instantiated). Again, the `__init__` method is simply inherited from `StateSpaceModel`. In fact, the `__init__` method of `StateSpaceModel` implements the following interesting behaviour. Try

```
my_2nd_sv_model = StochVol(mu=0.8, delta=3.)
my_2nd_sv_model.simulate(200)
```

What, no error? Missing parameter values (i.e. `rho` and `sigma`) are replaced by their default values, which are specified in the `default_params` dictionary. And extraneous parameters (here, `delta`) are simply ignored. This behaviour will be useful later, when we specify prior distributions for the parameters that may involve hyper-parameters.

The probability distributions available in module `distributions` are also defined as classes, with methods `rvs` (for random variable sampling), `logpdf` and so on. We follow the Python convention of naming classes using camel case (e.g. `StochVol`).

Bibliographical Notes

In Sect. 4.4 we developed probability kernels to work with Markov processes. Unavoidably, given the space we devoted, we have not gone into more detail than is necessary to proceed to the Feynman-Kac formalisation. For a complete mathematical treatment of conditional independence, see for example Chapter 5 of Kallenberg (1997). An excellent introduction to measure-theoretic probability is Williams (1991) and an accessible introduction to Markov chains is Grimmett and Stirzaker (1982).

Bibliography

Grimmett, G., & Stirzaker, D. (1982). *Probability and random processes. Oxford science publications*. New York: The Clarendon Press/Oxford University Press.

Kallenberg, O. (1997). *Foundations of modern probability. Probability and its applications (New York)*. New York: Springer.

Williams, D. (1991). *Probability with martingales. Cambridge mathematical textbooks*. Cambridge: Cambridge University Press.

Chapter 5
Feynman-Kac Models: Definition, Properties and Recursions

Summary Building on the concepts of the previous chapter we introduce in this chapter Feynman-Kac models. This concept plays a central role in this book, as all SMC algorithms may be viewed as a Monte Carlo approximation of some underlying Feynman-Kac model.

We start by giving the general definition of Feynman-Kac models, i.e., a sequence of distributions for variables $X_{0:t}$ obtained by applying a certain change of measure to the law of a Markov process. We highlight the connection between Feynman-Kac models and state-space models. We then work out various recursions that relate successive Feynman-Kac distributions. In particular, we cover the forward recursion, which underpins particle filtering (Chap. 10), and two types of forward-backward recursions, which underpin particle smoothing (Chap. 12).

5.1 Feynman-Kac Formalisation

5.1.1 Feynman-Kac Models

The starting point is a Markov probability law defined on a state space $\mathcal{X}$, with initial distribution $\mathbb{M}_0$ and transition kernels $M_{1:T}$:

$$\mathbb{M}_T(\mathrm{d}x_{0:T}) = \mathbb{M}_0(\mathrm{d}x_0) \prod_{t=1}^{T} M_t(x_{t-1}, \mathrm{d}x_t) \,.$$

Consider a sequence of so-called potential functions, $G_0 : \mathcal{X} \to \mathbb{R}^+$, and $G_t : \mathcal{X}^2 \to \mathbb{R}^+$, for $1 \leq t \leq T$. Then, a sequence, for $0 \leq t \leq T$, of Feynman-Kac models is given by probability measures on $(\mathcal{X}^{t+1}, \mathcal{B}(\mathcal{X})^{t+1})$, obtained as the following changes of measure from $\mathbb{M}_t$:

$$\mathbb{Q}_t(\mathrm{d}x_{0:t}) := \frac{1}{L_t} G_0(x_0) \left\{ \prod_{s=1}^{t} G_s(x_{s-1}, x_s) \right\} \mathbb{M}_t(\mathrm{d}x_{0:t}), \tag{5.1}$$

© Springer Nature Switzerland AG 2020
N. Chopin, O. Papaspiliopoulos, *An Introduction to Sequential Monte Carlo*, Springer Series in Statistics, https://doi.org/10.1007/978-3-030-47845-2_5

where L_t is the normalising constant needed for $\mathbb{Q}_t$ to be a probability measure,

$$L_t = \int_{\mathcal{X}^{t+1}} G_0(x_0) \prod_{s=1}^{t} G_s(x_{s-1}, x_s) \mathbb{M}_t(\mathrm{d}x_{0:t}) = \mathbb{E}_{\mathbb{M}_t}\left[G_0(X_0) \prod_{s=1}^{t} G_s(X_{s-1}, X_s) \right]. \tag{5.2}$$

Quantity L_t is also known by various names in different scientific fields, e.g. partition function and likelihood in Physics and Statistics respectively. Clearly, for this model to be well defined we require $0 < L_t < \infty$. Ratios of successive normalising constants will be denoted by $l_t := L_t/L_{t-1}$. Statistically, we might think of $\mathbb{Q}_t$ as an alternative model to $\mathbb{M}_t$ for the sequence of random variables $X_{0:t}$. The following terminology will be handy in the rest of the book.

Notation/Terminology The components of the Feynman-Kac model is a term that will be used to refer collectively to the kernels and functions that define it, i.e., T, $\mathbb{M}_0$, G_0 and $M_t(x_{t-1}, \mathrm{d}x_t)$, $G_t(x_{t-1}, x_t)$ for $1 \leq t \leq T$.

When $\mathbb{M}_0(\mathrm{d}x_0)$ and the Markov transition kernels $M_t(x_{t-1}, \mathrm{d}x_t)$ admit densities with respect to a common dominating measure, say $\nu(\cdot)$, we will denote those by $m_0(x_0)$ and $m_t(x_t|x_{t-1})$ respectively. In that case $\mathbb{Q}_t(\mathrm{d}x_{0:t})$ admits a density with respect to ν^{t+1}, which will be denoted by $q_t(x_{0:t})$ and is given by

$$q_t(x_{0:t}) = \frac{1}{L_t} G_0(x_0) m_0(x_0) \left\{ \prod_{s=1}^{t} G_s(x_{s-1}, x_s) m_s(x_s|x_{s-1}) \right\}.$$

Finally, we extend the definition of all the $\mathbb{Q}_t$'s to the same (possibly arbitrary) future horizon $T \geq t$:

$$\mathbb{Q}_t(\mathrm{d}x_{0:T}) = \frac{1}{L_t} G_0(x_0) \left\{ \prod_{s=1}^{t} G_s(x_{s-1}, x_s) \right\} \mathbb{M}_T(\mathrm{d}x_{0:T}), \tag{5.3}$$

so that $\mathbb{Q}_t(\mathrm{d}x_{0:t})$ becomes the marginal distribution of the $t+1$ first components relative to $\mathbb{Q}_t(\mathrm{d}x_{0:T})$. Other marginal distributions will be denoted in the same way; e.g. $\mathbb{Q}_t(\mathrm{d}x_t)$ is the marginal distribution of component x_t relative to $\mathbb{Q}_t(\mathrm{d}x_{0:T})$, and so on. Also let $\mathbb{Q}_{-1}(\mathrm{d}x_{0:T}) = \mathbb{M}_T(\mathrm{d}x_{0:T})$ for convenience.

Remark 5.1 A note of warning is due here: $\mathbb{Q}_t(\mathrm{d}x_{0:t})$ is a marginal of $\mathbb{Q}_t(\mathrm{d}x_{0:T})$, but not of $\mathbb{Q}_T(\mathrm{d}x_{0:T})$! Marginalising the Markov measures $\mathbb{M}_t$ is straightforward, as developed in Sect. 4.4. Marginalisation of $\mathbb{Q}_T$ is treated in Sect. 5.4.

5.1.2 Feynman-Kac Formalisms of a State-Space Model

Although a bit abstract, Feynman-Kac models are a very useful formalism for the sequential analysis of state-space models. We can relate state-space models to Feynman-Kac models in different ways. The main principle behind this book is to learn to work with Feynman-Kac models and find ways to relate them to state-space models in order to obtain mathematical results and algorithms. The following is one such important connection.

The "Bootstrap" Feynman-Kac Formalism of a State-Space Model
Consider a state-space model with initial distribution $\mathbb{P}_0(\mathrm{d}x_0)$, signal transition kernels $P_t(x_{t-1}, \mathrm{d}x_t)$ and observation densities $f_t(y_t|x_t)$. We define its "bootstrap" Feynman-Kac formalism to be the Feynman-Kac model with the following components

$$\mathbb{M}_0(\mathrm{d}x_0) = \mathbb{P}_0(\mathrm{d}x_0)\,, \qquad G_0(x_0) = f_0(y_0|x_0)\,,$$
$$M_t(x_{t-1}, \mathrm{d}x_t) = P_t(x_{t-1}, \mathrm{d}x_t)\,, \qquad G_t(x_{t-1}, x_t) = f_t(y_t|x_t)\,.$$

Then

- $\mathbb{Q}_{t-1}(\mathrm{d}x_{0:t}) = \mathbb{P}_t(X_{0:t} \in \mathrm{d}x_{0:t}|Y_{0:t-1} = y_{0:t-1})$,
- $\mathbb{Q}_t(\mathrm{d}x_{0:t}) = \mathbb{P}_t(X_{0:t} \in \mathrm{d}x_{0:t}|Y_{0:t} = y_{0:t})$,
- $L_t = p_t(y_{0:t})$ and $l_t := L_t/L_{t-1} = p_t(y_t|y_{0:t-1})$.

In this specification the potential functions G_t for $t \geq 1$ only depend on x_t, which is of course a possibility. Notice also how G_t depends implicitly on datapoint y_t, which is convenient since these datapoints are typically treated as fixed. We recommend the reader who would not be familiar with Feynman-Kac models to keep this particular interpretation in mind to follow more easily the forthcoming derivations. The correspondence between the $\mathbb{Q}$-laws above and the corresponding conditioned $\mathbb{P}$-laws follows immediately from the definition.

However, there are several other possible Feynman-Kac formalisms of a given state-space model. In this book we study two further families of formalisms. The first family is defined below.

Guided Feynman-Kac Formalisms of a State-Space Model
Consider a state-space model with initial distribution $\mathbb{P}_0(\mathrm{d}x_0)$, signal transition kernels $P_t(x_{t-1}, \mathrm{d}x_t)$ and observation densities $f_t(y_t|x_t)$. We call guided

(continued)

Feynman-Kac formalism any Feynman-Kac model such that

$$G_0(x_0)\mathbb{M}_0(\mathrm{d}x_0) = f_0(y_0|x_0)\mathbb{P}_0(\mathrm{d}x_0)\,, \tag{5.4}$$

$$G_t(x_{t-1}, x_t)M_t(x_{t-1}, \mathrm{d}x_t) = f_t(y_t|x_t)P_t(x_{t-1}, \mathrm{d}x_t)\,. \tag{5.5}$$

Then

- $\mathbb{Q}_t(\mathrm{d}x_{0:t}) = \mathbb{P}_t(X_{0:t} \in \mathrm{d}x_{0:t}|Y_{0:t} = y_{0:t})$,
- $L_t = p_t(y_{0:t})$, and $l_t = p_t(y_t|y_{0:t-1})$,
- $\mathbb{Q}_{t-1}(\mathrm{d}x_{0:t-1})P_t(x_{t-1}, \mathrm{d}x_t) = \mathbb{P}_t\left(X_{0:t} \in \mathrm{d}x_{0:t} \mid Y_{0:t-1} = y_{0:t-1}\right)$.

This family includes as a special case the bootstrap Feynman-Kac formalism: take $G_t(x_{t-1}, x_t) = f_t(y_t|x_t)$ and $M_t(x_{t-1}, \mathrm{d}x_t) = P_t(x_{t-1}, \mathrm{d}x_t)$, that is match each factor in (5.5).

Equalities (5.4) and (5.5) must be understood as equalities between two unnormalised probability measures, when $t = 0$, or between two unnormalised kernels, that is two kernels from $(\mathcal{X}, \mathcal{B}(\mathcal{X}))$ to itself that have not been normalised to integrate to one, when $t > 0$. In case both M_t and P_t admit probability densities $m_t(x_t|x_{t-1})$ and $p_t(x_t|x_{t-1})$, condition (5.5) may be rewritten as

$$G_t(x_{t-1}, x_t) = \frac{p_t(x_t|x_{t-1})f_t(y_t|x_t)}{m_t(x_t|x_{t-1})}$$

for x_{t-1}, x_t such that $m_t(x_t|x_{t-1}) > 0$.

We define below an even larger family of Feynman-Kac models that may be related to a given state-space model.

Auxiliary Feynman-Kac Formalisms of a State-Space Model
Consider a state-space model with initial distribution $\mathbb{P}_0(\mathrm{d}x_0)$, signal transition kernels $P_t(x_{t-1}, \mathrm{d}x_t)$ and observation densities $f_t(y_t|x_t)$. We call auxiliary Feynman-Kac formalism any Feynman-Kac model such that

$$G_0(x_0)\mathbb{M}_0(\mathrm{d}x_0) = f_0(y_0|x_0)\mathbb{P}_0(\mathrm{d}x_0)\eta_0(x_0)\,,$$

$$G_t(x_{t-1}, x_t)M_t(x_{t-1}, \mathrm{d}x_t) = f_t(y_t|x_t)P_t(x_{t-1}, \mathrm{d}x_t)\frac{\eta_t(x_t)}{\eta_{t-1}(x_{t-1})}\,,$$

for certain functions $\eta_t : \mathcal{X} \to \mathbb{R}^+$ such that $\mathbb{E}_{\mathbb{P}_t}[\eta_t(X_t)|Y_{0:t} = y_{0:t}] < \infty$ for all t.

We defer to Chap. 10 discussing in which way auxiliary Feynman-Kac models may be used, and how the $\mathbb{Q}$-distributions relate to the $\mathbb{P}$-distributions in this case. Again, guided Feynman-Kac models are recovered as a special case, by taking $\eta_t(x_t) = 1$ for all t.

The names we have assigned to the different formalisms are inspired by the names popular particle filter algorithms have in the literature; the bootstrap, guided and auxiliary particle filters, see Chap. 10. There is yet another degree of freedom for relating Feynman-Kac models to the laws of state-space models. We can choose a different state-space than that of the signal process. For example, we can allow it to change over time. This turns out to be useful when developing numerical methods to solve smoothing problems. We do not enter in details here, but postpone such analysis to Chap. 12.

5.1.3 Attractions of the Feynman-Kac Formalism

The following are some key reasons why we are interested in the Feynman-Kac formalism of state-space models:

- We can decouple a statistical model (the state-space model) from its mathematical representation that is used to study certain theoretical properties, such as recursions for marginal distributions, or to develop algorithms, such as particle filters. The multitude of Feynman-Kac formalisms for a given state-space model leads to a factory of numerical algorithms for learning and prediction with that model. Instead of dealing with each such variation as a separate algorithm, we can deal with all simultaneously as instances of the same framework with particular choice of a few "parameters", i.e. the components of the Feynman-Kac formalism.
- Feynman-Kac models share the same fundamental structure: the specific change of measure from a Markov measure. This structure can be exploited to obtain clean mathematical expressions and recursions for marginal distributions associated to a Feynman-Kac model. Thus, several results can be obtained without entering into a detailed specification of the components of the model. These results are contained in this chapter.
- The Feynman-Kac representation is ideal for the development of generic software code. The `particles` library developed to accompany this book (see the Python corners at the end of each chapter) capitalises on the modular structure of the Feynman-Kac representation to build object-oriented computing code for implementing particle filter algorithms.
- The Feynman-Kac representation can encode a variety of sequential simulation problems different from those that arise within the context of state-space models. All the problems described in Chap. 3 admit a Feynman-Kac representation.

5.2 Forward Recursion

An essential question is how to relate the successive $\mathbb{Q}_t$ measures. Recall from Sect. 4.4 (with notation appropriately adapted) that the Markov measures are related in a simple way:

$$\mathbb{M}_t(\mathrm{d}x_{0:t}) = \mathbb{M}_{0:t-1}(\mathrm{d}x_{t-1})M_t(x_{t-1}, \mathrm{d}x_t).$$

By the definition of $\mathbb{Q}_t$, one sees that $\mathbb{Q}_{t-1}$ may be related to $\mathbb{Q}_t$ through a simple change of measure:

$$\mathbb{Q}_t(\mathrm{d}x_{0:t}) = \frac{1}{\ell_t} G_t(x_{t-1}, x_t)\mathbb{Q}_{t-1}(\mathrm{d}x_{0:t-1})M_t(x_{t-1}, \mathrm{d}x_t)$$

where $\ell_t > 0$. However, what turns out to be even more useful, at least in order to derive e.g. filtering algorithms, is to relate the marginals $\mathbb{Q}_{t-1}(\mathrm{d}x_{t-1})$ and $\mathbb{Q}_t(\mathrm{d}x_t)$. In this way, one obtains the so-called forward recursion, which operates on the much smaller space $\mathcal{X}^2$. This recursion underpins particle filtering, as explained in Chap. 10.

Forward Recursion (Feynman-Kac Formalism)
Initialise with $\mathbb{Q}_{-1}(\mathrm{d}x_0) = \mathbb{M}_0(\mathrm{d}x_0)$, then, for $t = 0 : T$,

- Extension:

$$\mathbb{Q}_{t-1}(\mathrm{d}x_{t-1:t}) = \mathbb{Q}_{t-1}(\mathrm{d}x_{t-1})M_t(x_{t-1}, \mathrm{d}x_t)\,.$$

- Change of measure:

$$\mathbb{Q}_t(\mathrm{d}x_{t-1:t}) = \frac{1}{\ell_t} G_t(x_{t-1}, x_t)\mathbb{Q}_{t-1}(\mathrm{d}x_{t-1:t})\,,$$

with

$$l_0 = L_0 = \int_{\mathcal{X}} G_0(x_0)\mathbb{M}_0(\mathrm{d}x_0),$$

$$\ell_t = \frac{L_t}{L_{t-1}} = \int_{\mathcal{X}^2} G_t(x_{t-1}, x_t)\mathbb{Q}_{t-1}(\mathrm{d}x_{t-1:t}), \quad t \geq 1.$$

- Marginalisation:

$$\mathbb{Q}_t(\mathrm{d}x_t) = \int_{x_{t-1}\in\mathcal{X}} \mathbb{Q}_t(\mathrm{d}x_{t-1:t})\,.$$

To prove "Extension", note that

$$\mathbb{Q}_{t-1}(\mathrm{d}x_{0:t}) = \mathbb{Q}_{t-1}(\mathrm{d}x_{0:t-1})M_t(x_{t-1}, \mathrm{d}x_t)$$

and integrate both side with respect to $x_{0:t-2}$. For 'Change of measure', note that, for $t \geq 1$,

$$\mathbb{Q}_t(\mathrm{d}x_{0:t}) = \frac{L_{t-1}}{L_t} G_t(x_{t-1}, x_t)\mathbb{Q}_{t-1}(\mathrm{d}x_{0:t})$$

and integrate again with respect to $x_{0:t-2}$.

5.2.1 *Implications for State-Space Models: A Forward Recursion for the Filter, Prediction and the Likelihood*

The standard forward recursion for state-space models is easily deduced by applying the recursion of the previous section to the bootstrap formalism: for $t \geq 1$,

$$\begin{aligned} &\mathbb{P}_{t-1}(X_t \in \mathrm{d}x_t | Y_{0:t-1} = y_{0:t-1}) \\ &\quad = \int_{x_{t-1} \in \mathcal{X}} \mathbb{P}_{t-1}(X_{t-1} \in \mathrm{d}x_{t-1} | Y_{0:t-1} = y_{0:t-1}) P_t(x_{t-1}, \mathrm{d}x_t)\,, \end{aligned} \tag{5.6}$$

(apply 'Extension' and integrate with respect to x_{t-1}), and for $t \geq 0$

$$\mathbb{P}_t(X_t \in \mathrm{d}x_t | Y_{0:t} = y_{0:t}) = \frac{1}{p_t(y_t | y_{0:t-1})} f_t(y_t | x_t)\mathbb{P}_{t-1}(X_t \in \mathrm{d}x_t | Y_{0:t-1} = y_{0:t-1}) \tag{5.7}$$

where the predictive distribution must be replaced by $\mathbb{P}_0(\mathrm{d}x_0)$ at time $t = 0$ (apply "change of measure", and integrate with respect to x_{t-1}, noting that G_t does not depend on x_{t-1} in the bootstrap formalism).

Applying in turn (5.6) and (5.7), we are able to compute recursively the filtering and predictive distributions at times $0, 1, \ldots, T$. We also obtain at each time the likelihood factor as a by-product:

$$p_t(y_t | y_{0:t-1}) = \int_{\mathcal{X}} f_t(y_t | x_t)\mathbb{P}_{t-1}(X_t \in \mathrm{d}x_t | Y_{0:t-1} = y_{0:t-1}). \tag{5.8}$$

5.3 The Feynman-Kac Model as a Markov Measure

In this section we establish that the Feynman-Kac model is a Markov measure and identify its forward and backward kernels. We first define the following functions $H_{t:T}(x_t)$, which we will refer to as "cost-to-go" functions:

$$H_{T:T}(x_T) := 1, \quad H_{t:T}(x_t) := \int_{\mathcal{X}^{T-t}} \prod_{s=t+1}^{T} G_s(x_{s-1}, x_s) M_s(x_{s-1}, \mathrm{d}x_s), \quad t < T. \tag{5.9}$$

The terminology "cost-to-go" is inspired by dynamic programming, which is closely related to the computational paradigm we develop here. The definition leads to a backward recursion:

$$H_{t:T}(x_t) = \int_{\mathcal{X}} G_{t+1}(x_t, x_{t+1}) H_{t+1:T}(x_{t+1}) M_{t+1}(x_t, \mathrm{d}x_{t+1}), \quad t < T, \tag{5.10}$$

which is obtained by applying Fubini's theorem (to change the order of integration). We can also express the functions as conditional expectations. By making use of marginalisation properties of the Markov measure $\mathbb{M}_T$ (as in Proposition 4.5) we obtain:

$$\begin{aligned} H_{t:T}(x_t) &= \mathbb{E}_{\mathbb{M}_T}\left[\prod_{s=t+1}^{T} G_s(X_{s-1}, X_s) \,\middle|\, X_t = x_t \right] \\ &= \mathbb{E}_{\mathbb{M}_{t+1}}\left[G_{t+1}(x_t, X_{t+1}) H_{t+1:T}(X_{t+1}) \right], \end{aligned} \tag{5.11}$$

where each equation follows from the two alternative expressions for $H_{t:T}(x_t)$ obtained above.

Proposition 5.1 *$\mathbb{Q}_T$ is the law of a Markov process with state-space $\mathcal{X}$, initial distribution*

$$\mathbb{Q}_{0|T}(\mathrm{d}x_0) = \frac{H_{0:T}(x_0)}{L_T} G_0(x_0) \mathbb{M}_0(\mathrm{d}x_0), \tag{5.12}$$

forward transition kernels $Q_{t|T}(x_{t-1}, \mathrm{d}x_t)$ given by:

$$Q_{t|T}(x_{t-1}, \mathrm{d}x_t) = \frac{H_{t:T}(x_t)}{H_{t-1:T}(x_{t-1})} G_t(x_{t-1}, x_t) M_t(x_{t-1}, \mathrm{d}x_t), \tag{5.13}$$

and backward kernels given by:

$$\mathbb{Q}_T(\mathrm{d}x_t) \overleftarrow{Q}_{t-1|T}(x_t, \mathrm{d}x_{t-1}) = \mathbb{Q}_T(\mathrm{d}x_{t-1}) Q_{t|T}(x_{t-1}, \mathrm{d}x_t). \tag{5.14}$$

Proof Note first that due to the definition of the functions $H_{t:T}(x_t)$, $\mathbb{Q}_{0|T}(\mathrm{d}x_0)$ is a probability measure and the $Q_{t|T}(x_{t-1}, \mathrm{d}x_t)$'s are probability kernels, since the denominators in (5.12) and (5.13) are the integrals of the numerators. The expression for the transition kernel then follows by multiplying (5.1) by the telescoping product

$$H_0(x_0)\prod_{t=1}^{T}\frac{H_{t:T}(x_t)}{H_{t-1:T}(x_{t-1})}$$

which equals 1 because $H_{T:T}(x_T) = 1$, and by reorganising the terms. The expression of the backward kernel is a direct application of Proposition 4.6. □

Remark 5.2 See Corollary 5.2 in Sect. 5.4 for a more constructive expression for the backward kernel. In fact, what is not obvious in the above expression is that $\overleftarrow{Q}_{t-1|T}(x_t, \mathrm{d}x_{t-1}) = \overleftarrow{Q}_{t-1|t}(x_t, \mathrm{d}x_{t-1})$, which is made explicit in Corollary 5.2.

Notation/Terminology The notation $Q_{t|T}(x_{t-1}, \mathrm{d}x_t)$ recognises the dependence of the kernels on T. If we apply the result to find the corresponding parameters of the Markov representation of the Feynman-Kac model for some other time horizon, say $S < T$, then we obtain different initial distribution and kernels. The dependence of these parameters on the time horizon is due to the fact that each of the factors $G_t(x_{t-1}, x_t)M_t(x_{t-1}, \mathrm{d}x_t)$ in (5.1) is not properly normalised to be a transition kernel.

5.3.1 *Implications for State-Space Models: The Conditioned Markov Process*

It is easy to check that in the "bootstrap" formalism of a state-space model,

$$p_T(y_{t+1:T}|x_t) = H_{t:T}(x_t).$$

This quantity (introduced in Sect. 4.5) is the likelihood of future observations given the current value of the state. These functions may also be obtained via a backward recursion:

$$p_T(y_{t+1:T}|x_t) = \int_{\mathcal{X}} f_t(y_{t+1}|x_{t+1})p_T(y_{t+2:T}|x_{t+1})P_t(x_t, \mathrm{d}x_{t+1}) \tag{5.15}$$

with $p_T(y_{T+1:T}|x_T) := 1$.

Proposition 5.1 proves that in state-space models the process $\{X_t\}$ remains Markov when conditioned on data $Y_{0:T}$. The Proposition gives explicit expressions for the initial distribution and the transition kernels of the conditioned Markov process:

$$\mathbb{P}_{0|T}(\mathrm{d}x_0) = \frac{p_T(y_{1:T}|x_0)}{p_T(y_{0:T})} f_0(y_0|x_0)\mathbb{P}_0(\mathrm{d}x_0)\,,$$

$$P_{t|T}(x_{t-1}, \mathrm{d}x_t) = \frac{p_T(y_{t+1:T}|x_t)}{p_T(y_{t:T}|x_{t-1})} f_t(y_t|x_t)P_t(x_{t-1}, \mathrm{d}x_t)\,.$$

In principle, we can use $P_{t|T}(x_{t-1}, \mathrm{d}x_t)$ to simulate trajectories from the smoothing distribution: first run backwards a recursion to compute the cost-to-go functions, and then run a recursion forwards simulating from $P_{t|T}(x_{t-1}, \mathrm{d}x_t)$ at each step. Later in Sect. 5.4.4 we will see an alternative forward-backward scheme for simulating such trajectories, which is more appropriate in the context of particle approximations.

Notice that even if the Markov process $\{X_t\}$ is time-homogeneous according to its marginal law $\mathbb{P}$, it becomes time-inhomogeneous once conditioned upon $Y_{0:T} = y_{0:T}$. Each transition kernel $P_{t|T}(x_{t-1}, \mathrm{d}x_t)$ is a function of future observations $y_{t:T}$.

5.4 Forward-Backward Recursions in a Feynman-Kac Model

5.4.1 Forward-Backward Recursions Based on Cost-to-Go Functions

The representation of $\mathbb{Q}_T$ as a Markov process in Proposition 5.1 allows us to obtain its marginal distribution at time $t < T$, $\mathbb{Q}_T(\mathrm{d}x_{0:t})$, as a *change of measure* from $\mathbb{Q}_t(\mathrm{d}x_{0:t})$, effectuated by function $H_{t:T}(x_t)$.

Proposition 5.2 *For any $t < T$,*

$$\mathbb{Q}_T(\mathrm{d}x_{0:t}) = \frac{L_t}{L_T} H_{t:T}(x_t)\mathbb{Q}_t(\mathrm{d}x_{0:t}).$$

Proof From Proposition 4.5 and Proposition 5.1 we have that

$$\begin{aligned}
\mathbb{Q}_T(\mathrm{d}x_{0:t}) &= \mathbb{Q}_{0|T}(\mathrm{d}x_0)\prod_{s=1}^{t} Q_{s|T}(x_{s-1}, \mathrm{d}x_s) \\
&= \frac{1}{L_T} H_{t:T}(x_t)G_0(x_0)\mathbb{M}_0(\mathrm{d}x_0)\prod_{s=1}^{t} G_s(x_{s-1}, x_s)M_s(x_{s-1}, \mathrm{d}x_s) \\
&= \frac{L_t}{L_T} H_{t:T}(x_t)\mathbb{Q}_t(\mathrm{d}x_{0:t}).
\end{aligned}$$

□

By integrating out $x_{0:t-1}$ in the expression above, we obtain the following expression for marginals.

Corollary 5.1

$$\mathbb{Q}_T(\mathrm{d}x_t) = \frac{L_t}{L_T} H_{t:T}(x_t)\mathbb{Q}_t(\mathrm{d}x_t).$$

This corollary gives a recipe to compute all (or some of) the marginal distributions $\mathbb{Q}_T(\mathrm{d}x_t)$: use the forward recursion to compute recursively the $\mathbb{Q}_t(\mathrm{d}x_t)$'s; use backward recursion (5.10) to compute recursively the cost-to-go functions $H_{t:T}$; and then combine the two.

5.4.2 Implications for State-Space Models: Two-Filter Smoothing

Corollary 5.1 applied to the bootstrap formalism immediately implies the following representation for the marginal smoothing distributions of a state-space model:

$$\mathbb{P}(X_t \in \mathrm{d}x_t | Y_{0:T} = y_{0:T}) = \frac{1}{p_T(y_{t+1:T}|y_{0:t})} p_T(y_{t+1:T}|x_t)\mathbb{P}(X_t \in \mathrm{d}x_t | Y_{0:t} = y_{0:t}).$$

The derivation of this expression is simply a matter of replacing each factor in Corollary 5.1 by its equal in the state-space model. We see that the marginal smoothing distribution is a change of measure from the filtering distribution, with density proportional to $p_T(y_{t+1:T}|x_t)$, which can be computed via a backward recursion, according to (5.15).

In practice, this backward recursion may be implemented through a particle filter that corresponds to an artificial state-space model, where, basically, the data points are processed in reverse. The corresponding algorithm, known as two-filter smoothing, will be treated in Sect. 12.5 of Chap. 12 on particle smoothing.

5.4.3 Forward-Backward Recursions Based on Backward Kernels

Proposition 5.1 states that $\mathbb{Q}_T(\mathrm{d}x_{0:T})$ is the law of a Markov process, and gives the expression of the corresponding backward kernels. By replacing, in (5.14), $\mathbb{Q}_T(\mathrm{d}x_t)$ and $\mathbb{Q}_T(\mathrm{d}x_{t-1})$ with their equals in Corollary 5.1, we obtain the following alternative expression for these backward kernels.

Corollary 5.2

$$\mathbb{Q}_t(\mathrm{d}x_t)\overleftarrow{Q}_{t-1|T}(x_t, \mathrm{d}x_{t-1}) = \frac{1}{\ell_t} G_t(x_{t-1}, x_t)\mathbb{Q}_{t-1}(\mathrm{d}x_{t-1})M_t(x_{t-1}, \mathrm{d}x_t).$$

Notation/Terminology Since the backward kernel $\overleftarrow{Q}_{t-1|T}(x_t, \mathrm{d}x_{t-1})$ does not actually depend on T and on $G_{t+1}, \ldots, G_T$, we simplify its notation to $\overleftarrow{Q}_{t-1|t}(x_t, \mathrm{d}x_{t-1})$ from now on.

A much more constructive representation for the backward kernel is available if M_t admits a probability density: $M_t(x_{t-1}, \mathrm{d}x_t) = m_t(x_t|x_{t-1})\nu(\mathrm{d}x_t)$. This is of key importance to the development of smoothing algorithms in Chap. 12 and we highlight it in the following proposition.

Proposition 5.3 *Suppose that* $M_t(x_{t-1}, \mathrm{d}x_t) = m_t(x_t|x_{t-1})\nu(\mathrm{d}x_t)$. *Then,*

$$\overleftarrow{Q}_{t-1|t}(x_t, \mathrm{d}x_{t-1}) \propto G_t(x_{t-1}, x_t)m_t(x_t|x_{t-1})\mathbb{Q}_{t-1}(\mathrm{d}x_{t-1}) \tag{5.16}$$

where the proportionality constant is recovered by normalisation.

Proof By combining the steps of the forward recursion for Feynman-Kac models, one obtains that:

$$\mathbb{Q}_t(\mathrm{d}x_t) = \frac{1}{\ell_t} \int_{x_{t-1} \in \mathcal{X}} G_t(x_{t-1}, x_t)M_t(x_{t-1}, \mathrm{d}x_t)\mathbb{Q}_{t-1}(\mathrm{d}x_{t-1}).$$

Plugging this expression in the defining equation for the backward kernel, and using the assumption shows that the two sides are equal if and only if $\overleftarrow{Q}_{t-1|t}(x_t, \mathrm{d}x_{t-1})$ takes the form given in the Proposition. □

This proposition suggests a second type of forward-backward recursion, which may be used to simulate complete paths from $\mathbb{Q}_T(\mathrm{d}x_{0:T})$: first run the forward recursion until time T to compute all the $\mathbb{Q}_t(\mathrm{d}x_t)$'s; then simulate recursively a path as follows: sample X_T from $\mathbb{Q}_T(\mathrm{d}x_T)$, the initial distribution in the backward decomposition of $\mathbb{Q}_T(\mathrm{d}x_{0:T})$; then for $t = T, T-1, \ldots, 1$, simulate X_{t-1} given X_t according to kernel (5.16).

5.4.4 Implications for State-Space Models: FFBS Algorithms

For a given state-space model, we may deduce from Corollary 5.2 the backward kernel for process $\{X_t\}$ when conditioned on data $y_{0:T}$:

$$\mathbb{P}_t(X_{t-1} \in dx_{t-1}|Y_{0:t} = y_{0:t}) \overleftarrow{P}_{t-1|t}(x_t, dx_{t-1}) =$$

$$\frac{1}{p_t(y_t|y_{0:t-1})} f_t(y_t|x_t) P_t(x_{t-1}, dx_t) \mathbb{P}_{t-1}(X_{t-1} \in dx_{t-1}|Y_{0:t-1} = y_{0:t-1}). \tag{5.17}$$

Notation/Terminology We denote by $\overleftarrow{P}_{t-1|t}(x_t, dx_{t-1})$ the backward kernel of the state process conditionally on the data up to time t, to distinguish it from $\overleftarrow{P}_{t-1}(x_t, dx_{t-1})$, the backward kernel of the state process not conditioned on any data. The notation reflects that this kernel depends only on $y_{0:t}$ (and not on T and future data).

The backward kernel of the conditioned process simplifies when $P_t(x_{t-1}, dx_t) = p_t(x_t|x_{t-1})\nu(dx_t)$. By application of Proposition 5.3 we get that, in this case

$$\overleftarrow{P}_{t-1|t}(x_t, dx_{t-1}) \propto p_t(x_t|x_{t-1}) \mathbb{P}_{t-1}(X_{t-1} \in dx_{t-1}|Y_{0:t-1} = y_{0:t-1}). \tag{5.18}$$

Simulation algorithms that rely on backward kernels are called FFBS (forward filtering backward sampling) and will be covered in Sect. 12.3 of Chap. 12 (on particle smoothing).

Exercises

5.1 *Show that, if the potential functions G_t are constant, then $\mathbb{Q}_t(dx_{0:t}) = \mathbb{M}_t(dx_{0:t})$. Is this the only case where we have this equality?*

5.2 *Does Proposition 5.2 still hold for $t = T$?*

5.3 *Consider a stochastic process model for (X_t, Y_t) more general than the state-space model defined in Sect. 4.5, one where Y_t given $X_t = x_t$ also depends on Y_{t-1}. Identify two alternative "bootstrap" Feynman-Kac formalisms for this model, one based on modifying M_t and another on modifying G_t relative to their definitions in Sect. 5.1.2. Exercise 6.1 builds upon this idea to construct a filter for Markov switching time series models.*

Python Corner

In particles, Feynman-Kac models are, like state-space models, specified as objects of a custom class. For instance, the bootstrap Feynman-Kac models associated to stochastic volatility models (see Sect. 2.4.3), may be defined as follows:

```
import numpy as np
import particles
from particles import distributions as dists

class Bootstrap_SV(particles.FeynmanKac):
    """Bootstrap FK model associated to a stochastic volatility ssm. """

    def __init__(self, data=None, mu=0., sigma=1., rho=0.95):
        self.data = data
        self.T = len(data)  # number of time steps
        self.mu = mu
        self.sigma = sigma
        self.rho = rho
        self.sigma0 = self.sigma / np.sqrt(1. - self.rho**2)

    def M0(self, N):
        return dists.Normal(loc=self.mu, scale=self.sigma0).rvs(size=N)
    def M(self, t, xp):
        return dists.Normal(loc=self.mu + self.rho * (xp - self.mu),
                            scale=self.sigma).rvs(size=xp.shape[0])

    def logG(self, t, xp, x):
        return dists.Normal(scale=np.exp(0.5 * x)).logpdf(self.data[t])
```

Then we may define a particular Feynman-Kac model by instantiating the class:

```
y = dists.Normal().rvs(size=100)  # artificial data
fk_boot_sv = Bootstrap_SV(mu=-1., sigma=0.15, rho=0.9, data=y)
```

Method logG computes the logarithm of $G_t(x_{t-1}, x_t)$; the rationale behind the log is discussed in the next Python corner. Methods M0 and M do not really define distributions; rather, they implement *simulators* from these distributions. This is because the only operation a particle filter must perform with respect to $\mathbb{M}_0$ and M_t is precisely to simulate from it.

In the previous Python corner, we already defined the object my_sv_model (an instance of class StochVol), which represented a particular stochastic volatility model. It would be nice if we could generate automatically the corresponding Bootstrap Feynman-Kac model, without defining it manually as above. In particles, this may be done as follows:

```
from particles import state_space_models as ssm

fk_boot_sv = ssm.Bootstrap(ssm=my_sv_model, data=y)
```

This `Bootstrap` class defines in a generic way how `logG`, `M0` and `M` should depend on the components of the state-space model. Here is an extract from the definition of this class:

```
class Bootstrap(particles.FeynmanKac):

    # ... some bits missing

    def M0(self, N):
        return self.ssm.PX0().rvs(size=N)

    def M(self, t, xp):
        return self.ssm.PX(t, xp).rvs(size=xp.shape[0])

    def logG(self, t, xp, x):
        return self.ssm.PY(t, x).logpdf(self.data[t])
```

We may of course define other Feynman-Kac formalisms (such as the guided formalism). This will be explained in the Python corner of Chap. 10.

Bibliographical Notes

The Feynman-Kac formalism and its connections to algorithms based on interacting particle systems are developed in depth in the book of Del Moral (2004).

Bibliography

Del Moral, P. (2004). *Feynman-Kac formulae. Genealogical and interacting particle systems with applications. Probability and its applications.* New York: Springer.

Chapter 6
Finite State-Spaces and Hidden Markov Models

Summary This short chapter considers the special case where the state-space of the considered model is finite; $\mathcal{X} = \{1, \ldots, K\}$. In that case the integrals of the forward and backward recursions become sums over K terms, which can be computed exactly at a cost that is shown to be $\mathcal{O}(TK^2)$. State-space models with a finite state-space model are usually called hidden Markov models. Applying the generic algorithm to their "bootstrap" Feynman-Kac formalisation yields an exact algorithm known as the forward-backward algorithm.

6.1 Introduction: Recursions for Finite State-Spaces

Consider the Feynman-Kac model defined in Chap. 5, specifically Sect. 5.1, but specialised to $\mathcal{X} = \{1, \ldots, K\}$. In this setting $\{X_t\}$ is a Markov process with a finite state-space, defined through an initial distribution, given in terms of probabilities, and transition kernels represented in the form of transition matrices.

In this framework, the expressions that were obtained for the generic Feynman-Kac model and involved integration over measures on $\mathcal{X}$ become sums that can be computed explicitly. Therefore, the neat mathematical expressions for the marginal distributions in a Feynman-Kac model obtained in Chap. 5 convert immediately to numerical algorithms for their "derivation". In particular, marginal probabilities $\mathbb{Q}_T(x_t \in \{k\})$ can be computed for all $t \in 0 : T$ and $k \in 1 : K$. In this short chapter we only address two further points. First, for the sake of providing a quick reference, we state the recursions for the filtering, smoothing and prediction probabilities for finite state-space models. Second, we discuss the numerical complexity of the operations involved.

© Springer Nature Switzerland AG 2020
N. Chopin, O. Papaspiliopoulos, *An Introduction to Sequential Monte Carlo*, Springer Series in Statistics, https://doi.org/10.1007/978-3-030-47845-2_6

6.2 Hidden Markov Models, Recursions and Simulation

Typically, although there is large discrepancy in the literature, a state-space model with a finite state-space is called a hidden Markov model, see also the discussion in Sect. 2.4.8. Using the "bootstrap" Feynman-Kac formalism of such models and exploiting the nature of the state-space we obtain the following recursions that may be used to perform sequential inference with respect to a hidden Markov model.

Recursions for Hidden Markov Models

Consider a hidden Markov model with initial distribution $p_0(k)$ (the probability that $X_0 = k$ for $k = 1 : K$), transition probabilities $p_t(k|l)$ (i.e. the probability that $X_t = k$ given $X_{t-1} = l$, for $k, l = 1 : K$) and observation (emission) densities $f_t(y_t|k)$ (the density of $Y_t|X_t = k$). The following are understood for all $k = 1 : K$.

- **Prediction:** At time 0, let $\mathbb{P}_{-1}(X_0 = k) := p_0(k)$. At time $t > 0$,

$$\mathbb{P}_{t-1}(X_t = k|Y_{0:t-1} = y_{0:t-1}) = \sum_{l=1}^{K} \mathbb{P}_{t-1}(X_{t-1} = l|Y_{0:t-1} = y_{0:t-1}) p_t(k|l) .$$

- **Filter:**

$$\mathbb{P}_t(X_t = k|Y_{0:t} = y_{0:t}) = \frac{1}{p(y_t|y_{0:t-1})} \mathbb{P}_{t-1}(X_t = k|Y_{0:t-1} = y_{0:t-1}) f_t(y_t|k) .$$

- **Likelihood factors:**

$$p_t(y_t|y_{0:t-1}) = \sum_k \mathbb{P}_{t-1}(X_t = k|Y_{0:t-1} = y_{0:t-1}) f_t(y_t|k) .$$

- **Likelihood of future observations given current state:**

$$p_T(y_{t+1:T}|x_t = k) = \sum_l f_{t+1}(y_{t+1}|l) p(y_{t+2:T}|l) p_t(l|k) .$$

- **Marginal smoother:**

$$\mathbb{P}(X_t = k|Y_{0:T} = y_{0:T}) = \frac{1}{p(y_{t+1:T}|y_{0:t})} p(y_{t+1:T}|x_t = k) \mathbb{P}(X_t = k|Y_{0:t} = y_{0:t}) .$$

The resultant algorithm for the computation of the filters is known as the forward-backward algorithm. It is also possible to sample complete trajectories $X_{0:T}$ conditional on $Y_{0:T} = y_{0:T}$; see Exercise 6.3.

6.3 Numerical Complexity

Hidden Markov models give a first illustration of the power of forward-backward recursions related to the Feynman-Kac models. A brute-force approach to computing, e.g., the likelihood would require computing sums over $\mathcal{O}(K^{T+1})$ terms, where $\mathcal{O}$ is the "big O" notation for characterising asymptotic behaviour. On the other hand, the complexity of the forward-backward algorithm is $\mathcal{O}(TK^2)$ for the same task. Computing the predictive probabilities requires a summation of K terms for each probability to be computed, and there are K such probabilities; and once these probabilities are computed, the filtering probabilities and the likelihood factor may be computed in $\mathcal{O}(K)$ time. It may be possible to reduce this complexity in specific cases, for instance when the transition matrix is sparse (as in change-point modelling, see Exercise 6.2). When K is very large, the forward-backward algorithm might still become too expensive; in that case, one might consider implementing a particle filter: see Exercise 10.12 in Chap. 10.

Exercises

6.1 *Show how to adapt (slightly) the forward-backward algorithm to deal with a hidden Markov model such that Y_t given $X_t = k$ also depends on Y_{t-1}.*

6.2 *Consider a hidden Markov model such that $X_0 = 1$, and $X_t|X_{t-1} = k$ may only take values k and $k+1$ (say with probabilities τ_k, and $1 - \tau_k$, respectively). Explain how such a model relates to change-point modelling, determine the distribution between two changes, and discuss the complexity of the forward-backward algorithm when applied to this particular model.*

6.3 *As a way to prepare yourself to particle smoothing (see in particular Sect. 12.3 on the FFBS algorithm), determine how you may adapt the forward-backward recursion based on backward kernels (Sect. 5.4.3) to sample paths $X_{0:T}$ conditional on $Y_{0:T} = y_{0:T}$ in a hidden Markov model. Discuss the complexity of the obtained algorithm.*

6.4 *Consider a hidden Markov model, with Markov process $\{X_t\}$ which is homogeneous, and $\mathcal{X} = \{1, \ldots, K\}$. For simplicity, assume first that $K = 2$ and $X_0 = 1$ with probability one. Define $M_t = \max_{s \in 0:t} X_s$. Show that $X'_t = (M_t, X_t)$ is Markov, and give its transition matrix. What is the advantage of hidden Markov model $\{X'_t, Y_t\}$ in terms of parameter estimation? What happens if we drop the assumption that $X_0 = 1$? To deal with that case, consider a Bayesian approach, where the parameters (in particular the transition probabilities $p(k|l) = \mathbb{P}(X_t = l|X_{t-1} = k)$, but also the parameters that may be attached to the emission distribution) are assigned an exchangeable prior. Consider now the process $X'_t = (M_t, \sigma(X_t))$, where $\sigma = (1, 2)$ (resp. $(2, 1)$) if $X_0 = 1$ (resp. $X_0 = 2$). Construct a transition matrix for X'_t, such that the posterior distributions of the parameters (up* *to a certain re-ordering) are equivalent for both hidden Markov models,* $\{(X_t, Y_t)\}$, *and* $\{(X'_t, Y_t)\}$. *Generalise to* $K \geq 2$. *Give the complexity in* K *of the forward-backward algorithm when applied to the alternate hidden Markov model where states are ordered by order of appearance. (Note first that the transition matrix is sparse for this alternate model.)*

Python Corner

Package `particles` has a `hmm` module which implements a basic version of the forward-backward algorithm. If you look at the source code, you will see that all the computations are performed on the log scale; see the Python corner of Chap. 8 for the rationale behind this. Other noteworthy Python libraries that implement the forward-backward algorithm are: `hmmlearn` (formerly a module of `scikit-learn`, now an independent package at https://github.com/hmmlearn/hmmlearn), and `pomegranate` (see https://github.com/jmschrei/pomegranate). See also `ghmm` (ghmm.org), a C library with Python bindings.

Bibliographical Notes

Hidden Markov models and forward-backward recursions were developed in the late sixties for speech processing; see the tutorial of Rabiner (1989), and its references to the pioneering work of Leonard E. Baum; e.g. Baum et al. (1970). In speech processing, $\{Y_t\}$ typically stands for an acoustic signal (a speech recording), and $\{X_t\}$ for a sequence of words, or phonemes. Other important areas of application are bioinformatics (where $\{X_t\}$ typically stands for a genetic sequence), and natural language processing. In the latter area, hidden Markov models are used in particular for part-of-speech tagging, that is, the process of associating words (the Y_t's) to a grammatical category X_t such as noun, verb, adjective, etc.

Standard approaches for parameter estimation in hidden Markov models are: maximum likelihood through the EM algorithm (see Chap. 14) and Bayes through Gibbs sampling (see Chap. 16). These two topics are also covered respectively in e.g. the books of McLachlan and Krishnan (2008) and Frühwirth-Schnatter (2001). The standard EM algorithm for hidden Markov models is often called the Baum-Welch algorithm. See also Johansen et al. (2008) for how to use SMC to perform Bayesian inference and model choice for hidden Markov models.

Hidden Markov models are closely related to mixture models. In particular, like mixture models, they are invariant to state relabelling. One way to deal with this difficulty is to label the states by order of appearance; see Exercise 6.4 and Chopin (2007b) for more details.

Since the forward-backward algorithm has cost $\mathcal{O}(K^2)$ per time step, it becomes unfeasible when K is too large; in that case it makes sense to use a particle algorithm instead. Change-point models may be cast as hidden Markov model with a value of K which increases quickly with time; see Exercise 6.2. For more details on particle algorithms for change-point models, see Fearnhead (2006), Chopin (2007a), and Whiteley et al. (2010).

Bibliography

Baum, L. E., Petrie, T., Soules, G., & Weiss, N. (1970). A maximization technique occurring in the statistical analysis of probabilistic functions of Markov chains. *Annals of Mathematical Statistics, 41*, 164–171.

Chopin, N. (2007a). Dynamic detection of change points in long time series. *Annals of the Institute of Statistical Mathematics, 59*(2), 349–366.

Chopin, N. (2007b). Inference and model choice for sequentially ordered hidden Markov models. *Journal of the Royal Statistical Society: Series B (Statistical Methodology), 69*(2), 269–284.

Fearnhead, P. (2006). Exact and efficient Bayesian inference for multiple changepoint problems. *Statistics and Computing, 16*(2), 203–213.

Frühwirth-Schnatter, S. (2001). Markov chain Monte Carlo estimation of classical and dynamic switching and mixture models. *Journal of the American Statistical Association, 96*(453), 194–209.

Johansen, A. M., Doucet, A., & Davy, M. (2008). Particle methods for maximum likelihood estimation in latent variable models. *Statistics and Computing, 18*(1), 47–57.

McLachlan, G. J. & Krishnan, T. (2008). *The EM algorithm and extensions. Wiley series in probability and statistics* (2nd ed.). Hoboken, NJ: Wiley.

Rabiner, L. R. (1989). A tutorial on hidden Markov models and selected applications in speech recognition. *Proceedings of the IEEE, 77*, 257–284.

Whiteley, N., Andrieu, C., & Doucet, A. (2010). Efficient Bayesian inference for switching state-space models using discrete particle Markov chain Monte Carlo methods. *ArXiv preprints 1011.2437*.

Chapter 7
Linear-Gaussian State-Space Models

Summary Another special case where the forward and backward recursions developed in Chap. 5 may be implemented exactly is when the considered state-space model is linear and Gaussian. The corresponding algorithms are commonly known as the Kalman filter and the Kalman smoother. The recursions follow immediately from the generic formulae of Chap. 5, but in this setting they become linear algebra calculations. Various alternative, mathematically equivalently but computationally different, recursions can be obtained. This chapter provides insights into these possibilities and touches upon the practical implementation of such recursions.

7.1 Linear Gaussian State-Space Models

We return to the linear Gaussian state-space model first introduced in Sect. 2.4.7, and extend the model as originally presented in Eqs. (2.5)–(2.6) to allow for time-varying coefficients:

$$X_t = A_t X_{t-1} + U_t \,, \tag{7.1}$$

$$Y_t = B_t X_t + V_t \,, \tag{7.2}$$

where U_t and V_t are independent Gaussian innovations, with $U_t \sim \mathcal{N}(0, \Sigma_t)$ and $V_t \sim \mathcal{N}(0, R_t)$. At time 0, (7.1) becomes $X_0 \sim \mathcal{N}(0, \Sigma_0)$.

Matrices Σ_t and R_t are not necessarily invertible: e.g. one may have $\lambda \in \mathbb{R}^{d_x}$, $\lambda \neq 0$, such that $\lambda^T \Sigma_t \lambda = 0$, and thus $\lambda^T U_t = 0$ almost surely. In that case, the transition distribution does not admit a Lebesgue density. (For the sake of simplicity, we will actually assume that R_t is invertible in order to simplify the proof of some intermediate results, but the expressions we will obtain are valid even when R_t is not invertible.)

© Springer Nature Switzerland AG 2020

N. Chopin, O. Papaspiliopoulos, *An Introduction to Sequential Monte Carlo*, Springer Series in Statistics, https://doi.org/10.1007/978-3-030-47845-2_7

For this family of models the filtering and smoothing distributions evolve within the family of Gaussian distributions, which we will denote by $\mathcal{N}(m_t, Q_t)$ for the filter and $\mathcal{N}(\overleftarrow{m}_t, \overleftarrow{Q}_t)$ for the smoother. The cost-to-go functions also have a parametric representation,

$$H_{t:T}(x) = c_t \exp\left\{-\frac{1}{2}x^T C_t x + u_t^T x\right\}$$

where c_t is constant in x, C_t is a positive semi-definite matrix and u_t a vector. In this section we obtain forward and backward recursions for the computation of the parameters in these distributions and functions. A key result that we will repeatedly use is the following Lemma.

Lemma 7.1 *Suppose that $X \sim \mathcal{N}(m_0, Q_0)$ and $Y|X \sim \mathcal{N}(BX, R)$ where R and Q_0 are positive semi-definite matrices. Then $X|Y = y \sim \mathcal{N}(m_1, Q_1)$, with*

$$\begin{aligned} Q_1 &= (I_{d_x} + Q_0 B^T R^{-1} B)^{-1} Q_0 \\ m_1 &= (I_{d_x} + Q_0 B^T R^{-1} B)^{-1} m_0 + Q_1 B^T R^{-1} y \,. \end{aligned} \tag{7.3}$$

The following are alternative expressions for the same quantities:

$$\begin{aligned} Q_1 &= Q_0[I_{d_x} - B^T (B Q_0 B^T + R)^{-1} B Q_0] \\ m_1 &= [I_{d_x} - Q_0 B^T (B Q_0 B^T + R)^{-1} B] m_0 + Q_0 B^T (B Q_0 B^T + R)^{-1} y \,. \end{aligned} \tag{7.4}$$

This lemma can be proved in a variety of ways; one is by properties of Gaussian distribution, by writing down the joint distribution of X, Y and then conditioning on Y; another (that we prefer) is by a formal calculation and application of Schur's complement (see Exercise 7.1); a third is to parameterise the problem in terms of the innovation $Z \sim \mathcal{N}(0, I_{d_x})$, $X = m_0 + Q_0^{1/2} Z$, and then use Bayes theorem by completing the square and exploiting the fact that now the covariance matrices in the prior and likelihood are invertible. The method that is more suitable and easier to work with depends on the context (and one's previous training).

Additionally, by applying Schur's complement and matrix algebra creatively one can obtain various alternative expressions that may or may not require Q_0 and R to be invertible: in fact, expression (7.3) requires R to be invertible, while (7.4) does not require Q_0 or R to be invertible. This machinery can also be used to obtain expressions that allow matrix inversion at the lowest possible dimension. Note in particular that (7.3) involves inverses of matrices of size $d_y \times d_y$ and $d_x \times d_x$, while (7.4) involves only inverses of size $d_y \times d_y$. In many applications, $d_x \geq d_y$, i.e. the dimension of X_t is larger than that of Y_t, hence the latter set of expressions is computationally preferable.

7.2 Kalman Forward and Backward Recursions

We first obtain the filtering recursions. Since we rely on the recursions obtained in Chap. 5, we use again the notations corresponding to the bootstrap formalism, i.e. $\mathbb{Q}_t(dx_t)$ stands for the filtering distribution $\mathbb{P}_t(dx_t|Y_{0:t} = y_{0:t})$, $\mathbb{Q}_{t-1}(dx_t)$ stands for the predictive distribution $\mathbb{P}_t(dx_t|Y_{0:t-1} = y_{0:t-1})$, and so on.

Suppose that $\mathbb{Q}_{t-1}(dx_{t-1}) = \mathcal{N}(m_{t-1}, Q_{t-1})$. Since $X_t = A_t X_{t-1} + U_t$, $U_t \sim \mathcal{N}(0, \Sigma_t)$, straightforward Gaussian calculation yields that

$$\mathbb{Q}_{t-1}(dx_t) = \mathcal{N}(A_t m_{t-1}, A_t Q_{t-1} A_t^T + \Sigma_t)\,.$$

The variance of the predictive distribution will reappear on various formulae, hence it is convenient to define

$$E_t = A_t Q_{t-1} A_t^T + \Sigma_t\,. \tag{7.5}$$

Appealing directly to Lemma 7.1 we obtain that

$$\begin{aligned}
&\mathbb{Q}_t(dx_t) = \mathcal{N}(m_t, Q_t)\\
&\quad Q_t = E_t[I_{d_x} - B_t^T(B_t E_t B_t^T + R_t)^{-1} B_t E_t]\\
&\quad m_t = [I_{d_x} - E_t B_t^T(B_t E_t B_t^T + R_t)^{-1} B_t] A_t m_{t-1} + E_t B_t^T(B_t E_t B_t^T + R_t)^{-1} y_t
\end{aligned} \tag{7.6}$$

where neither R_t nor Q_t are assumed to be invertible, and the size of matrices being inverted is $d_y \times d_y$.

We can obtain the likelihood factors working directly from the expression of Sect. 5.2 and using an elementary Gaussian calculation: ℓ_t is the density at point y_t of the Gaussian $\mathcal{N}\left(B_t A_t m_{t-1}, B_t E_t B_t^T + R_t\right)$ hence

$$\ell_t = \frac{\exp\{-\frac{1}{2}(y_t - B_t A_t m_{t-1})^T(B_t E_t B_t^T + R_t)^{-1}(y_t - B_t A_t m_{t-1})\}}{(2\pi)^{d_y/2}|B_t E_t B_t^T + R_t|^{1/2}}\,, \tag{7.7}$$

where for a matrix A, $|A|$ is its determinant.

Working from the basic recursions of Sect. 5.3, we observe that the cost-to-go functions have the following parametric representation:

$$H_{t:T}(x) = c_t \exp\left\{-\frac{1}{2}x^T C_t x + u_t^T x\right\}$$

where c_t is constant in x, C_t is a positive semi-definite matrix and u_t a vector; these quantities are obtained by a backward recursion, given below:

$$\begin{aligned}
C_t &:= A_{t+1}\left(C_{t+1} + B_{t+1}^T R_{t+1}^{-1} B_{t+1}\right)\Gamma_t A_{t+1} \\
u_t &:= A_{t+1}^T \Gamma_t^T \gamma_t \\
c_t &:= (2\pi)^{-d_y/2}|R_t|^{-1/2}|\Gamma|^{1/2}\exp\left\{-\frac{1}{2}y_{t+1}^T R_{t+1}^{-1} y_{t+1} + \gamma_t \Gamma_t \Sigma_{t+1}\gamma_t\right\} c_{t+1} \\
\gamma_t &:= u_{t+1} + B_{t+1}^T R_{t+1}^{-1} y_{t+1} \\
\Gamma_t &:= \left[\Sigma_{t+1}(C_{t+1} + B_{t+1}^T R_{t+1}^{-1} B_{t+1}) + I_{d_x}\right]^{-1}
\end{aligned} \tag{7.8}$$

with $C_T = 0$, $u_T = 0$ and $c_T = 1$. The only assumption in the recursions is that R_t^{-1} exists.

The cost-to-go function is proportional to a likelihood arising from the model $U_t \sim \mathcal{N}(C_t X_t, C_t)$. Therefore, appealing again to Lemma 7.1, and using Corollary 5.1 we get that the smoothing distributions are Gaussian,

$$\begin{aligned}
\mathbb{Q}_T(\mathrm{d}x_t) &= \mathcal{N}(\overleftarrow{m}_t, \overleftarrow{Q}_t) \\
\overleftarrow{m}_t &= m_t + Q_t(C_t Q_t + I_{d_x})^{-1}(u_t - C_t m_t) \\
\overleftarrow{Q}_t &= Q_t(C_t Q_t + I_{d_x})^{-1}.
\end{aligned}$$

The resulting formulae are not those commonly encountered for the smoothing distributions in linear Gaussian state-space models. We may obtain those from the following two results:

1. The smoothing distributions are Gaussian (this is already evident from the previous analysis);
2. $p(x_t|y_{0:T}, x_{t+1}) = p(x_t|y_{0:t}, x_{t+1}) \propto p(x_t|y_{1:t})p(x_{t+1}|x_t)$,

where $\propto$ means that the density (in x_t) on the left equals that on the right up to multiplicative terms (that can depend on anything but x_t, e.g., on $y_{0:t}$). Using the second result, and appealing to Lemma 7.1 we directly get that

$$X_t|Y_{0:t} = y_{0,t}, X_{t+1} \sim \mathcal{N}(m_t + D_t(X_{t+1} - A_{t+1}m_t), Q_t - D_t A_{t+1} Q_t),$$

for

$$D_t = Q_t A_{t+1}^T (A_{t+1} Q_t A_{t+1}^T + \Sigma_{t+1})^{-1}.$$

Combined with the first result above we obtain an alternative backward recursion for the parameters of the Gaussian smoothing distributions. Combined with the identity

$$A_{t+1}Q_t = (A_{t+1}Q_tA_{t+1}^T + \Sigma_{t+1})D_t^T$$

we obtain a backward recursion that it is commonly referred to as the Rauch-Tung-Striebel smoother,

$$\begin{aligned}\overleftarrow{m}_t &= m_t + D_t(\overleftarrow{m}_{t+1} - A_{t+1}m_t) \\ \overleftarrow{Q}_t &= Q_t + D_t(\overleftarrow{Q}_{t+1} - A_{t+1}Q_tA_{t+1}^T - \Sigma_{t+1})D_t^T .\end{aligned} \tag{7.9}$$

The following frame summarises the key recursions for linear Gaussian state-space models.

Recursions for Linear Gaussian State-Space Models

- **Prediction:**

$$X_t | Y_{0:t-1} = y_{0:t-1} \sim \mathcal{N}(A_t m_{t-1}, A_t Q_{t-1} A_t^T + \Sigma_t) .$$

- **Filter:**

$$X_t | Y_{0:t} = y_{0:t} \sim \mathcal{N}(m_t, Q_t) .$$

- **Likelihood factors:**

$$p_t(y_t | y_{0:t-1}) = \ell_t .$$

- **Likelihood of future observations given current state:**

$$p_T(y_{t+1:T} | x) = c_t \exp\{-\frac{1}{2} x^T C_t x + u_t^T x\} .$$

- **Smoother:**

$$X_t | Y_{0:T} = y_{0:T} \sim \mathcal{N}(\overleftarrow{m}_t, \overleftarrow{Q}_t) .$$

In the above, m_t, Q_t are given in (7.6), ℓ_t in (7.7), c_t, C_t, u_t in (7.8), and $\overleftarrow{m}_t$, $\overleftarrow{Q}_t$ in (7.9).

7.3 Numerical Complexity and Stability

For concreteness we assume that $d_x \geq d_y$, which is the most common situation in applications of these models. We carefully avoided writing inverses of $d_x \times d_x$ matrices in the forward recursion of the Kalman filter. Unfortunately the complexity of the Kalman filter remains $\mathcal{O}(d_x^3)$ per time step, as it involves matrix-matrix multiplications of size $d_x \times d_x$, see e.g. (7.5). Therefore, the overall complexity of the recursions is $\mathcal{O}(d_x^3)$ per time step. On the other hand, avoiding these matrix inverses should significantly reduce the constant in front of the d_x^3 and in practice leads to a more efficient and numerically stable implementation.

As the practical implementation of the Kalman filter relies on numerical linear algebra, a potential issue is that round-off errors accumulate and render invertible matrices non-invertible. So-called square root filters (which store and propagate square roots of the matrices involved in the computation) have been developed to address this issue. We return to this point in the Python corner.

Exercises

7.1 *Prove Lemma 7.1.* Hint and a guide to a practical way to deriving complicated-looking formulae easily! *It is easy to verify that the formula for $X|Y = y$ is correct in the following way. Using the "forward" definition of the model, in terms of $Y|X$, work out the marginal moments of Y and $\mathbb{E}[XY^T]$. By construction, the joint distribution of (X, Y) is Gaussian. Then, using the "backward" transition density for $X|Y$ given in* (7.1)*, work out the marginal moments of X and the implied $\mathbb{E}[XY^T]$. What you will find in this way is that the marginal moments and the covariance of (X, Y) are the same, computed either "forwards" or "backwards", hence we obtain the correct joint distribution for (X, Y) using $X|Y$ as given in* (7.1)*. The point of the exercise is to establish a direct way to obtain the formula. One such way is to work* formally*, in the sense of the term as used in Applied Mathematics, see for example https://en.wikipedia.org/wiki/Formal_calculation. For example, assuming that both R and Q_0 are invertible, it is easy to obtain (e.g. by Bayes theorem using the Gaussian densities) that $Q_1^{-1} = Q_0^{-1} + A^T R^{-1} A$, from which we have $Q_1 = (Q_0^{-1} + A^T R^{-1} A)^{-1}$. This formula relies on Q_0 being positive definite. Basic properties of matrix inverse and multiplication yields the equal expression $Q_1 = (I + Q_0 A^T R^{-1} A)^{-1} Q_0$, where now Q_0^{-1} does not appear. Although we have derived the formula "cheating" that Q_0 is positive definite, the final answer can directly be verified that it is a valid one even when Q_0 is semi-definite. Hence, the "cheating" was only a convenient way to obtain a useful expression. Another tool indispensable in obtaining matrix identities is Schur's complement, see e.g. https://en.wikipedia.org/wiki/Schur_complement; this is so central in the derivation of matrix identities that it is often called the matrix inversion lemma. One version of this result is the following identity, for matrices A, B, C of appropriate dimensions* *and where inverses are assumed to exist where they appear:*

$$A(BA + C)^{-1}B = I - (I + AC^{-1}B)^{-1}.$$

(This identity is also referred to as Woodbury matrix identity.)

Python Corner

`NumPy` and related numerical libraries do not currently implement Kalman filtering. We thus decided to implement it in `particles`; see module `kalman`.

One benefit of developing our own implementation is that we can re-use parts of it to compute proposal distributions for state-space models that are not linear and Gaussian; we shall return to this point in Chap. 10.

Another way to run a Kalman filter in Python is to call through `rpy2` one of the several R libraries that implement Kalman filtering. Tusell (2011) reviews and compares five such libraries. In particular, the paper mentions concerns in the literature regarding the numerical stability of the Kalman filter, when transcribed directly from its equations. To address this concern, one may propagate (one version of) the square root of the covariance matrix of the filter. The resulting algorithm (called square root covariance filter) is typically 2–3 times slower, but it has the advantage of always producing filtering covariances that are definite positive.

Surprisingly, Tusell (2011) did not observe any significant differences (in the considered examples) between the results obtained from square root filtering (as implemented in package `dlm`) and from standard Kalman filtering (as implemented in the other considered packages). Still, it is worth keeping in mind that the standard Kalman filter might be numerically unstable in difficult problems (e.g. high-dimensional state-spaces or near singular matrix parameters).

Bibliographical Notes

The historical papers on Kalman filtering are Kalman (1960) and Kalman and Bucy (1961). Stratonovitch derived around the same time the Kalman filter as a special case of a more general theory on filtering. Because of this, some authors call the method the Kalman-Bucy filter, or the Stratonovitch-Kalman-Bucy filter. The first numerical implementation of the Kalman filter was used in the navigation system of the Appolo project, which accomplished landing human beings on the Moon.

In certain applications (e.g. weather forecasting) d_x and d_y are so large that the update formulae of the Kalman filter become computationally infeasible. One may then use an approximate algorithm known as the ensemble Kalman filter; see e.g. the books of Reich and Cotter (2015) and Law et al. (2015) for more details.

It is worth having in mind that the recursions we develop in this chapter hold more broadly for processes on trees. The models we consider throughout the book involve a latent stochastic process indexed by time. Another perspective on this family of models is that each latent variable generates a single offspring. One can consider more general stochastic processes where each latent variable generates a number of offsprings, whereas still each latent variable has a single parent. The corresponding graphical models are trees. If all the distributional links among the latent variables are linear-Gaussian, computations, such as computing smoothing distributions and the likelihood, can be carried out by a little variation of the formulae we obtain in this chapter. The whole framework turns out to be a version of what is known as belief propagation; see Papaspiliopoulos and Zanella (2017).

Bibliography

Kalman, R. E. (1960). A new approach to linear filtering and prediction problems. *Transactions of the ASME–Journal of Basic Engineering, 82*(Series D), 35–45.

Kalman, R. E., & Bucy, R. S. (1961). New results in linear filtering and prediction theory. *Transactions of the ASME. Series D, Journal of Basic Engineering, 83*, 95–108.

Law, K., Stuart, A., & Zygalakis, K. (2015). *Data assimilation. Texts in applied mathematics* (Vol. 62). Cham: Springer. A mathematical introduction.

Papaspiliopoulos, O., & Zanella, G. (2017). A note on MCMC for nested multilevel regression models via belief propagation. *arXiv e-prints 1704.06064*.

Reich, S., & Cotter, C. (2015). *Probabilistic forecasting and Bayesian data assimilation*. New York: Cambridge University Press.

Tusell, F. (2011). Kalman filtering in R. *Journal of Statistical Software, 39*(2), 1–27.

Chapter 8
Importance Sampling

Summary Roughly speaking, a particle filter is an algorithm that iterates importance sampling and resampling steps, in order to approximate a sequence of filtering (or related) distributions. This chapter covers the basics of importance sampling; resampling will be treated in the following chapter. We describe in particular the two versions of importance sampling (normalised and auto-normalised), we make a connection with the probabilistic notion of change of measure, we derive the variance of importance sampling estimators, we describe different empirical measures of efficiency such as effective sample size, we discuss the inherent curse of dimensionality of importance sampling, and we extend importance sampling to situations where importance weights are randomised.

8.1 Monte Carlo

Monte Carlo is a well known and very popular numerical method, which offers for any quantity that may written as an expectation

$$\mathbb{E}_q[\varphi(X)] = \int_{\mathcal{X}} \varphi(x) q(x)\, \mathrm{d}x$$

with respect to some probability density q (relative to measure $\mathrm{d}x$) the following approximation

$$\frac{1}{N}\sum_{n=1}^{N} \varphi(X^n), \qquad X^n \sim q, \tag{8.1}$$

where the second part above means that the X^n's are N i.i.d. random variables with common density $q(x)$. For the sake of exposition, we assume that $\varphi : \mathcal{X} \to \mathbb{R}$, but extension to vectors (or other spaces) of the discussion below should be trivial by working component-wise.

© Springer Nature Switzerland AG 2020

N. Chopin, O. Papaspiliopoulos, *An Introduction to Sequential Monte Carlo*, Springer Series in Statistics, https://doi.org/10.1007/978-3-030-47845-2_8

The approximation error may be assessed in various ways. For instance, assuming that $\mathbb{E}_q[\varphi(X)^2] < +\infty$, the mean square error (MSE) is

$$\text{MSE}\left[\frac{1}{N}\sum_{n=1}^{N}\varphi(X^n)\right] = \mathbb{E}\left[\left\{\frac{1}{N}\sum_{n=1}^{N}\varphi(X^n) - \mathbb{E}_q[\varphi(X)]\right\}^2\right] = \frac{1}{N}\text{Var}_q[\varphi(X)]$$

since (8.1) is an unbiased estimator of $\mathbb{E}_q[\varphi(X)]$; note $\text{Var}_q[\varphi(X)] = \mathbb{E}_q[\varphi(X)^2] - \{\mathbb{E}_q[\varphi(X)]\}^2$ is simply the variance of $\varphi(X)$ (relative to density q). This leads to two remarks. First, by Markov's inequality, the approximation error vanishes at rate $\mathcal{O}_P(N^{-1/2})$, which is rather slow: to divide the error by 10, you need 100 more samples. Second, the P in the $\mathcal{O}_P$ symbol means that the bound on the approximation error is stochastic: it is possible to state that the error is below a certain threshold with some large probability, but not with complete certainty.

For instance, applying the law of large numbers and the central limit theorem, one gets that, as $N \to +\infty$,

$$\left\{\frac{1}{N}\sum_{n=1}^{N}\varphi(X^n) - \mathbb{E}_q[\varphi(X)]\right\} \to 0 \quad \text{a.s.}$$

$$N^{1/2}\left\{\frac{1}{N}\sum_{n=1}^{N}\varphi(X^n) - \mathbb{E}_q[\varphi(X)]\right\} \Rightarrow \mathcal{N}\left(0, \text{Var}_q[\varphi(X)]\right),$$

where $\Rightarrow$ means convergence in distribution, and assuming that $\mathbb{E}_q[\varphi(X)^2] < +\infty$. The latter result implies that the confidence interval

$$\left[\frac{1}{N}\sum_{n=1}^{N}\varphi(X^n) \pm \frac{z_{1-\alpha/2}}{N^{1/2}}\sqrt{\text{Var}_q[\varphi(X)]}\right]$$

where $z_{1-\alpha/2}$ is the $(1-\alpha/2)$-quantile of the $\mathcal{N}(0, 1)$ distribution has limiting (as $N \to +\infty$) probability $1-\alpha$ to contain the true value of $\mathbb{E}_q[\varphi(X)]$. It is good practice to report such a confidence interval alongside the value of the estimator; one may take e.g. $z_{1-\alpha/2} = 3$ to have $\alpha \approx 2.5 \times 10^{-3}$. (Of course, this is an arbitrary choice, but in the context of scientific calculations this seems more sensible that the conventional value $\alpha = 0.05$, giving $z_{1-\alpha/2} \approx 1.96$). When $\text{Var}_q[\varphi(X)]$ cannot be computed (which is most of the time), it may be replaced by the (N^{-1} as opposed to the $(N-1)^{-1}$ version of the) empirical variance of the $\varphi(X^n)$

$$\frac{1}{N}\sum_{n=1}^{N}\left\{\varphi(X^n) - \frac{1}{N}\sum_{m=1}^{N}\varphi(X^m)\right\}^2 = \frac{1}{N}\sum_{n=1}^{N}\{\varphi(X^n)\}^2 - \left\{\frac{1}{N}\sum_{n=1}^{N}\varphi(X^n)\right\}^2$$

which is also a Monte Carlo estimate, this time of $\mathrm{Var}_q[\varphi(X)] = \mathbb{E}_q[\varphi(X)^2] - \{\mathbb{E}_q[\varphi(X)]\}^2$.

Notation/Terminology We will broadly refer to a function of Monte Carlo estimators of certain quantities as the Monte Carlo estimator of the function of the quantities. In the example above, $\mathrm{Var}_q[\varphi(X)] = f(\mathbb{E}_q[\varphi(X)^2], \mathbb{E}_q[\varphi(X)])$ is estimated by plugging-in Monte Carlo estimators of $\mathbb{E}_q[\varphi(X)^2]$ and $\mathbb{E}_q[\varphi(X)]$.

So far we have assumed that $\mathbb{E}_q[\varphi(X)^2] < +\infty$. It is safe to say that the Monte Carlo method is mostly useless when this condition is not met. In such a case, the approximation error may either not converge, or converge at too slow a rate for practical use. Ideally one should always check that $\mathbb{E}_q[\varphi(X)^2] < +\infty$ before using Monte Carlo. We will assume that this condition holds in the rest of this chapter.

8.2 Basics of Random Variable Generation

We now briefly outline some bare essentials on random variable generation that any Monte Carlo user should know.

Computers cannot generate true randomness, because they are deterministic machines. However, they can generate pseudo-random sequences, that is, deterministic sequences that behave for most practical purposes like a sequence of independent random variables.

In particular, basic PRNGs (pseudo-random number generators) usually consist of an iterative sequence of the form $u^n = \Psi(u^{n-1})$, $n > 0$, taking values in $0 : (k - 1)$ (e.g. $k = 2^{32}$ or 2^{64}). Once the "seed" (starting value) u^0 is chosen, the sequence is completely fixed and periodic, with a period smaller than or equal to k (as it is bound to repeat itself at least once before iteration $k + 1$). Yet, for a good choice of Ψ, this sequence will approximate reasonably well a string of independent variables from the uniform distribution on the set $0 : (k - 1)$.

From such a discrete generator, one may construct a generator for the continuous distribution $\mathcal{U}([0, 1])$ by taking $v^n = u^n/k$. To generate non-uniform random variables, various tricks may be used. The simplest one is based on the inverse CDF (cumulative distribution function).

Proposition 8.1 *Let X a real-valued random variable, and let F^{-1} its inverse CDF function, i.e. $F^{-1}(u) = \inf\{x : F(x) \geq u\}$, then $F^{-1}(U)$, $U \sim \mathcal{U}([0, 1])$, has the same distribution as X.*

For a proof, see Exercise 8.1.

The trick above works only for scalar random variables. Another, more general trick is rejection sampling. This latter approach requires to be able to simulate from a proposal distribution, with density m, such that the ratio $q(x)/m(x)$ is always defined, and may be evaluated up to a constant.

Algorithm 8.1: Rejection sampling

Input: A function $w(x)$ such that $w(x) \propto q(x)/m(x)$, and a constant C such that $w(x) \leq C$.

Output: A draw $X \sim q$.

repeat

  $X \sim m$

  $U \sim \mathcal{U}([0, 1])$

until $CU \leq w(X)$

Exercise 8.2 establishes that this algorithm indeed yields draws from q. Exercise 8.4 illustrates rejection sampling through two simple examples.

This overview is awfully brief, and glosses over important subtleties and pitfalls of random variable generation; see the bibliography for references to more in-depth materials. For uniform sampling, we strongly recommend non-experts to rely on well tested implementations of modern, advanced generators (such as the Mersenne twister, which is the default generator in, e.g., R, Python or Matlab) rather than trying to implement manually more basic (but obsolete) generators, as this might introduce undesired artefacts in numerical experiments.

For non-uniform sampling, one thing to remember is that random variables are always obtained as some (possibly non-trivial) transform of uniform variables; we shall return to this point when discussing quasi-Monte Carlo in Chap. 13.

8.3 Importance Sampling

8.3.1 Normalised Importance Sampling

Standard Monte Carlo requires simulating from target density q. Unfortunately, this is sometimes difficult or even impossible. When this happens, one may instead use *importance sampling*, a technique that approximates expectations with respect to q, using simulations from some other density m.

Importance sampling is based on the following trivial identity for any pair (m, q) of probability densities defined with respect to the same measure dx:

$$\int_{\mathcal{X}} \varphi(x)q(x)\mathrm{d}x = \int_{\mathcal{X}} \varphi(x)\frac{q(x)}{m(x)}m(x)\mathrm{d}x$$

where the ratio $w = q/m$ is called the importance weight function, $q(x)$ the target density, and $m(x)$ the proposal density. The identity above is correct provided $m(x) > 0$ whenever $q(x)\varphi(x) > 0$. A sufficient condition is that the support of m contains the support of q; see Sect. 8.4 for a more rigorous discussion of this point.

The identity above leads to the Monte Carlo estimator

$$\frac{1}{N}\sum_{n=1}^{N} w(X^n)\varphi(X^n), \qquad w(x) = \frac{q(x)}{m(x)}, \qquad X^n \sim m$$

which is unbiased with

$$\mathrm{MSE}\left[\frac{1}{N}\sum_{n=1}^{N} w(X^n)\varphi(X^n)\right] = \frac{1}{N}\mathrm{Var}_m[w(X)\varphi(X)]$$

assuming $\mathbb{E}_m[w(X)^2\varphi(X)^2] < \infty$; we return to this result in Proposition 8.2 and Sect. 8.5. We call the above the normalised importance sampling estimator from now on since the weight is properly normalised to satisfy $\mathbb{E}_m[w(X)] = 1$; this will differentiate it from another introduced in the next section, which does not assume such normalisation.

As discussed in the previous section, this estimator is pretty much useless if the condition $\mathbb{E}_m[w(X)^2\varphi(X)^2] < \infty$ does not hold. Provided that $\mathbb{E}_q[\varphi(X)^2] < \infty$, a sufficient condition for $\mathbb{E}_m[w(X)^2\varphi(X)^2] < \infty$ is that $w(x)$ is upper bounded in x. Note that bounded weights cannot be achieved if there are directions in the space along which the tail of the target $q(x)$ dies out slower than that of the proposal $m(x)$.

Example 8.1 Assume $\mathcal{X} = \mathbb{R}$ and $q(x) = \phi(x)\mathbb{1}\{x > c\}/\Phi(-c)$, where ϕ is the PDF of a $\mathcal{N}(0, 1)$ random variable function, $\phi(x) = (2\pi)^{-1/2}\exp(-x^2/2)$, and Φ is the corresponding CDF. In words, q is the density of the distribution of $X \sim \mathcal{N}(0, 1)$, conditional on $X > c$, which is called a truncated Gaussian distribution. For c large, the support of q becomes small, which might suggest to do importance sampling with proposal $\mathcal{N}(\mu, \sigma^2)$, with $\mu = c$ (or a bit above), and σ small. Unfortunately, in this case,

$$\mathbb{E}_m[w(X)^2] = \frac{\sigma}{\Phi(-c)^2\sqrt{2\pi}}\int_c^{\infty} \exp\left\{-x^2 + \frac{(x-\mu)^2}{2\sigma^2}\right\} \mathrm{d}x = +\infty$$

(continued)

Example 8.1 (continued)
as soon as $\sigma^2 < 1/2$. That is, if σ is too small, the tails of the proposal distribution become too light relative to the tails of the target distribution. A better proposal is $m(x) = \lambda \exp\{-\lambda(x-c)\}\mathbb{1}\{x \geq c\}$, an exponential shifted to start at c; then the variance of the weights is finite whatever $\lambda > 0$, and q clearly has lighter tails than m. See Exercise 8.5 for a discussion on how to choose λ.

8.3.2 Auto-Normalised Importance Sampling

In many practical settings, and in particular those this book is concerned with, the weight function $w = q/m$ may be computed *only up to a multiplicative constant*. This happens typically when $q(x) = q_u(x)/L_q$, $m(x) = m_u(x)/L_m$, where both q_u and m_u may be evaluated point-wise, but either L_q or L_m, or both are intractable. This is why the first importance sampling estimator introduced above is dubbed as *normalised*, as it cannot be used in this situation. Instead, one replaces the first importance sampling identity, (8.3.1), by the following one:

$$\int_{\mathcal{X}} \varphi(x)q(x)\,\mathrm{d}x = \frac{\int_{\mathcal{X}} \varphi(x)\frac{q(x)}{m(x)}m(x)\,\mathrm{d}x}{\int_{\mathcal{X}} \frac{q(x)}{m(x)}m(x)\,\mathrm{d}x} = \frac{\int_{\mathcal{X}} \varphi(x)\frac{q_u(x)}{m_u(x)}m(x)\,\mathrm{d}x}{\int_{\mathcal{X}} \frac{q_u(x)}{m_u(x)}m(x)\,\mathrm{d}x}$$

where the intractable normalising constants conveniently cancel out. (Again, this identity is valid provided that the support of m contains the support of q, but see next section for a more formal statement.) This second identity leads to the auto-normalised importance sampling estimator:

$$\frac{\sum_{n=1}^N w(X^n)\varphi(X^n)}{\sum_{n=1}^N w(X^n)}, \qquad w(x) = \frac{q_u(x)}{m_u(x)}, \qquad X^n \sim m\,.$$

Contrary to the normalised estimator, the auto-normalised estimator is biased, because it is a ratio of unbiased estimators. Its bias and its variance are hard to compute; we return to these considerations in Sect. 8.5. In the meantime, the reader might wish to think which of the two importance sampling estimators is preferable in terms of MSE; see Exercise 8.7.

8.3.3 How to Choose the Proposal Density?

Given $\mathbb{E}_q[\varphi(X)]$, a question of practical interest is how to choose the proposal distribution m. Consider first normalised importance sampling, where it is assumed that $w = q/m$ can be computed. Of course, a first requirement is that it is reasonably easy to sample random variables from m. A second natural requirement is that the Monte Carlo error is small. One then may ask how to choose m so that the MSE is as small as possible.

Proposition 8.2 *The MSE of the normalised IS estimator of* $\mathbb{E}_q[\varphi(X)]$, *is minimised with respect to m by taking* $m = m^\star$,

$$m^\star(x) = \frac{q(x)\,|\varphi(x)|}{\int_{\mathcal{X}} |\varphi(x')|\, q(x')\,\mathrm{d}x'}.$$

Proof This is a direct consequence of Jensen's inequality. First, note that since the estimator is unbiased its MSE equals its variance. Then,

$$\begin{aligned}\mathrm{Var}_m\left[q(X)\varphi(X)/m(X)\right] &= \mathbb{E}_m\left[q(X)^2\varphi(X)^2/m(X)^2\right] - \mathbb{E}_q[\varphi(X)]^2 \\ &\geq \left\{\mathbb{E}_q\left[|\varphi(X)|\right]\right\}^2 - \mathbb{E}_q[\varphi(X)]^2 \\ &= \mathrm{Var}_{m^\star}\left[q(X)\varphi(X)/m^\star(X)\right].\end{aligned}$$

□

For $\varphi \geq 0$ and $m = m^\star$, the estimator equals the estimated quantity

$$\frac{1}{N}\sum_{n=1}^{N} w(X^n)\varphi(X^n) = \mathbb{E}_q[\varphi(X)] \quad \text{a.s.},$$

and has variance zero. More generally, for $m = m^\star$,

$$\frac{1}{N}\sum_{n=1}^{N} w(X^n)\varphi(X^n) = \mathbb{E}_q\left[|\varphi(X)|\right]\left\{\frac{1}{N}\sum_{n=1}^{N}\mathrm{sign}(\varphi(X^n))\right\}.$$

The applicability of this optimality result is therefore quite limited: to compute our estimator we need to compute $\mathbb{E}_q[|\varphi(X)|]$, which will typically be unavailable, and is precisely the quantity of interest when $\varphi \geq 0$. A more practical recommendation is that the proposal distribution should be chosen as close as possible to $m^\star$ (or more simply to q, since $m^\star(x) \propto q(x)|\varphi(x)|$) within a class of tractable distributions (i.e. distributions we are able to simulate from).

An analogous result can be obtained for the proposal density that minimises the *asymptotic* variance of the auto-normalised importance sampling estimator. The result is obtained using again Jensen's inequality on an expression for the asymptotic variance obtained in Sect. 8.5. The result is that the optimal choice is

$$m^\star(x) = \frac{q(x)|\overline{\varphi}(x)|}{\int_{\mathcal{X}} q(x')|\overline{\varphi}(x')|\,\mathrm{d}x'}$$

with $\overline{\varphi}(x) = \varphi(x) - \mathbb{E}_q[\varphi(X)]$, and minimised variance given by $\left\{\mathbb{E}_q[|\overline{\varphi}(X)|]\right\}^2$; see Exercise 8.9. Note that this is not zero unless φ is (almost surely) a constant. The same remarks apply regarding the usefulness of this result.

There is yet another criticism of the usefulness of this type of optimality results within the context of statistical inference. Typically, we are interested in approximating the expectations under q of several different test functions φ. Therefore, it does not make sense to try and tune the proposal to one of them. We wish instead to find a proposal that works fairly well for a broad family of test functions. This idea leads to a different perspective on importance sampling, that of a particle approximation, developed next.

8.4 A More Formal Perspective on Importance Sampling

So far our treatment of Monte Carlo and importance sampling have been rather elementary, but in this section we would like to adopt a more formal perspective, with a view towards gaining some insights that will be particularly useful in the context of this book. We focus on the auto-normalised importance sampling estimator, as it is more generally applicable, and forms the backbone of all particle filter algorithms. Our new more formal framework is achieved via three steps.

The first step—in complete analogy to the previous chapters in this book—is to consider expectations of test functions φ with respect to probability measure $\mathbb{Q}$, rather than with respect to density q; we shall use the standard (in Probability) notation $\mathbb{Q}(\varphi)$ for such expectations; see the note on this notation in Sect. 4.2.

The second step is casting importance sampling in the more formal context of absolute continuity and Radon-Nikodym derivatives of Sect. 4.2. In particular, we re-interpret the importance sampling identity (8.3.1) as a change of measure, from $\mathbb{M}$ to $\mathbb{Q}$.

Notation/Terminology In the rest of this chapter and this book more broadly, the potentially un-normalised Radon-Nikodym derivative between $\mathbb{Q}$ and $\mathbb{M}$ will be denoted by $w(x)$, i.e.

$$w(x) \propto \frac{\mathbb{Q}(\mathrm{d}x)}{\mathbb{M}(\mathrm{d}x)},$$

with $\mathbb{M}(w)$ being the necessary constant to obtain the normalised derivative. In previous Sections we wrote $w(x)$ for the normalised derivative, but, since this is hardly ever available in non-toy problems, we reserve the notation for what is typically available, the un-normalised version and avoid a special notation for the normalised one.

Our main focus is the auto-normalised importance sampling estimator, whose validity follows from Lemma 4.3 and the identity:

$$\mathbb{M}(\varphi w) = \mathbb{Q}(\varphi)\mathbb{M}(w),$$

for any un-normalised weight function w. This identity suggests the following estimator of $\mathbb{Q}(\varphi)$,

$$\sum_{n=1}^{N} W^n \varphi(X^n), \qquad W^n = \frac{w(X^n)}{\sum_{m=1}^{N} w(X^m)}, \qquad X^n \sim \mathbb{M}.$$

Notation/Terminology W^n will denote the auto-normalised weights as defined above. We will occasionally write w^n as a shorthand notation for $w(X^n)$.

The third step is switching attention from thinking about this as an estimator for *one* test function φ to interpreting it as an approximation for a *class* of test functions φ. In particular, we interpret the estimator as the expectation of φ with respect to the random probability measure

$$\mathbb{Q}^N(\mathrm{d}x) := \sum_{n=1}^{N} W^n \delta_{X^n}(\mathrm{d}x), \qquad X^n \sim \mathbb{M}.$$

This probability measure is a mixture of Dirac masses at points X^n. Thus, and in accordance to the terminology already established, the auto-normalised estimator of $\mathbb{Q}(\varphi)$ is precisely $\mathbb{Q}^N(\varphi)$, that is

$$\mathbb{Q}^N(\varphi) = \sum_{n=1}^{N} W^n \varphi(X^n).$$

From this new perspective, importance sampling effectively amounts to approximating a probability measure $\mathbb{Q}$ by another probability measure $\mathbb{Q}^N$, which will be referred to as the particle approximation of $\mathbb{Q}$. We will refer to $(X^{1:N}, W^{1:N})$ as a weighted sample, and we will occasionally use the same term even when the weights are not auto-normalised, e.g. for $(X^{1:N}, w^{1:N})$. Note that the notation does not explicitly show the dependence of the particle approximation on the proposal measure $\mathbb{M}$—a choice made to avoid dealing with horrific formulae. Going further, one may consider the notion of *weak convergence of measures*, i.e. we say that

$$\mathbb{Q}^N \Rightarrow \mathbb{Q}$$

as $N \to +\infty$ provided that $\mathbb{Q}^N(\varphi) \to \mathbb{Q}(\varphi)$ for any function $\varphi : \mathcal{X} \to \mathbb{R}$ that is continuous and bounded. The concept of weak convergence will not be central in this book, but the more general idea of certain random measures being approximations of certain (fixed) probability measures will prove very useful. Next section establishes this weak convergence.

8.5 The MSE of Importance Sampling Estimators

We study the MSE of importance sampling estimators in the following framework: target $\mathbb{Q}$, proposal $\mathbb{M}$, un-normalised weight $w(x)$ and normalised weight $w(x)/\mathbb{M}(w)$. In accordance to the notation already established, the auto-normalised importance sampling estimator of a test function φ is denoted by $\mathbb{Q}^N(\varphi)$, whereas we will write explicitly the expression for the normalised estimator.

For the normalised importance sampling estimator we already have a simple expression for its MSE. We repeat it here for completeness:

$$\text{MSE}\left[\frac{1}{N}\sum_{n=1}^{N}\frac{w(X^n)}{\mathbb{M}(w)}\varphi(X^n)\right] = \frac{1}{N}\frac{\text{Var}_{\mathbb{M}}[w\varphi]}{\mathbb{M}(w)^2} = \frac{1}{N}\frac{\left\{\mathbb{M}(w^2\varphi^2) - \mathbb{Q}(\varphi)^2\right\}}{\mathbb{M}(w)^2}\,.$$

The analysis of the MSE of the auto-normalised importance sampling estimator is more complicated since it involves a ratio of random variables. However, useful results can be obtained either by appealing to asymptotics in N or by giving upper bounds. We discuss both possibilities below. Both approaches are based on the following simple identity:

$$\mathbb{Q}^N(\varphi) - \mathbb{Q}(\varphi) = \frac{N^{-1}\sum_{n=1}^{N} w(X^n)[\varphi(X^n) - \mathbb{Q}(\varphi)]}{N^{-1}\sum_{n=1}^{N} w(X^n)}\,. \tag{8.2}$$

8.5.1 Asymptotic Results

The right hand side of (8.2) is a ratio of two quantities that may be interpreted as *normalised* importance sampling estimators, corresponding to function $\overline{\varphi} = \varphi - \mathbb{Q}(\varphi)$ for the numerator, and to the constant function for the denominator.

Regarding the numerator, provided

$$\mathbb{M}(w^2\overline{\varphi}^2) < \infty\,, \tag{8.3}$$

by a standard application of the CLT for averages

$$\sqrt{N}\left\{\frac{1}{N}\sum_{n=1}^{N} w(X^n)\overline{\varphi}(X^n)\right\} \Rightarrow \mathcal{N}\left(0, \mathbb{M}(w^2\overline{\varphi}^2)\right) .$$

Regarding the denominator, by the strong law of large numbers we get that

$$\frac{1}{N}\sum_{n=1}^{N} w(X^n) \to \mathbb{M}(w) \quad \text{a.s.} .$$

Then, a direct application of what is typically known as Slutsky's theorem shows that

$$\sqrt{N}\left\{\mathbb{Q}^N(\varphi) - \mathbb{Q}(\varphi)\right\} \Rightarrow \mathcal{N}\left(0, \frac{\mathbb{M}(w^2\overline{\varphi}^2)}{\mathbb{M}(w)^2}\right) \tag{8.4}$$

from which we get that for large N,

$$\text{MSE}\left\{\mathbb{Q}^N(\varphi)\right\} \approx \frac{1}{N}\frac{\mathbb{M}(w^2\overline{\varphi}^2)}{\mathbb{M}(w)^2} .$$

The following result gives practical conditions under which the asymptotic variance of the auto-normalised importance sampling estimator is finite for a certain class of functions φ.

Proposition 8.3 *Assuming that $\mathbb{Q}$ is not a trivial probability measure, the condition $\mathbb{M}(w^2\overline{\varphi}^2) < +\infty$, with $\overline{\varphi} = \varphi - \mathbb{Q}(\varphi)$, holds for all bounded functions φ if and only if $\mathbb{M}(w^2) < \infty$.*

Proof Sufficient condition: assume $\mathbb{M}(w^2) < \infty$, and take φ such that $|\varphi| < C$. Then $|\overline{\varphi}| \leq 2C$ and $\mathbb{M}(w^2\overline{\varphi}^2) \leq 4C^2\mathbb{M}(w^2) < +\infty$. Necessary condition: if $\mathbb{Q}$ is not a trivial probability measure, we can choose some set $A \in \mathcal{B}(\mathcal{X})$ such that $0 < \mathbb{Q}(A) \leq 1/2$. Then, taking $\varphi = \mathbb{1}_A$, note that

$$\mathbb{M}(w^2\overline{\varphi}^2) = \mathbb{M}(w^2\mathbb{1}_A)(1 - 2\mathbb{Q}(A)) + \mathbb{Q}(A)^2\mathbb{M}(w^2) .$$

Therefore, if $\mathbb{M}(w^2) = \infty$ then $\mathbb{M}(w^2\overline{\varphi}^2) = \infty$ for some bounded function φ. □

The proposition gives a nice justification on why the variance of the weights relative to $\mathbb{M}$, which equals $\mathbb{M}(w^2) - \mathbb{M}(w)^2$, is a good criterion for the performance of importance sampling when applied to a class of test functions, rather than to a specific test function φ. We return to this point in Sect. 8.6.

8.5.2 Non-asymptotic Results

Non-asymptotic bounds on the MSE of the auto-normalised importance sampling estimator may be obtained from concentration inequalities. The following result is obtained in Agapiou et al. (2017), and makes very strong assumptions on the class of test functions, i.e., bounded, but very weak ones on the weights.

Theorem 8.4

$$\sup_{|\varphi|\leq 1} \text{MSE}\{\mathbb{Q}^N(\varphi)\} \leq \frac{4}{N}\frac{\mathbb{M}(w^2)}{\mathbb{M}(w)^2}\,.$$

Note that this is precisely the upper bound on the MSE obtained by using the asymptotic expression and assuming that $|\overline{\varphi}| \leq 2$; in this sense the bound is not too conservative.

This result is related to the following distance (Rebeschini and van Handel 2015) between two random measures

$$d(\mu, \nu) := \sup_{|\varphi|\leq 1} \left[\mathbb{E}\big(\mu(\varphi) - \nu(\varphi)\big)^2\right]^{1/2}. \tag{8.5}$$

Therefore,

$$d(\mathbb{Q}^N, \mathbb{Q})^2 = \text{MSE}\{\mathbb{Q}^N(\varphi)\}\,,$$

and we also have $N^{-\frac{1}{2}}$ convergence of $\mathbb{Q}^N$ to $\mathbb{Q}$ under this norm provided $\mathbb{M}(w^2) < \infty$. An alternative approach in Agapiou et al. (2017) yields a different bound, obtained under weaker assumptions on the test functions and stronger on the weights.

Theorem 8.5 *For test functions φ and importance sampling weights w with the obvious sufficient regularity as determined by the right hand side, we have the following bound:*

$$\begin{aligned}\text{MSE}\{\mathbb{Q}^N(\varphi)\} \leq & \frac{1}{N}\left(\frac{3}{\mathbb{M}(w)^2}m_2[\varphi w] + \frac{3}{\mathbb{M}(w)^4}\mathbb{M}(|\varphi w|^{2d})^{\frac{1}{d}} C_{2e}^{\frac{1}{e}} m_{2e}[w]^{\frac{1}{e}}\right.\\ &\left. + \frac{3}{\mathbb{M}(w)^{2(1+\frac{1}{p})}}\mathbb{M}(|\varphi|^{2p})^{\frac{1}{p}} C_{2q(1+\frac{1}{p})}^{\frac{1}{q}} m_{2q(1+\frac{1}{p})}^{\frac{1}{q}}\right),\end{aligned}$$

where

$$m_t[h] = \mathbb{M}(|h - \mathbb{M}(h)|^t),$$

the constants $C_t > 0$, $t \geq 2$ *satisfy* $C_t^{\frac{1}{t}} \leq t - 1$ *and the pairs of parameters* d, e, *and* p, q *are conjugate indices:* $1/d + 1/e = 1/p + 1/q = 1$.

The main message from this result is that non-asymptotically the MSE of the auto-normalised importance sampling converges to 0 at rate N^{-1} for a much broader family of test functions than simply bounded functions. In particular, the result implies (in conjunction with the Hölder inequality) that if the weights admit exponential moments, then we get this rate for all test functions φ such that $\mathbb{M}(\varphi^{2+\epsilon}) < \infty$.

8.6 The Effective Sample Size

A popular measure of efficiency in importance sampling is the so-called *effective sample size*:

$$\mathrm{ESS}(W^{1:N}) := \frac{1}{\sum_{n=1}^{N}(W^n)^2} = \frac{\{\sum_{n=1}^{N} w(X^n)\}^2}{\sum_{n=1}^{N} w(X^n)^2}.$$

The interpretation of this quantity as a sample size comes from the facts that $\mathrm{ESS}(W^{1:N}) \in [1, N]$, and that if k weights equal one, and $N - k$ weights equal zero, then $\mathrm{ESS}(W^{1:N}) = k$; see Exercise 8.10. The particular case $\mathrm{ESS}(W^{1:N}) = N$ occurs when $\mathbb{Q} = \mathbb{M}$, i.e. importance sampling collapses to standard Monte Carlo.

In addition, note that

$$\frac{N}{\mathrm{ESS}(W^{1:N})} = \frac{N^{-1}\sum_{n=1}^{N} w(X^n)^2}{\{N^{-1}\sum_{n=1}^{N} w(X^n)\}^2} \tag{8.6}$$

$$= 1 + \frac{N^{-1}\sum_{n=1}^{N} w(X^n)^2 - \{N^{-1}\sum_{n=1}^{N} w(X^n)\}^2}{\{N^{-1}\sum_{n=1}^{N} w(X^n)\}^2}. \tag{8.7}$$

Both equations re-express the ESS as a simple transformation of (empirical versions of) quantities we have seen previously. The second term in (8.7) is the square of the coefficient of variation of the weights, and thus is directly related to their variance. (The coefficient of variation is the standard deviation divided by the average.) And the ratio in (8.6) is the empirical version of the ratio $\mathbb{M}(w^2)/(\mathbb{M}(w)^2)$ that appears in Theorem 8.4, Sect. 8.5, for the MSE of auto-normalised importance sampling estimators of bounded functions. In both cases we have the simple interpretation that a small ESS should translate into a large variance for bounded functions, and presumably for other functions as well.

Something more formal to say about the ESS is that the second term in (8.7) converges (as $N \to +\infty$) to the chi-squared pseudo-distance between $\mathbb{M}$ and $\mathbb{Q}$:

$$\mathbb{M}\left(\left\{\frac{d\mathbb{Q}}{\mathrm{d}\mathbb{M}} - 1\right\}^2\right) = \mathrm{Var}_{\mathbb{M}}\left[\frac{w}{\mathbb{M}(w)}\right]$$

which is non-negative, and equals 0 only if $\mathbb{M} = \mathbb{Q}$. Hence a small ESS indicates that proposal $\mathbb{M}$ may be too far away from target $\mathbb{Q}$ to allow for reasonable performance for importance sampling.

This remark suggests alternative criteria for assessing the quality of a proposal and comparing alternatives, by considering other pseudo-distances between $\mathbb{M}$ and $\mathbb{Q}$. One that we consider here is a simple transformation of the entropy of the discrete probability measure $W^{1:N}$. Recall that the entropy of such a discrete measure is defined as

$$\mathrm{Ent}(W^{1:N}) := -\sum_{n=1}^{N} W^n \log(W^n)\,.$$

We can work instead with

$$\sum_{n=1}^{N} W^n \log(NW^n) = \log N - \mathrm{Ent}(W^{1:N})\,.$$

This is a non-negative function of the weights, that is minimised for $W^n = 1/N$, which is precisely the set of weights with maximum entropy; see Exercise 8.11. Additionally, Exercise 8.11 establishes that $\sum_{n=1}^{N} W^n \log(NW^n)$ is a consistent estimator of

$$D_{\mathrm{KL}}(\mathbb{Q}||\mathbb{M}) := \int \log\left(\frac{\mathrm{d}\mathbb{Q}}{\mathrm{d}\mathbb{M}}\right) \mathrm{d}\mathbb{Q}\,,$$

the Kullback-Leibler divergence of $\mathbb{M}$ relative to $\mathbb{Q}$ (see Exercise 8.12 for more insights into this divergence). We find again the idea of measuring a (pseudo-)distance between $\mathbb{M}$ and $\mathbb{Q}$ in order to assess whether importance sampling from $\mathbb{M}$ to $\mathbb{Q}$ may perform well (produce low variance estimates).

8.7 Curse of Dimensionality

It is a commonplace that importance sampling suffers from the curse of dimensionality. A very concrete demonstration of this curse is afforded by what turns out to be one of the hardest importance sampling problems: that of sampling jointly independent random variables! In particular, consider distributions $\mathbb{M}$ and $\mathbb{Q}$ over $\mathcal{X}$ such that $\mathbb{M} \gg \mathbb{Q}$, and let

$$\mathbb{Q}_t(\mathrm{d}x_{0:t}) = \prod_{s=0}^{t} \mathbb{Q}(\mathrm{d}x_s)\,, \qquad \mathbb{M}_t(\mathrm{d}x_{0:t}) = \prod_{s=0}^{t} \mathbb{M}(\mathrm{d}x_s)$$

that is, both $\mathbb{Q}_t$ and $\mathbb{M}_t$ are the law of $(t+1)$ i.i.d. copies from respectively $\mathbb{Q}$ and $\mathbb{M}$. Let $w(x) = \mathrm{d}\mathbb{Q}/\mathrm{d}\mathbb{M}$, and

$$w_t(x_{0:t}) = \frac{\mathrm{d}\mathbb{Q}_t}{\mathrm{d}\mathbb{M}_t} = \prod_{s=0}^{t} w(x_s).$$

(Both w and w_t are normalised in this section.) Then

$$\mathrm{Var}_{\mathbb{M}_t}[w_t(X_{0:t})] = \mathbb{M}_t(w_t^2) - 1 = \mathbb{M}(w^2)^{t+1} - 1\,.$$

As a function of t this is exponentially increasing since $\mathbb{M}(w^2) \geq 1$ with $\mathbb{M}(w^2) = 1$ only if $\mathbb{M} = \mathbb{Q}$ (by Jensen's inequality). Therefore, the variance of the (normalised) weights grows exponentially with t whenever the proposal differs from the target. This is a manifestation of the curse of dimensionality.

This example is not entirely artificial: if we are interested in test functions that involve the x_s's in functional forms other than sums or products, there is no trivial way to reduce the simulation problem to one of t decoupled simulations. Therefore, although the structure of the target distribution is simple, the form of test functions can still make this a non-trivial simulation problem suffering from the curse of dimensionality.

Although the increase might not be that drastic in other situations, there is *always* going to be an increase of the variance with dimension. To see this, consider the general case where $\mathbb{M}_t$ and $\mathbb{Q}_t$ admit probability densities $m_t(x_{0:t})$ and $q_t(x_{0:t})$, respectively, with respect to a common dominating measure over $\mathcal{X}^{t+1}$. Then, from the standard definition of conditional densities

$$q_t(x_t|x_{0:t-1}) := \frac{q_t(x_{0:t})}{q_{t-1}(x_{0:t-1})}$$

we obtain the following representation of the weight function:

$$w_t(x_{0:t}) = w_{t-1}(x_{0:t-1}) \frac{q_t(x_t|x_{0:t-1})}{m_t(x_t|x_{0:t-1})}\,, \tag{8.8}$$

from which we deduce that $w_t(X_{0:t})$ is a martingale, for $X_{0:t} \sim \mathbb{M}_t$ (Exercise 8.13). This directly implies that, while the expectation of $w_t(X_{0:t})$ stays constant (and equals one), its variance increases over time (again, see Exercise 8.13).

8.8 Random Weight Importance Sampling

8.8.1 General Methodology

Importance sampling can be cast in yet more general terms, which are helpful in devising, analysing and understanding a variety of Monte Carlo schemes that turn out to be related. Rather than requiring the computation of the unnormalised weight $w(x)$, we can derive an algorithm that only requires an *unbiased estimator* of $w(x)$.

Concretely, let $\mathbb{Q}$ be the target measure, $\mathbb{M}$ the proposal and $w(x)$ proportional to $\mathrm{d}\mathbb{Q}/\mathrm{d}\mathbb{M}$. Then, the following input

$$(X^n, \mathcal{W}^n), \quad X^n \sim \mathbb{M}, \quad \mathbb{E}[\mathcal{W}^n | X^n] = cw(X^n),$$

can be used to construct the auto-normalised importance sampling estimator

$$\frac{\sum_{n=1}^N \mathcal{W}^n \varphi(X^n)}{\sum_{n=1}^N \mathcal{W}^n};$$

the proportionality constant $c > 0$ is arbitrary, and its effect is cancelled by the re-normalisation. Note that by the iterated property of conditional expectation we trivially have that

$$\frac{\mathbb{E}[\mathcal{W}\varphi(X)]}{\mathbb{E}[\mathcal{W}]} = \frac{\mathbb{M}(w\varphi)}{\mathbb{M}(w)} = \mathbb{Q}(\varphi).$$

The $\mathcal{W}$'s are referred to as *random weights* since they are allowed to be non-trivial random variables conditionally on the X^n's.

8.8.2 Variance Increase Relative to Standard Importance Sampling

The MSE analysis of Sect. 8.5 can be replicated for the more generic framework of random weight auto-normalised importance samplers. Assuming that the pairs $(X^n, \mathcal{W}^n)$ are i.i.d., we immediately obtain that

$$\sqrt{N}\left\{\frac{\sum_{n=1}^N \mathcal{W}^n \varphi(X^n)}{\sum_{n=1}^N \mathcal{W}^n} - \mathbb{Q}(\varphi)\right\} \Rightarrow \mathcal{N}\left(0, \frac{\mathbb{E}[\mathcal{W}^2 \overline{\varphi}(X)^2]}{\mathbb{E}[\mathcal{W}]^2}\right). \tag{8.9}$$

What is interesting to notice is that this asymptotic variance is always larger than with non-random weights. Indeed, $\mathbb{E}[\mathcal{W}] = c\mathbb{M}(w)$ and by Jensen's inequality (assuming that the random weights are not degenerate random variables):

$$\mathbb{E}[\mathcal{W}^2\bar{\varphi}(X)^2] = \mathbb{E}\left[\mathbb{E}\left[\mathcal{W}^2 \,\middle|\, X\right]\bar{\varphi}^2(X)\right] > \mathbb{E}\left[\mathbb{E}\left[\mathcal{W} \mid X\right]^2 \bar{\varphi}(X)^2\right] = \mathbb{M}(w^2\bar{\varphi}^2)\,.$$

This inequality shows the price to pay for randomising the weights: the variance of the weights increases. The principle behind the above argument is what is typically known as *Rao-Blackwellization*.

Therefore, when random weights are employed it is either because the actual weights $w(x)$ are intractable, or because the importance sampling step is part of a larger simulation exercise for which the randomisation of the weights has outweighing advantages.

8.8.3 Connection with Rejection Sampling

As a (slightly artificial) illustration of random weight importance sampling, consider rejection sampling, which was described in Sect. 8.2.

Both importance sampling and rejection sampling sample from a proposal distribution $\mathbb{M}$ in order to approximate a target distribution $\mathbb{Q}$. Which approach performs best in practice? To answer this question, one must modify slightly rejection sampling so as to ensure it has exactly the same CPU budget as importance sampling, meaning N draws from the proposal. Using our random weight notations, $X^n \sim \mathbb{M}$, and $\mathcal{W}^n|X^n \sim \mathcal{B}er(w(X^n)/C)$, so that $\mathbb{E}[\mathcal{W}^n|X^n] = cw(X^n)$ with $c = 1/C$, as requested. Then the output of the algorithm can be represented via the weighted sample $(X^{1:N}, \mathcal{W}^{1:N})$ and Monte Carlo averages with respect to a test function φ using the accepted draws can be equivalently written as

$$\frac{\sum_{n=1}^{N} \mathcal{W}^n \varphi(X^n)}{\sum_{n=1}^{N} \mathcal{W}^n}\,, \tag{8.10}$$

provided not all draws were rejected.

From the discussion in the previous section, we can immediately conclude that importance sampling outperforms rejection sampling, since (8.10) has larger asymptotic variance than the importance sampling estimator

$$\frac{\sum_{n=1}^{N} w(X^n)\varphi(X^n)}{\sum_{n=1}^{N} w(X^n)}\,.$$

That said, there are of course many practical situations where rejection sampling is useful, e.g., because one may need to sample exactly from target $\mathbb{Q}$.

8.8.4 Non-negative Weights and Auxiliary Variables

In the generic construction the random weights do not have to be non-negative valued. The validity of the method is based on the property that $\mathbb{E}_{\mathbb{M}}[\mathcal{W}|X = x] \propto w(x)$ were the proportionality is with respect to x.

Nevertheless, non-negative weights have the advantage that the resultant weighted sample $(X^{1:N}, \mathcal{W}^{1:N})$ can be used to build a particle approximation of $\mathbb{Q}$ as with the standard importance sampling output:

$$\mathbb{Q}^N(\mathrm{d}x) := \sum_{n=1}^{N} W^n \delta_{X^n}(\mathrm{d}x), \qquad X^n \sim \mathbb{M}, \qquad W^n = \frac{\mathcal{W}^n}{\sum_{m=1}^{N} \mathcal{W}^m}.$$

When the weights are non-negative by construction, random weight importance sampling has an interesting interpretation. Let Z denote the random variable used to generate a random weight, where

$$Z|X = x \sim M(x, \mathrm{d}z)$$

for a probability kernel M, and $\mathcal{W} = f(Z, X)$, for some function $f \geq 0$; the random weight assumption may be reformulated as:

$$\int f(z, x) M(x, \mathrm{d}z) = c w(x).$$

Then, consider the extended target distribution defined as the following change of measure from $\mathbb{M}(\mathrm{d}x) M(x, \mathrm{d}z)$:

$$\begin{aligned} \mathbb{Q}(\mathrm{d}x, \mathrm{d}z) &= \frac{f(z, x)}{c\mathbb{M}(w)} \mathbb{M}(\mathrm{d}x) M(x, \mathrm{d}z) \\ &\propto f(z, x) \mathbb{M}(\mathrm{d}x) M(x, \mathrm{d}z). \end{aligned}$$

By construction, marginalising over z this measure we obtain the original target $\mathbb{Q}(\mathrm{d}x)$. Therefore, $\mathbb{Q}(\mathrm{d}x, \mathrm{d}z)$ is a distribution constructed on a larger space that includes the distribution of interest as a marginal. It is now possible to carry out plain-vanilla importance sampling for $\mathbb{Q}(\mathrm{d}x, \mathrm{d}z)$ with proposals from $\mathbb{M}(\mathrm{d}x) M(x, \mathrm{d}z)$, in which case the weights are precisely $\mathcal{W} = f(Z, X)$. This simple observation links random weight importance sampling with the notion of *auxiliary variables*. The basic principle behind auxiliary variable methods is the apparent paradox that it is often easier to simulate from a higher dimensional probability measure that admits the one of interest as a marginal, than from the one of interest directly. This is arguably one of the most productive ideas in stochastic simulation that results in a multitude of practically useful algorithms. In this context, if w cannot be computed importance sampling for target $\mathbb{Q}(\mathrm{d}x)$ and proposal $\mathbb{M}$

cannot be carried out, but provided non-negative unbiased estimators of w can be generated, importance sampling from $\mathbb{Q}(dx, dz)$ as defined before is feasible.

Auxiliary variables will re-appear in various places in this book, notably Chaps. 16 and 18.

Exercises

8.1 *Prove Proposition 8.1. To do so, the simplest approach is to compute the CDF of $F^{-1}(U)$. For simplicity, you may consider first the case where F is bijective (i.e. F is continuous and increasing). Then you may generalise to the case where F is not bijective, and its pseudo-inverse is defined as in Proposition 8.1.*

A related property is that $F(X)$ is distributed as a $\mathcal{U}([0, 1])$ distribution. Derive conditions on F under which this statement holds. (Hint: does this property holds when X has a finite support, such as e.g. a Bernoulli variable?)

8.2 *The aim of this exercise is to prove that Algorithm 8.1 samples from $\mathbb{Q}$. First, give the probability that a draw is accepted, and the distribution of the number of draws required have one draw accepted. Then, let $\tilde{q}$ denote the density of the distribution generated by this Algorithm. Use a recursive argument to show that $\tilde{q}(x) = m(x)\{q(x)/Cm(x)\} + \tilde{q}(x)(1 - 1/C)$, and deduce the result. See also Exercise 8.3 for an alternative proof.*

8.3 *Consider the graph of a given probability density q: $\mathcal{G} = \{(x, y) \in \mathcal{X} \times \mathbb{R}^+ : 0 \leq y \leq q(x)\}$, and let (X, Y) be uniformly distributed over $\mathcal{G}$. Determine the marginal distribution of X, and the conditional distribution $Y|X = x$. Use this property to derive an alternative proof for the fact that Algorithm 8.1 samples from distribution $\mathbb{Q}$.*

8.4 *We wish to sample from the truncated Gaussian distribution defined in Example 8.1. To do so, we consider the following algorithm: Sample $X \sim \mathcal{N}(0, 1)$ until $X > c$. Explain why this is a special case of rejection sampling (Algorithm 8.1) and compute the acceptance probability. Clearly, this approach becomes inefficient when c is too large. Instead, we may use a rejection sampler based on a shifted exponential distribution, with density $m(x) = \lambda \exp\{\lambda(c - x)\}\mathbb{1}\{x \geq c\}$. Explain why we should set constant C as follows: $C = \sup\{q(x)/m(x)\}$, and then choose λ so as to minimise C. Compute the optimal λ. In practice, it is sometimes recommended to take $\lambda = c$. Discuss.*

8.5 *In Example 8.1 in Sect. 8.3.1, compute the variance of the weights when the proposal density is $m(x) = \lambda \exp\{-\lambda(x - c)\}\mathbb{1}\{x \geq c\}$, and find the value of λ that minimises this quantity. Discuss what happens when $c \to +\infty$. Compute the value of λ that minimises the variance of the estimator of the expectation (with respect to q) of test function $\varphi(x) = \exp(\tau x)$, and comment on how much it differs from the value that minimises the variance of the weights.*

8.6 *Consider the settings of normalised importance sampling, see Sect. 8.3.1, where both the proposal density m and the target density q are known exactly, and $w(x) = q(x)/m(x)$. Show that a generalisation of the normalised importance sampling estimator is:*

$$\frac{1}{N}\sum_{n=1}^{N} w(X^n)\varphi(X^n) + \lambda\left\{\frac{1}{N}\sum_{n=1}^{N} w(X^n) - 1\right\}$$

and compute the value of λ that minimises the MSE of this estimator. The weight $w(X^n)$ acts as a control variate, that is, a random variable with known expectation that may be used to reduce the variance of a MC estimator.

8.7 *Following the previous exercise on control variates and normalised importance sampling, another way to use the fact that $\mathbb{M}(w) = 1$ to possibly achieve variance reduction is to divide the normalised importance sampling estimator by the average of the weights:*

$$\frac{\sum_{n=1}^{N} w(X^n)\varphi(X^n)}{\sum_{n=1}^{N} w(X^n)} \tag{8.11}$$

by which we recover of course auto-normalised importance sampling. This suggests that, in situations where we can use both (i.e. when both q and m are known exactly), auto-normalised importance sampling may perform better when there is a positive correlation between the numerator and the denominator of (8.11)*. Show that a necessary and sufficient condition for the asymptotic variance of auto-normalised importance sampler to be smaller than the MSE of the normalised importance sampler is:*

$$2\mathbb{Q}(w\varphi) > \mathbb{Q}(\varphi)\{\mathbb{Q}(w) + 1\}\,.$$

In particular, show that if w is an indicator function, $w(x) = \mathbb{1}_{\mathcal{A}}(x)$ for some set $\mathcal{A}$, then this condition is always met. Show that, when $\mathbb{Q}(\varphi) > 0$, the condition above implies that $w(X^n)\varphi(X^n)$ and $w(X^n)$ are positively correlated. (But the converse does not hold, i.e. the condition above is actually stronger than assuming positive correlation when $\mathbb{Q}(\varphi) > 0$, or negative correlation when $\mathbb{Q}(\varphi) < 0$.)

8.8 *The optimal normalised importance sampling estimator has variance zero, when $\varphi \geq 0$, as discussed in Sect. 8.3.3. More generally, how can we construct a zero-variance estimator for an arbitrary φ, by summing two such normalised IS estimators? Discuss the applicability of this result.*

8.9 *Adapt the proof of Proposition 8.2 to obtain the density $m^\star$ that minimises the asymptotic variance of a normalised importance sampling estimator, see* (8.4)*, when both $\mathbb{Q}$ and $\mathbb{M}$ admit probability densities q and m, respectively.*

8.10 *Show that the ESS criterion (defined in Sect. 8.6) equals k whenever k particles have un-normalised weight* 1*, and* $N - k$ *particles have weight 0. Show that the ESS always takes values in* $[1, N]$.

8.11 *Show that the quantity* $\sum_{n=1}^{N} W^n \log(NW^n)$*, defined in Sect. 8.6 for normalised weights,* $\sum_{n=1}^{N} W^n = 1$*, is minimised by weights* $W^n = 1/N$*. Deduce that this quantity is non-negative. In addition, show that this quantity converges as* $N \to +\infty$ *to the Kullback-Leibler divergence of* $\mathbb{M}$ *relative to* $\mathbb{Q}$*.*

8.12 *Show that the Kullback-Leibler divergence between* $\mathbb{Q}$ *and* $\mathbb{M}$ *is non-negative and it is 0 if and only if* $\mathbb{Q} = \mathbb{M}$*. (Hint: to show the non-negative property, simply apply Jensen's inequality. The "if" part is trivial but the "only if" is tricky—and in fact it has little to do with the details of the Kullback-Leibler divergence. It boils down to showing that for a convex function* $c(\cdot)$ *and a random variable* X*,* $\mathbb{E}_{\mathbb{Q}}[c(X)] = c(\mathbb{E}_{\mathbb{Q}}[X])$ *if and only if* $X = \mathbb{E}_{\mathbb{Q}}[X]$ *a.s., i.e., X is a trivial random variable.)*

8.13 *A process* $\{Z_t\}_{t\geq 0}$ *is a martingale if and only* $\mathbb{E}[Z_t|Z_{0:t-1}] = Z_{t-1}$*. Show that the weight function* $w_t(X_{0:t})$*, as defined in (8.8), is a martingale. Establish that the expectation of a martingale is constant over time, and its variance is non-decreasing over time; as an intermediate result, you may establish first that* Z_{t-1} *and* $Z_t - Z_{t-1}$ *are uncorrelated.*

Python Corner

Something every data scientist discovers sooner or later, sometimes at her own expense, is that probability densities (and related quantities) should always be computed on the log scale. If you are not already convinced, try the following in Python:

```
import numpy as np
from scipy import stats

x = stats.norm.rvs(size=1000)
L = np.prod(stats.norm.pdf(x))
if L == 0.:
    print('told you!')
```

This is a typical example of numerical underflow: quantity L is smaller than $\approx 10^{-300}$, the smallest number that may be represented using double precision arithmetic. (Overflow is also possible: find how to modify the above example to that effect.). One should do instead:

```
l = np.sum(stats.norm.logpdf(x))
print(l)
```

Multiplying two numbers stored on the log-scale is easy: simply add their logs. Adding two such numbers is a bit more tricky. One may use so-called Gaussian logarithms:

```
def logplus(la, lb):
    """Sum of two numbers stored on log-scale."""
    if la > lb:
        return la + np.log1p(np.exp(lb - la))  # log1p(x) = log(1 + x)
    else:
        return lb + np.log1p(np.exp(la - lb))
```

Function `logplus` exponentiates only negative numbers, and thus will not produce any numerical overflow.

How do these remarks apply to importance sampling? importance weights should of course also be computed on the log scale. To obtain normalised weights, one may adapt the `logplus` function as follows:

```
def exp_and_normalise(lw):
    """Compute normalised weigths W, from log weights lw."""
    w = np.exp(lw - lw.max())
    return w / w.sum()
```

With these remarks in mind, here is a generic function to compute importance sampling estimates:

```
def importance_sampling(target, proposal, phi, N=1000):
    x = proposal.rvs(size=N)
    lw = target.logpdf(x) - proposal.logpdf(x)
    W = exp_and_normalise(lw)
    return np.average(phi(x), weights=W)
```

and here is a quick example:

```
f = lambda x: x  # function f(x)=x
est = importance_sampling(stats.norm(), stats.norm(scale=2.), f)
```

One last remark about log scale computation: there is apparently some push to develop hardware (microchips) based on LNS (logarithmic number systems) rather than the standard floating-point number systems (Coleman et al. 2008). Such hardware would perform any type of numerical computation on the log scale, invisibly from the user. If that ever happens, we will not have to worry any more about numerical overflow when dealing with importance sampling. But, for now, ignore this piece of advice at your own peril.

Bibliographical Notes

A classical reference on random variable generation is the book of Devroye (1986). Good books on Monte Carlo include Robert and Casella (2004) and Owen (2013) (in progress, the ten first chapters are available on the author's web-site). The

ESS criterion seems to have appeared from the first time in Kong et al. (1994). More background on, and practical applications of, random weight importance sampling may be found in Fearnhead et al. (2010). See also Agapiou et al. (2017) and Chatterjee and Diaconis (2018) on some interesting theoretical insights on importance sampling.

Bibliography

Agapiou, S., Papaspiliopoulos, O., Sanz-Alonso, D., & Stuart, A. M. (2017). Importance sampling: intrinsic dimension and computational cost. *Statistical Science, 32*(3), 405–431.

Chatterjee, S., & Diaconis, P. (2018). The sample size required in importance sampling. *Annals of Applied Probability, 28*(2), 1099–1135.

Coleman, J. N., Softley, C. I., Kadlec, J., Matousek, R., Tichy, M., Pohl, Z., et al. (2008). The European logarithmic microprocesor. *IEEE Transactions on Computers, 57*(4), 532–546.

Devroye, L. (1986). *Non-uniform random variate generation*. New York: Springer.

Fearnhead, P., Papaspiliopoulos, O., Roberts, G. O., & Stuart, A. (2010). Random-weight particle filtering of continuous time processes. *Journal of the Royal Statistical Society: Series B (Statistical Methodology), 72*(4), 497–512.

Kong, A., Liu, J. S., & Wong, W. H. (1994). Sequential imputation and Bayesian missing data problems. *Journal of the American Statistical Association, 89*, 278–288.

Owen, A. B. (2013). *Monte Carlo theory, methods and examples*. (in progress). https://statweb.stanford.edu/~owen/mc/

Rebeschini, P., & van Handel, R. (2015). Can local particle filters beat the curse of dimensionality? *Annals of Applied Probability, 25*(5), 2809–2866.

Robert, C. P., & Casella, G. (2004). *Monte Carlo statistical methods* (2nd ed.). New York: Springer.

Chapter 9
Importance Resampling

Summary Resampling is the action of drawing randomly from a weighted sample, so as to obtain an unweighted sample. Resampling may be viewed as a random weight importance sampling technique. However it deserves a separate chapter because it plays a central role in particle filtering. In particular, we explain that resampling has the curious property of potentially reducing the variance at a later stage, provided that this later stage corresponds to a Markov update that forgets its past in some way. This point is crucial for the good performance of particle algorithms.

This chapter explains informally this particular property, formalises importance resampling, describes several algorithms to perform resampling, and compares these algorithms numerically.

9.1 Motivation

Consider the following apparently innocuous problem: say we have at our disposal the following particle approximation of measure $\mathbb{Q}_0(\mathrm{d}x_0)$,

$$\mathbb{Q}_0^N(\mathrm{d}x_0) = \sum_{n=1}^N W_0^n \delta_{X_0^n}, \qquad X_0^n \sim \mathbb{M}_0, \qquad W_0^n = \frac{w_0(X_0^n)}{\sum_{m=1}^N w_0(X_0^m)},$$

obtained through importance sampling based on proposal $\mathbb{M}_0$, and weight function w_0 proportional to $\mathrm{d}\mathbb{Q}_0/\mathrm{d}\mathbb{M}_0$. How can we recycle this so as to approximate the following extended probability measure?

$$(\mathbb{Q}_0 M_1)(\mathrm{d}x_{0:1}) = \mathbb{Q}_0(\mathrm{d}x_0) M_1(x_0, \mathrm{d}x_1). \tag{9.1}$$

Our notations are now very close to those we have used when presenting Feynman-Kac models, which is of course on purpose. Interestingly, and importantly as far as importance sampling is concerned, there are two solutions to this problem.

© Springer Nature Switzerland AG 2020

N. Chopin, O. Papaspiliopoulos, *An Introduction to Sequential Monte Carlo*, Springer Series in Statistics, https://doi.org/10.1007/978-3-030-47845-2_9

The first one is to recognise that importance sampling from $\mathbb{M}_1 = \mathbb{M}_0 M_1$ to $\mathbb{Q}_0 M_1$ requires (a) to sample (X_0^n, X_1^n) from $\mathbb{M}_0 M_1$, and (b) to compute weights, which in this case are precisely $w_0(X_0^n)$, since X_1^n conditionally on X_0^n is sampled from the correct distribution. Thus, if we indeed want to follow this particular importance sampling strategy, the only thing left is to sample $X_1^n \sim M_1(X_0^n, \mathrm{d}x_1)$, for $n = 1 : N$. This first strategy belongs to the broader framework known as *sequential importance sampling*. As far as inference relative to $(\mathbb{Q}_0 M_1)$ is concerned, it may be analysed as standard importance sampling, without paying attention to the intermediate step which allowed for inference relative to $\mathbb{Q}_0$.

The second one is to consider a two-step approximation, where first we replace $\mathbb{Q}_0$ by $\mathbb{Q}_0^N$ in (9.1):

$$\mathbb{Q}_0^N(\mathrm{d}x_0) M_1(x_0, \mathrm{d}x_1) = \sum_{n=1}^{N} W_0^n M_1(X_0^n, \mathrm{d}x_1) \delta_{X_0^n}(\mathrm{d}x_0) \tag{9.2}$$

then, in a second step, we sample N times from this intermediate approximation so as to form:

$$\frac{1}{N} \sum_{n=1}^{N} \delta_{\widetilde{X}_{0:1}^n}, \qquad \widetilde{X}_{0:1}^n \sim \mathbb{Q}_0^N(\mathrm{d}x_0) M_1(x_0, \mathrm{d}x_1).$$

There are several ways to sample from (9.2), but here we shall consider the simplest one: for $n \in 1 : N$, we sample independently the pairs $(A_1^n, \widetilde{X}_{0:1}^n)$ as

$$A_1^n \sim \mathcal{M}(W_0^{1:N}), \qquad \widetilde{X}_{0:1}^n = (X_0^{A_1^n}, X_1^n), \qquad X_1^n \sim M_1(X_0^{A_1^n}, \mathrm{d}x_1),$$

where $\mathcal{M}(W_0^{1:N})$ is the multinomial distribution that generates value n with probability W_0^n, for $n \in 1 : N$.

This second approach is called *importance resampling*, because of its connection with resampling techniques such as the bootstrap, where one draws samples with replacement.

At first sight, importance resampling is unappealing, because the two approximation steps above suggest that, relative to sequential importance sampling, (a) it should lead to a larger Monte Carlo error (introduction of variance at the resampling stage); and (b) it should be more expensive computationally (because of the extra step to draw the A_1^n variables).

Intuition behind (a) turns out to be incorrect: if one is interested in the marginal distribution with respect to X_1 (i.e., to compute $\mathbb{Q}_1(\varphi)$ for functions that depend only on x_1), and if in addition M_1 in some sense 'forgets its past' (i.e., the measure $M_1(x_0, \mathrm{d}x_1)$ does not depend so much on x_0) then importance resampling leads to smaller Monte Carlo error, sometimes dramatically. The following example illustrates this point.

9.2 A Toy Example Demonstrating the Appeal of Resampling

Assume $\mathcal{X} = \mathbb{R}$, $\mathbb{M}_0$ is $\mathcal{N}(0, 1)$, and $w_0(x) = \mathbb{1}_{[-\tau,\tau]}(x)$. $\mathbb{Q}_0$ is then a truncated Gaussian distribution, i.e., the distribution of $X_0 \sim \mathcal{N}(0, 1)$ conditional on $X_0 \in [-\tau, \tau]$. Let $M_1(x_0, \mathrm{d}x_1)$ be the Markov kernel such that $X_1 = \rho X_0 + \sqrt{1-\rho^2}U$, with $U \sim \mathcal{N}(0, 1)$, independently of X_0.

Consider the test function $\varphi(x_1) = x_1$; note that $(\mathbb{Q}_0 M_1)(\varphi) = 0$ (because the density of $\mathcal{N}(0, 1)$ truncated to $[-\tau, \tau]$ is symmetric). We have two alternative estimators:

$$\text{importance sampling:} \quad \widehat{\varphi}_{\mathrm{IS}} = \sum_{n=1}^{N} W_0^n X_1^n \,, \qquad (X_0^n, X_1^n) \sim \mathbb{M}_0 M_1 \,,$$

$$\text{importance resampling:} \quad \widehat{\varphi}_{\mathrm{IR}} = \frac{1}{N} \sum_{n=1}^{N} X_1^n \,, \qquad X_1^n \sim \mathbb{Q}_0^N M_1 \,,$$

where $X_1^n \sim \mathbb{Q}_0^N M_1$ means a draw from the corresponding marginal distribution of $\mathbb{Q}_0^N M_1$ defined in (9.2).

Figure 9.1 gives some intuition on the respective merits of each estimator. If τ is small, only a few particles 'survive' (get non-zero weight) at time 0. Without resampling, the number of alive particles remains small at time 1; in particular, generating a particle X_1^n from a 'dead' ancestor is a waste of time. On the other hand, with resampling, we get N alive particles at time 1, but they are correlated, because they originate from the few survivors at time 0.

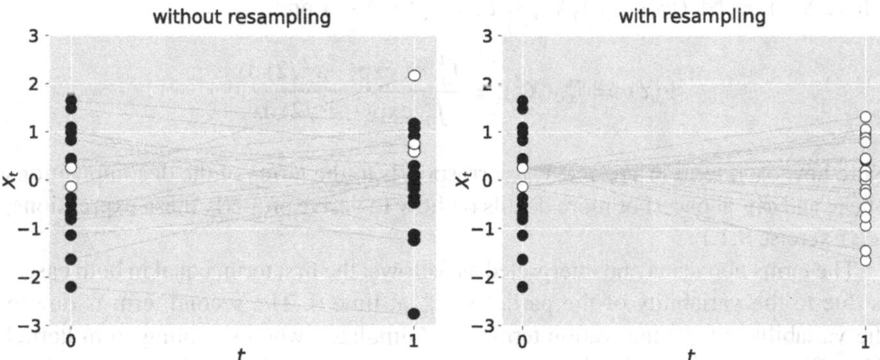

Fig. 9.1 Pictorial representation of the effect of resampling for the toy example discussed in the text; $N = 20$, $\tau = 0.3$, $\rho = 0.5$, white (resp. black) dots represent particles with non-zero (resp. zero) weight. A light grey line connects each particle at time 1 with its ancestor at time 0

It is possible in this particular example to compare formally both estimators, by computing their asymptotic variances. To that aim, we decompose these estimators as follows:

$$\widehat{\varphi}_{\mathrm{IS}} = \rho \sum_{n=1}^{N} W_0^n X_0^n + \sqrt{1-\rho^2} \sum_{n=1}^{N} W_0^n U^n$$

and

$$\widehat{\varphi}_{\mathrm{IR}} = \rho \sum_{n=1}^{N} W_0^n X_0^n + \sqrt{1-\rho^2} \left(\frac{1}{N} \sum_{n=1}^{N} U^n \right) + \rho \left(\frac{1}{N} \sum_{n=1}^{N} X_0^{A_0^n} - \sum_{n=1}^{N} W_0^n X_0^n \right). \tag{9.3}$$

From this, one may deduce the following asymptotic results:

$$\sqrt{N} \hat{\varphi}_{\mathrm{IS}} \Rightarrow \mathcal{N}(0, V_{\mathrm{IS}})$$
$$\sqrt{N} \hat{\varphi}_{\mathrm{IR}} \Rightarrow \mathcal{N}(0, V_{\mathrm{IR}})$$

with asymptotic variances:

$$V_{\mathrm{IS}} = \rho^2 \frac{\gamma(\tau)}{S(\tau)} + (1-\rho^2) \frac{1}{S(\tau)}$$
$$V_{\mathrm{IR}} = \rho^2 \frac{\gamma(\tau)}{S(\tau)} + (1-\rho^2) + \rho^2 \gamma(\tau)$$

where $S(\tau) := \mathbb{M}_0(w_0) = \mathbb{P}(|X_0| \leq \tau) = 2\Phi(\tau) - 1$ and

$$\gamma(\tau) := \mathbb{Q}_0(X_0^2) = \frac{\int_0^\tau x^2 \exp(-x^2/2)\, \mathrm{d}x}{\int_0^\tau \exp(-x^2/2)\, \mathrm{d}x}.$$

Note how each term in V_{IS} and V_{IR} corresponds to the terms of the decompositions of $\widehat{\varphi}_{\mathrm{IS}}$ and $\widehat{\varphi}_{\mathrm{IR}}$ above. (For more details on how to derive properly these expressions, see Exercise 9.1.)

The terms above may be interpreted as follows: the first term, equal in both cases, is due to the variability of the particles X_0^n at time 0. The second term is due to the variability of the innovation terms U^n (simulated when sampling from kernel M_1). Observe it is smaller for importance resampling ($S(\tau) < 1$), because these innovation terms are not affected by the weights of the previous step. The third term appears only in V_{IR}, and comes from the randomness introduced by the resampling step.

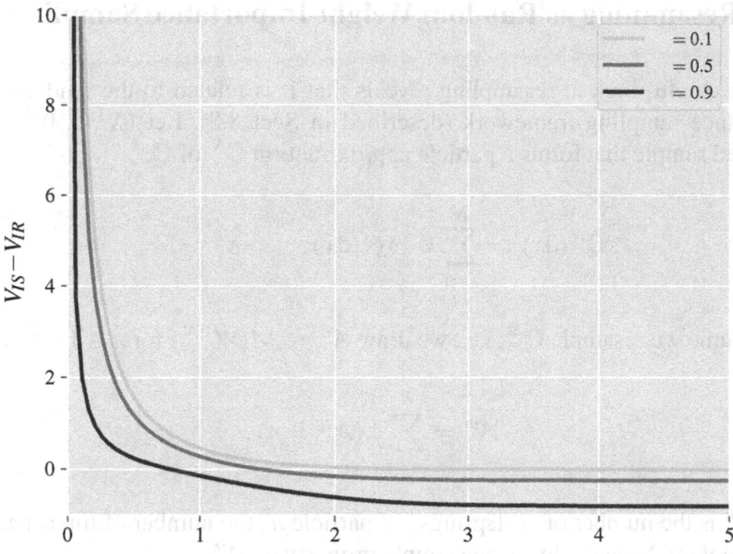

Fig. 9.2 Quantity $V_{\mathrm{IS}} - V_{\mathrm{IR}}$ as a function of τ for selected values of ρ

Figure 9.2 plots the quantity

$$V_{\mathrm{IS}} - V_{\mathrm{IR}} = (1-\rho^2)\left\{\frac{1}{S(\tau)} - 1\right\} - \rho^2\gamma(\tau)$$

as a function of τ, for various values of ρ. It is easy to check that, for fixed ρ, this quantity diverges to $+\infty$ when $\tau \to 0$, while it converges to $-\rho^2$ when $\tau \to +\infty$. Parameter τ determines how restrictive is the support of w_0 (and therefore how far away is $\mathbb{Q}_0$ from $\mathbb{M}_0$). We see that importance resampling reduces the variance by a large amount in challenging cases (τ small), and increases it slightly in simple cases (τ large). In doubt, that is for realistic models where the calculations above are not tractable, resampling seems to be the "safer" option.

Furthermore, parameter ρ determines how much X_1 depends on X_0. It is easy to check that $V_{\mathrm{IR}} \leq V_{\mathrm{IS}}$ as soon as:

$$\rho^2 \leq \frac{1 - S(\tau)}{1 - S(\tau) + S(\tau)\gamma(\tau)} \in [0, 1].$$

In particular, whatever τ, we always have $V_{\mathrm{IR}} < V_{\mathrm{IS}}$ for $\rho = 0$ (X_1 is independent from X_0), and $V_{\mathrm{IR}} > V_{\mathrm{IS}}$ for $|\rho| = 1$ ($X_1 = X_0$ or $X_1 = -X_0$). The decomposition of V_{IS} and V_{IR} above reveals the subtle double effect of resampling: resampling adds noise to the "past", while on the other hand reduces the variability of the "present". Provided the Markov kernel M_1 forgets its past sufficiently (ρ is small enough), resampling is beneficial.

9.3 Resampling as Random Weight Importance Sampling

A more careful look at resampling reveals that it is related to the random weight importance sampling framework (described in Sect. 8.8). Let $(X^{1:N}, W^{1:N})$ be a weighted sample that forms a particle approximation $\mathbb{Q}^N$ of $\mathbb{Q}$:

$$\mathbb{Q}^N(\mathrm{d}x) := \sum_{n=1}^{N} W^n \delta_{X^n}(\mathrm{d}x), \qquad X^n \sim \mathbb{M},$$

and assume we resample $\mathbb{Q}^N$, i.e., we draw $A^n \sim \mathcal{M}(W^{1:N})$ for $n \in 1:N$. Let

$$\mathcal{W}^n = \sum_{m=1}^{N} \mathbb{1}(A^m = n),$$

i.e., $\mathcal{W}^n$ is the number of "offsprings" of particle n (the number of times particle n is resampled). Note the following simple property of $\mathcal{W}^n$:

$$\mathbb{E}\left[\mathcal{W}^n \,\middle|\, X^{1:N}\right] = N W^n .$$

Clearly, for any function φ, we have that

$$\frac{1}{N}\sum_{n=1}^{N} \varphi(X^{A^n}) = \frac{1}{N}\sum_{n=1}^{N} \mathcal{W}^n \varphi(X^n) = \frac{\sum_{n=1}^{N} \mathcal{W}^n \varphi(X^n)}{\sum_{n=1}^{N} \mathcal{W}^n} \tag{9.4}$$

and thus our resampled estimator may be viewed as a random weight importance sampling estimator, with random weights $\mathcal{W}^n$, such that $\mathbb{E}(\mathcal{W}^n | X^{1:N}) = N W^n$.

Does it mean that this resampled estimate has larger variance than the initial estimate $\sum_{n=1}^{N} W^n \varphi(X^n)$, as is often the case with random weights? To answer this question, we cannot apply the CLT derived in the previous chapter, (8.9): that result assumes that the $\mathcal{W}^n$'s are independent, which is not the case here. Instead, we may apply directly the law of total variance to (9.4):

$$\begin{aligned}
\mathrm{Var}\left[\frac{1}{N}\sum_{n=1}^{N} \mathcal{W}^n \varphi(X^n)\right] &= \mathrm{Var}\left[\mathbb{E}\left\{\frac{1}{N}\sum_{n=1}^{N} \mathcal{W}^n \varphi(X^n) \,\middle|\, X^{1:N}\right\}\right] \\
&\quad + \mathbb{E}\left[\mathrm{Var}\left\{\frac{1}{N}\sum_{n=1}^{N} \mathcal{W}^n \varphi(X^n) \,\middle|\, X^{1:N}\right\}\right] \\
&\geq \mathrm{Var}\left[\sum_{n=1}^{N} W^n \varphi(X^n)\right].
\end{aligned}$$

Indeed, our resampled estimator has always larger variance than the initial estimator $\sum_{n=1}^{N} W^n \varphi(X^n)$ (although it may decrease the variance at a later step, as discussed in the previous section.)

9.4 Multinomial Resampling

We now explain how to simulate efficiently the ancestor variables A^n from the multinomial distribution $\mathcal{M}(W^{1:N})$. The corresponding algorithm is aptly named multinomial resampling.

The standard method to sample from such a discrete distribution is the inverse CDF transformation method: generate N uniform variates U^m, $m \in 1 : N$, and set A^m according to:

$$C^{n-1} \leq U^m \leq C^n \quad \Leftrightarrow \quad A^m = n \tag{9.5}$$

where the C^n's are the cumulative weights:

$$C^0 = 0, \qquad C^n = \sum_{l=1}^{n} W^l \quad \text{for } n \in 1 : N.$$

Figure 9.3 illustrates the idea.

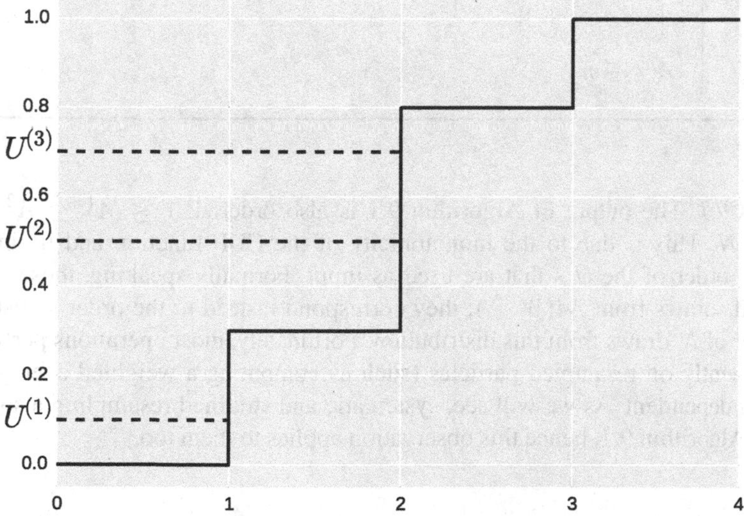

Fig. 9.3 A pictorial representation of the inverse CDF algorithm: the black line represents the empirical CDF $x \to \sum_{n=1}^{N} W^n \mathbb{1}(x \geq n)$. The height of step n is W^n. If $U^{(m)}$ 'picks' step n, then $A^m = n$; $A^{1:3} = (1, 2, 2)$ in this particular case

Equation (9.5) suggests that $\mathcal{O}(N^2)$ comparisons should be performed to generate N draws, i.e., each individual simulation requires $\mathcal{O}(N)$ comparisons to be made and we need N draws.

Fortunately, a nice short-cut exists if the U^m's are ordered in a preliminary step; denote $U^{(1)}, \dots, U^{(N)}$ the order statistics of $U^{1:N}$, $0 < U^{(1)} < \dots < U^{(N)} < 1$. (The probability of a tie is zero.) Again, this idea is illustrated by Fig. 9.3: once we have determined that $A^m = n$, and thus that $C^{n-1} \leq U^{(m)} \leq C^n$, then by construction $U^{(m+1)} \geq C^{n-1}$, and one may skip $(n-1)$ comparisons. This remark leads to Algorithm 9.1, which has $\mathcal{O}(N)$ complexity.

Algorithm 9.1: Function `icdf` for computing the inverse of CDF $x \to \sum_{n=1}^{N} W^n \mathbb{1}\{n \leq x\}$ at ordered points $U^{(1)}, \dots U^{(N)}$

Input: Normalised weights $W^{1:N}$, and $0 < U^{(1)} < \dots < U^{(N)} < 1$.

Output: $A^{1:N}$ (N indices in $1:N$).

Function `icdf(`$W^{1:N}$, $U^{(1:N)}$`):`

```
    s ← W^1
    m ← 1
    for n = 1 to N do
        while s < U^(n) do
            m ← m + 1
            s ← s + W^m
        A^n ← m
```

Remark 9.1 The output of Algorithm 9.1 is also ordered: $1 \leq A^1 \leq A^2 \dots \leq A^n \leq N$. This is due to the monotonicity of the CDF function and its inverse, and the order of the U's that are used as input. Formally speaking, these A^n are not i.i.d. draws from $\mathcal{M}(W^{1:N})$; they correspond instead to the order statistics of a vector of N draws from this distribution. Fortunately, most operations performed subsequently on resampled particles (such as computing a weighted average) are order-independent. As we will see, systematic and stratified resampling also make use of Algorithm 9.1, hence this observation applies to them too.

How do we obtain the order statistics of N uniform variates? A naive approach is to generate these N uniform variates, and to sort them. The complexity of this approach is $\mathcal{O}(N \log(N))$. The following well-known property of uniform spacings provides a $\mathcal{O}(N)$ algorithm.

Proposition 9.1 *Let $E^1, \ldots, E^{N+1}$ be i.i.d. draws from the exponential distribution $\mathcal{E}(1)$, and let $S^n = \sum_{m=1}^{n} E^m$. Then the vector*

$$\left(\frac{S^1}{S^{N+1}}, \ldots, \frac{S^N}{S^{N+1}} \right)$$

has the same distribution as the joint distribution of the order statistics of N random variables from $\mathcal{U}([0, 1])$.

See Exercise 9.3 for how to prove this result from first principles. A more elegant proof is to recognise this result as a simple corollary of known properties of the Poisson process on $\mathbb{R}^+$ with rate 1: the E^n's represent the inter-arrival times between events (which are exponentially distributed), and conditional on the $(N+1)$-th event occurring at time t, the dates of the N previous events are uniformly distributed over $[0, t]$.

Algorithm 9.2: Generation of uniform spacings

Input: Integer N.

Output: An ordered sequence $0 < U^{(1)} < \ldots < U^{(N)} < 1$ in $[0, 1]$.

```
Function uniform_spacings(N):
    S^0 ← 0
    for n = 1 to (N + 1) do
        E^n ~ ℰ(1)
        S^n ← S^{n-1} + E^n
    for n = 1 to N do
        U^(n) ← S^n / S^{N+1}
```

Algorithm 9.3 summarises the overall approach for multinomial resampling.

Algorithm 9.3: Multinomial resampling

Input: Normalised weights $W^{1:N}$ ($\sum_{n=1}^{N} W^n = 1$, $W^n \geq 0$).

Output: N draws from $\mathcal{M}(W^{1:N})$.

```
U^(1:N) ← uniform_spacings(N)                  ▷ using Algorithm 9.2
A^1:N ← icdf(W^1:N, U^(1:N))                   ▷ using Algorithm 9.1
```

Multinomial resampling may introduce quite a lot of extra noise in certain cases, as the following example shows.

Example 9.1 Consider a situation where the importance weights W^n are all equal, i.e,. $W^n = 1/N$. Then $\mathcal{W}^n \sim \mathcal{B}in(N, 1/N)$, and $\mathbb{P}(\mathcal{W}^n = 0) = (1 - 1/N)^N \approx e^{-1} \approx 0.37$ for large N. In words, when the weights are all equal, about 37% of the initial particles disappear after the resampling step, although they had the same weight initially.

In Sect. 9.3, we interpreted multinomial resampling as a random weight importance technique, with random weight $\mathcal{W}^n$ such that (a) $\mathcal{W}^n$ is integer-valued (it represents the number of offsprings of particle n); (b) $\mathcal{W}^n$ has expectation NW^n; and (c) $\sum_{n=1}^{N} \mathcal{W}^n = N$. In fact, any method that generates ancestor variables $A^{1:N}$ in such a way that the number of copies $\mathcal{W}^n = \sum_{m=1}^{n} \mathbb{1}\{A^m = n\}$ fulfils these three properties will be an *unbiased resampling scheme*, i.e,. it will satisfy

$$\mathbb{E}\left[\frac{1}{N}\sum_{n=1}^{N} \varphi(X^{A^n}) \,\middle|\, X^{1:N}\right] = \mathbb{E}\left[\frac{1}{N}\sum_{n=1}^{N} \mathcal{W}^n \varphi(X^n) \,\middle|\, X^{1:N}\right] = \sum_{n=1}^{N} W^n \varphi(X^n),$$

or, in words, it will generate an unweighted sample that provides estimates with the same expectation (but with increased variance) as the original weighted sample.

We now consider alternative resampling schemes that satisfy these three constraints, while introducing less noise than multinomial resampling.

9.5 Residual Resampling

Let $\text{frac}(x) = x - \lfloor x \rfloor$, where $\lfloor x \rfloor$ is the integer part of x, and thus $\text{frac}(x)$ is the fractional part of x. To reduce the variance of $\mathcal{W}^n$, subject to $\mathbb{E}(\mathcal{W}^n) = NW^n$, a simple solution is to take $\mathcal{W}^n = \lfloor NW^n \rfloor + \widetilde{\mathcal{W}}^n$, where $\widetilde{\mathcal{W}}^n$ is a $\mathbb{Z}^+$-valued random variable such that $\mathbb{E}(\widetilde{\mathcal{W}}^n) = \text{frac}(NW^n)$. This is the main idea behind residual resampling.

In practice, one first creates $\lfloor NW^n \rfloor$ "deterministic copies" of label n, for each $n \in 1:N$. In this way, one obtains a vector of labels of length $N - R := \sum_{n=1}^{N} \lfloor NW^n \rfloor$. The R remaining labels are generated randomly, by sampling R times from $\mathcal{M}(r^{1:N})$, with $r^n = \text{frac}(NW^n)/R$. Thus, the number $\widetilde{\mathcal{W}}^n$ of "random copies" of label n follows the binomial distribution $\mathcal{B}in(R, r^n)$, with expectation $Rr^n = \text{frac}(NW^n)$. An algorithmic description of residual resampling is given by Algorithm 9.4.

Note that we again exploit the property that any permutation of the sample labels is permitted, hence identify the first indices particles with offsprings that are generated "deterministically" and the later those by residual sampling.

Algorithm 9.4: Residual resampling

Input: Normalised weights $W^{1:N}$.

Output: $A^{1:N} \in 1:N$.

Compute $r^n = \text{frac}(NW^n)$ (for each $n \in 1:N$) and $R = \sum_{n=1}^{N} r^n$.

Construct $A^{1:(N-R)}$ as a vector of size $(N-R)$ that contains $\lfloor NW^n \rfloor$ copies of value n for each $n \in 1:N$.

Sample $A^{N-R+1:N} \sim \mathcal{M}(r^{1:N}/R)$ ▷ `using Algorithm 9.3`

9.6 Stratified and Systematic Resampling

Multinomial resampling applies some transformation (the inverse CDF) to i.i.d. uniform variates. A well-known strategy for variance reduction is to replace i.i.d. uniforms by values that cover [0, 1] more regularly. For instance, one may take $U^n \sim \mathcal{U}((n-1)/N, n/N)$ so that each sub-interval $[n - 1/N, n/N]$ contains exactly one point. This approach is known as stratification, and may be justified by the following basic result.

Lemma 9.2 *Let* $\varphi : [0,1] \to \mathbb{R}$ *such that* $\int_0^1 \varphi^2(u)\,\mathrm{d}u < \infty$. *Consider the following approximations of* $I := \int_0^1 \varphi(u)\,\mathrm{d}u$ *(based on independent random variables):*

$$I_N = \frac{1}{N}\sum_{n=1}^{N} \varphi(U^n)\,, \qquad U^n \sim \mathcal{U}([0,1])\,,$$

$$J_N = \frac{1}{N}\sum_{n=1}^{N} \varphi(V^n)\,, \qquad V_n \sim \mathcal{U}\left(\left[\frac{n-1}{N}, \frac{n}{N}\right]\right)\,.$$

Both estimates are unbiased, and

$$\mathrm{Var}[J_N] \leq \mathrm{Var}[I_N]\,.$$

Proof Without loss of generality, take $I = 0$ (i.e., we may replace φ by $\varphi - I$). Then

$$\mathrm{Var}[I_N] = \frac{1}{N}\mathrm{Var}[\varphi(U^1)] = \frac{1}{N}\int_0^1 \varphi^2(u)\,\mathrm{d}u\,,$$

and

$$\begin{aligned}\mathrm{Var}[J_N] = \frac{1}{N^2}\sum_{n=1}^{N}\mathrm{Var}[\varphi(V_n)] &\leq \frac{1}{N^2}\sum_{n=1}^{N}\mathbb{E}[\varphi^2(V_n)]\\ &= \frac{1}{N}\sum_{n=1}^{N}\int_{(n-1)/N}^{n/N}\varphi^2(u)\,\mathrm{d}u\\ &= \frac{1}{N}\int_0^1 \varphi^2(u)\,\mathrm{d}u\,.\end{aligned}$$

□

If φ is sufficiently regular, we may even show that the variance of the stratified estimate is $\mathcal{O}(N^{-3})$ (Exercise 9.4). This strong result does not apply in our case however: in resampling, the function that maps the uniform variates into ancestor variables (i.e., the inverse CDF of Fig. 9.3) is discontinuous.

Still, Lemma 9.2 gives already some motivation for using stratification within resampling. The corresponding algorithm, called stratified resampling, is described as Algorithm 9.5.

Algorithm 9.5: Stratified resampling

Input: Normalised weights $W^{1:N}$ ($\sum_{n=1}^{N} W^n = 1$, $W^n \geq 0$).

Output: N random indices $A^{1:N}$, taking values in $1:N$.

```
for n = 1 to N do
    U^(n) ~ U([(n − 1)/N, n/N])
A^{1:N} ← icdf(W^{1:N}, U^{(1:N)})        ▷ using Algorithm 9.1
```

One may reduce further the randomness of the procedure by simulating a *single* uniform, $U \sim \mathcal{U}([0, 1])$, and taking $U^{(n)} = (n - 1 + U)/N$. Using this input for Algorithm 9.1 gives systematic resampling. See Algorithm 9.6 for an algorithmic description of systematic resampling, which is essentially Algorithm 9.1 specialised to the particular structure of the $U^{(n)}$'s.

Algorithm 9.6: Systematic resampling

Input: Normalised weights $W^{1:N}$, and $U \in [0, 1]$.

Output: N random indices $A^{1:N}$, taking values in $1:N$

Compute the cumulative weights as $v^n = \sum_{m=1}^{n} N W^m$ for $n \in 1:N$.

```
s ← U
m ← 1
for n = 1 to N do
    while v^m < s do
        m ← m + 1
    A^n ← m
    s ← s + 1
```

To see that systematic resampling is unbiased, i.e., $\mathbb{E}[\mathcal{W}^n] = NW^n$, we make the following observations. First, both schemes have the property that each $U^{(n)} \sim \mathcal{U}((n-1)/N, n/N)$ marginally. Second, per (9.5), we know that

$$\mathbb{P}(A^m = n) = \mathbb{P}(U^{(m)} \in [C_{n-1}, C_n]) = \lambda\left([C^n - C^{n-1}] \cap [\frac{n-1}{N}, \frac{n}{N}]\right)$$

where $\lambda(J)$ is the length of interval J (i.e., its Lebesgue measure). Thus

$$\begin{aligned}\mathbb{E}[\mathcal{W}^n] = \mathbb{E}\left[\sum_{m=1}^{N} \mathbb{1}(A^m = n)\right] &= \sum_{m=1}^{N} \mathbb{P}(A^m = n)\\ &= \sum_{n=1}^{N} \lambda\left([C^n - C^{n-1}] \cap [\frac{n-1}{N}, \frac{n}{N}]\right)\\ &= (C^n - C^{n-1}) = W^n .\end{aligned}$$

9.7 Which Resampling Scheme to Use in Practice?

Systematic resampling is often recommended by practitioners, as it is fast, and seems to work better than other schemes empirically (i.e., it yields lower-variance estimates). The following properties give some support to this recommendation.

First, we note that systematic resampling is such that each $\mathcal{W}^n$ may only take two values: $\lfloor NW^n \rfloor$, or $\lfloor NW^n \rfloor + 1$ (Exercise 9.11). Thus, $\mathcal{W}^n$ has (marginally) minimal variance, under the constraint that $\mathbb{E}[\mathcal{W}^n] = NW^n$, as shown by the following lemma.

Lemma 9.3 *For given $x \geq 0$ the random variable $\mathcal{W}$ that takes the value $\lfloor x \rfloor$ with probability $1 - \text{frac}(x)$, and $\lfloor x \rfloor + 1$ with probability $\text{frac}(x)$, is the integer-valued random variable with the smallest variance that has expectation x.*

For a proof, see Exercise 9.10.

Another interesting criterion is the 'body count':

$$\mathbb{E}\left[\sum_{n=1}^{N} \mathbb{1}\{\mathcal{W}^n = 0\}\right] = \sum_{n=1}^{N} \mathbb{P}(\mathcal{W}^n = 0)$$

that is, the expected number of particles that disappear during the resampling step. For systematic resampling, we have

$$\sum_{n=1}^{N} \mathbb{P}(\mathcal{W}^n = 0) = \sum_{n=1}^{N} (1 - NW^n)^+$$

i.e., a particle such that $NW^n \geq 1$ has at least one offspring by construction, while a particle such that $NW^n < 1$, may have one offspring with probability NW^n (and zero otherwise). This quantity turns out to be a lower bound over all resampling schemes.

Proposition 9.4 *Let $\mathcal{W}$ a non-negative integer-valued random variable. Then,*

$$\mathbb{P}(\mathcal{W} = 0) \geq (1 - \mathbb{E}[\mathcal{W}])^+ .$$

Proof One has:

$$\mathbb{E}[\mathcal{W}] = \sum_{k \geq 1} k\mathbb{P}(\mathcal{W}^n = k) \geq \sum_{k \geq 1} \mathbb{P}(\mathcal{W} = k) = 1 - \mathbb{P}(\mathcal{W} = 0) ,$$

from which the result follows directly. □

Finally, we consider the following criterion:

$$\frac{1}{2N} \sum_{n=1}^{N} |NW^n - \mathcal{W}^n| \tag{9.6}$$

the total variation distance between the two discrete distributions

$$\sum_{n=1}^{N} W^n \delta_{X^n}(\mathrm{d}x) , \quad \text{and} \quad \frac{1}{N} \sum_{n=1}^{N} \mathcal{W}^n \delta_{X^n}(\mathrm{d}x) .$$

This quantity evaluates how much resampling degrades the approximation of $\mathbb{Q}$ brought by importance sampling; see Sect. 11.4.1 for formal definitions.

We conduct the following numerical experiment: the proposal distribution is set to $\mathcal{N}(0, 1)$, the weight function to $w(x) = \exp\{-\tau(x - 1)^2/2\}$, and the sample size to $N = 10^4$. Figure 9.4 plots the average over 100 independent runs of criterion (9.6) as a function of τ, and for the four resampling schemes discussed in this chapter.

We observe that systematic resampling dominates the other resampling schemes in this particular example. Note also the interesting behaviour of these curves at $\tau = 0$; this relates to Example 9.1, i.e., $W^n \to (1/N)$. For multinomial resampling, our TV criterion stays strictly positive even when the weights are constants, i.e., even when $W^n = 1/N$, about 37% of particles are not resampled, as discussed in Example 9.1. But under the other resampling schemes, each particle gets one and only one offspring when $W^n = 1/N$, hence our TV criterion goes to zero when $\tau \to 0$.

Note however that systematic resampling enjoys little theoretical support, relative to other schemes. For instance, we know that estimates obtained by stratified resampling have lower variance than estimated obtained by multinomial resampling (Lemma 9.2). There is no such result for systematic resampling; see also the bibliography for recent results on resampling schemes.

What matters most in practice is to avoid using multinomial resampling, as it is significantly outperformed by other schemes.

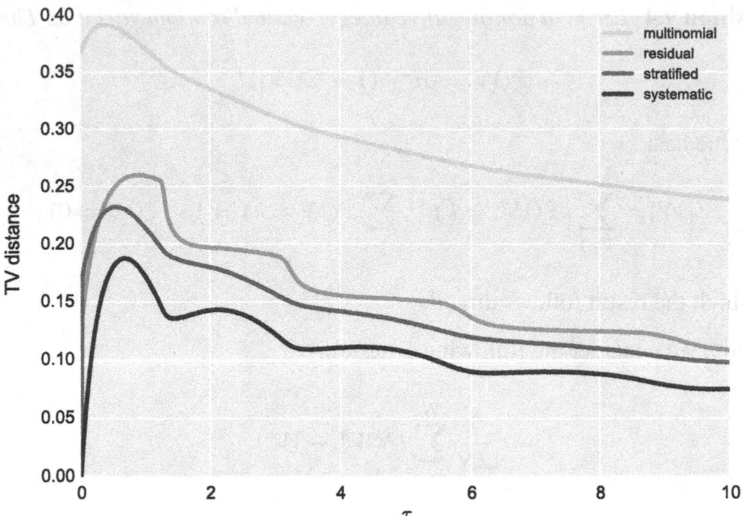

Fig. 9.4 Comparison of resampling schemes: average (over 100 runs) of total variation distance $\sum_{n=1}^{N} |W^n - \mathcal{W}^n/N|/2$ as a function of τ, where proposal is $\mathcal{N}(0, 1)$, weight function is $w(x) = \exp(-\tau(x-1)^2/2)$, and for different resampling schemes

Exercises

9.1 *The aim of this exercise is to justify the expression of* V_{IS} *and* V_{IR} *obtained in the toy example of Sect. 9.2 (appeal of resampling), using only basic probabilistic tools. For* V_{IS}*, simply apply the CLT for auto-normalised importance sampling obtained in the previous chapter, (8.4), and use that* $X_1^2 = \rho^2 X_0^2 + (1-\rho^2)U^2 + 2\rho\sqrt{1-\rho^2}X_0 U$ *to obtain the desired result. For* V_{IR}*, we will show instead the weaker result that* $\text{MSE}(\sqrt{N}\widehat{\varphi}_{\text{IS}}) \to V_{\text{IS}}$ *as* $N \to +\infty$*. (For a proper CLT for* $\widehat{\varphi}_{\text{IS}}$*, see Chap. 11.) To this aim, we shall use decomposition (9.3). Obviously* $N^{-1}\sum_{n=1}^{N} U^n$ *is independent of the two other terms, and has variance 1. For* $\sum_{n=1}^{N} W_0^n X_0^n$*, simply re-use the corresponding result obtained for* $\widehat{\varphi}_{\text{IS}}$*. (Technical point: convergence in law implies* L^p *convergence for bounded random variables. Explain why.) Use simple properties of the multinomial distribution to compute the expectation and variance of the third term, conditional on the random variables* $X_0^{1:N}$*. Deduce the limit of the MSE of the third term as* $N \to +\infty$*. (Recall the technical point above.) Use similar ideas to show that the first and the third term are not correlated. Conclude.*

9.2 *Justify the following expression for* $\gamma(\tau)$ *that appears in the toy example of Sect. 9.2 (appeal of resampling):*

$$\gamma(\tau) = \frac{\int_0^\tau x^2 \exp(-x^2/2)\,\mathrm{d}x}{\int_0^\tau \exp(-x^2/2)\,\mathrm{d}x} = \frac{2}{\sqrt{\pi}} \times \frac{\Gamma(3/2, \tau^2/2)}{2\Phi(\tau) - 1}$$

where $\Gamma(s,t)$ *denotes the incomplete gamma function,* $\Gamma(a,t) = \int_0^t x^{a-1}$ $\exp(-x)\mathrm{d}x$. *Differentiate the first expression to show that* $\gamma(\tau)$ *is an increasing function. Use the second expression to plot* γ *as a function of* τ. *Show that* $\gamma(\tau) \to 1$ *as* $\tau \to +\infty$, *and* $\gamma(\tau) \to 0$ *as* $\tau \to 0$. *(To this effect, you may show first that* $\gamma(\tau) = \mathbb{Q}_0(X_0^2) \leq \tau^2$.*)*

9.3 *Establish Proposition 9.1. As an intermediate step, you may derive the joint PDF of* $(S_1, \ldots, S_{N+1})$.

9.4 *In the settings of Lemma 9.2, assume furthermore that* φ *is Lipschitz. Show that the variance of the stratified estimate is* $\mathcal{O}(N^{-3})$ *in that case. (Hint: as in Lemma 9.2, you may take* $\int_0^1 \varphi(u)\,\mathrm{d}u = 0$*; also note that* $2\mathrm{Var}[X] = E[(X - X')^2]$ *where* X' *is a independent 'copy' of* X *(i.e., a random variable with the same distribution).*

9.5 *The resampling algorithms described in this chapter draw* N *indices from a collection of size* N *(the* N *particles). Generalise these algorithms to the generation of* M *indices from a collection of size* N. *Explain why the complexity of these generalised resampling schemes is* $\mathcal{O}(N + M)$. *Compare your solution with those in the Python corner.*

9.6 *The alias algorithm draws a random integer in the range* $1 : N$ *as follows: given vectors* $q \in [0, 1]^N$ *and* $m \in (1 : N)^N$, *draw* A *uniformly in* $1 : N$, *with probability* $q(A)$ *return* A, *otherwise return* $m(A)$. *The aim of this exercise is to find how to set up* q *and* m *so that this algorithm draws from a given multinomial distribution* $\mathcal{M}(W^{1:N})$. *Say a particle* n *is small (resp. large) if* $NW^n \leq 1$ *(resp.* > 1*). Explain how, by selecting two particles, one small and one large, one may set* $q(n)$ *and* $m(n)$ *for a single* n. *Deduce a recursive algorithm that will set up all these values in at most* N *steps. Is the alias algorithm competitive with Algorithm 9.3 for sampling* N *times from* $\mathcal{M}(W^{1:N})$*?*

9.7 *Consider the multinomial distribution* $\mathcal{M}(W^{1:N})$. *Using Algorithm 9.3, we are able to generate* N *draws from this distribution at cost* $\mathcal{O}(N)$. *It might be tempting to say that each draw costs* $\mathcal{O}(1)$. *Consider however a situation where one needs to generate such draws 'on the fly', without knowing any advance how many draws* M *will be required. Using the generalised multinomial resampling derived in Exercise 9.5 (with* $M = 1$*) would cost* $\mathcal{O}(N)$ *per draw, leading to an overall* $\mathcal{O}(MN)$ *cost. Think of how to adapt this algorithm to obtain a much lower computational cost. (There are quite a few ways to answer this question; think for instance of binary search, or the algorithm described in 9.6, or a queue algorithm. See the Python corner for a discussion.)*

9.8 *Show that criterion* (9.7) *equals*

$$\sum_{n=1}^{N} \mathbb{P}(\mathcal{W}^n = 0) = \sum_{n=1}^{N} (1 - W^n)^N$$

for multinomial resampling. Show that this quantity is always above $N(1 - 1/N)^N \approx Ne^{-1}$.

9.9 *Show that the* $\mathcal{W}^n$*'s (where* $\mathcal{W}^n$ *is the number of times particle n is resampled) are pairwise negatively correlated under multinomial resampling. Extend this result to residual resampling.*

9.10 *Prove Lemma 9.3. To do so, first check that the distribution defined in the lemma has a support of size at most two, and that it has the correct expectation. Then derive the Lagrangian that corresponds to the problem of maximising* $\mathbb{E}[\mathcal{W}^2]$ *under the constraints that* $\mathbb{E}[\mathcal{W}]$ *and that* $\sum_k p_k = 1$*, where* $p_k := \mathbb{P}(\mathcal{W} = k)$*. Show that any distribution with a support of cardinal 3 or more cannot be a solution of this Lagrangian. Conclude.*

9.11 *Show that systematic resampling is such that* $\mathcal{W}^n$ *is either* $\lfloor NW^n \rfloor$ *or* $\lfloor NW^n \rfloor + 1$.

Python Corner

All the resampling algorithms discussed in this chapter are implemented in the `resampling` module of `particles`. They take as arguments `M` (the desired number of draws) and `W` (a vector of normalised weights of size N). Note that we may have $N \neq M$, i.e. the number of resampled particles may be different from the number of initial particles (as discussed in Exercise 9.5).

The main routine in the resampling module is the inverse CDF algorithm, which we described in Algorithm 9.1 in this chapter:

```
def inverse_cdf(su, W):
    """Inverse CDF algorithm for a finite distribution.

        Parameters
        ----------
        su: (M,) ndarray
            M sorted variates (i.e. M ordered points in [0,1]).
        W: (N,) ndarray
            a vector of N normalized weights (>=0 and sum to one)

        Returns
        -------
        A: (M,) ndarray
            a vector of M indexes in range 0, ..., N-1
    """
    j = 0
    s = W[0]
    M = su.shape[0]
    A = np.empty(M, 'int')
    for n in range(M):
        while su[n] > s:
            j += 1
```

```
            s += W[j]
        A[n] = j
    return A
```

Once `inverse_cdf` is defined, the resampling algorithms become one-liners. For stratified and systematic, we have:

```
from numpy import random

def stratified(M, W):
    su = (random.rand(M) + np.arange(M)) / M
    return inverse_cdf(su, W)

def systematic(M, W):
    su = (random.rand(1) + np.arange(M)) / M
    return inverse_cdf(su, W)
```

For multinomial resampling, we write separately the generation of uniform spacings:

```
def uniform_spacings(N):
    z = np.cumsum(-np.log(random.rand(N + 1)))
    return z[:-1] / z[-1]

def multinomial(M, W):
    return inverse_cdf(uniform_spacings(M), W)
```

Residual resampling is left as an exercise (or alternatively check directly the source code of `particles`).

Function `inverse_cdf` involves a loop, and as such may not be very fast in an interpreted language such as Python. A bright student of the first author came up with the following elegant one-liner for multinomial resampling:

```
def multinomial(M, W):
    return np.searchsorted(np.cumsum(W), random.rand(M))
```

`searchsorted` is a `NumPy` function that finds, for each component of the second argument, the position where this component should be inserted in the first argument to maintain order. It is based on binary search, hence its complexity is $\mathcal{O}(M \log N)$. Yet it is faster than the $\mathcal{O}(N + M)$ implementation above, provided M and N are not too large, again because it does not loop (in Python).

However, as discussed in Chap. 1, the loop above may be made to run much faster using `Numba`. Figure 9.5 compares the running times of the three implementations for different values of N. (We took $M = N$.) The `searchsorted` implementation is faster than the naive (pure Python) implementation until $N \approx 10^6$. Then it becomes more expensive than the two other implementations, because of its greater complexity. The `Numba` implementation is consistently faster than the two other implementations.

Finally, consider the problem of simulating a long stream of draws from the multinomial distribution $\mathcal{M}(W^{1:N})$, without knowing in advance how many draws M will be required (Exercise 9.7). This may be useful for instance in rejection

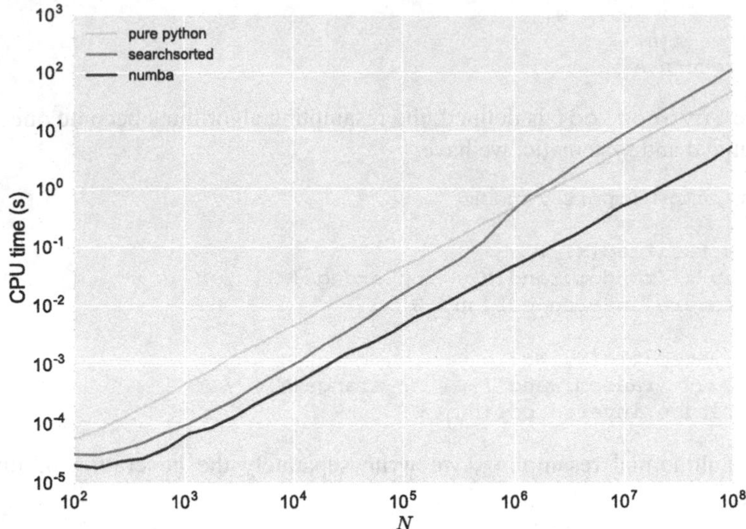

Fig. 9.5 CPU time (in seconds) as a function of N for the three implementations of multinomial resampling discussed in the text

sampling, in case the proposal is a multinomial distribution. Certain smoothing algorithms presented in Chap. 12 will be based on such a rejection strategy.

A possible solution is to implement a queue. At creation, the queue is populated with N draws. Each time a draw is required, it is 'dequeued' (removed from the queue). When the queue is empty, we regenerate N draws. At most $N - 1$ draws may be generated without being used, so the overall complexity remains $\mathcal{O}(M+N)$.

The following Python class implements such a queue. As per Remark 9.1, we must randomly permute the N draws whenever the queue is filled; note that generating a random permutation is a $\mathcal{O}(N)$ operation.

```
class MulinomialQueue(object):
    def __init__(self, W):
        self.W = W
        self.N = W.size
        self.j = 0
        self.enqueue()

    def enqueue(self):
        perm = random.permutation(self.N)
        self.A = multinomial(self.N, self.W)[perm]

    def dequeue(self, k):
        """Outputs *k* draws from the multinomial distribution."""
        if self.j + k <= self.N:
            out = self.A[self.j:(self.j+k)]
            self.j += k
        elif k <= self.N:
            out = np.empty(k, 'int')
```

```
                nextra = self.j + k - self.M
                out[:(k - nextra)] = self.A[self.j:]
                self.enqueue()
                out[(k - nextra):] = self.A[:nextra]
                self.j = nextra
            else:
                raise ValueError('MultinomialQueue: k must be <= N')
            return out
```

Another, perhaps more elegant, solution to this problem is the alias algorithm (discussed in Exercise 9.6). For the record, here is a succinct implementation of the alias algorithm, based on lists. Note again how relevant is object orientation for this type of algorithm.

```
class AliasTable(object):
    def __init__(self, W):
        self.N = W.size
        self.q = W * self.N
        self.big_bro = np.empty(self.N, 'int')
        small_ones = [n for n, q in enumerate(self.q) if q<=1.]
        big_ones = [n for n, q in enumerate(self.q) if q>1.]
        while small_ones and big_ones:
            small = small_ones.pop()
            big = big_ones.pop()
            self.big_bro[small] = big
            self.q[big] -= (1.-self.q[small])
            if self.q[big] > 1.:
                big_ones.append(big)
            else:
                small_ones.append(big)

    def sample(self, size=1):
        ns = random.randint(self.N, size=size)
        u = random.rand(size)
        return np.where(u < self.q[ns], ns, self.big_bro[ns])
```

As an exercise, you may rewrite this class by using `NumPy` arrays, and compare your implementation with the one in `particles`.

Bibliographical Notes

An early reference for Algorithm 9.2 (uniform spacings), which may be used to perform multinomial resampling, is the book of Devroye (1986, p. 208).

Residual resampling appeared in the genetic algorithms literature (Baker 1985) under the name of "stochastic remainder sampling" and then was rediscovered by Higuchi (1997) and Liu and Chen (1998) in the particle filtering literature.

Both stratified and systematic resampling were introduced in the appendix of Kitagawa (1996). That early version of systematic resampling was purely deterministic however. The version presented here (with a random U) appeared in Carpenter et al. (1999). Similar algorithms appeared previously in the genetic algorithms literature (Whitley 1994), and in the survey sampling literature, under the name of "unequal probability systematic sampling" (e.g. Hartley and Rao 1962).

We also mention in passing the tree-based resampling algorithm of Crisan and Lyons (2002), which also ensures that $\mathcal{W}^n$ has minimal variance (like systematic resampling), and in addition can be shown to dominate multinomial resampling in terms of asymptotic variance (Künsch 2005). However, although it also has $\mathcal{O}(N)$ complexity, it is a bit more difficult to implement than systematic resampling. Other recently proposed resampling schemes are the transport-based algorithm of Reich (2013) and the chop-thin algorithm of Gandy and Lau (2016). Finally, Murray et al. (2016) and Guldas et al. (2017) present and discuss resampling schemes that are better suited for parallel computation; we return to this point in Chap. 19.

Up to recently, there were few formal results on resampling schemes (besides multinomial resampling, which is easier to study). Gerber et al. (2019) provide some such results. For instance, this paper shows that estimates obtained by stratified resampling are consistent, but those obtained by systematic resampling may not be (if particles are ordered in a certain artificial way). This paper also describes a resampling scheme (called SSP) which is a bit more complicated to implement, but which enjoys the best properties of both stratified and systematic resampling (consistency, has lower variance than multinomial, minimises the variance of $\mathcal{W}^n$ and the body count).

Bibliography

Baker, J. L. (1985). Adaptive selection methods for genetic algorithms. In J. Grefenstette (Ed.), *Proceedings of the International Conference on Genetic Algorithms and Their Applications* (pp. 101–111). Hillsdale, NJ: Lawrence Erlbaum Associates.

Carpenter, J., Clifford, P., & Fearnhead, P. (1999). Improved particle filter for nonlinear problems. *IEE Proceedings - Radar, Sonar Navigation, 146*(1), 2–7.

Crisan, D., & Lyons, T. (2002). Minimal entropy approximations and optimal algorithms for the filtering problem. *Monte Carlo Methods and Applications, 8*(4), 343–356.

Devroye, L. (1986). *Non-uniform random variate generation*. New York: Springer.

Gandy, A., & Lau, F. D.-H. (2016). The chopthin algorithm for resampling. *IEEE Transactions on Signal Processing, 64*(16), 4273–4281.

Gerber, M., Chopin, N., & Whiteley, N. (2019). Negative association, ordering and convergence of resampling methods. *Annals of Statistics, 47*(4), 2236–2260.

Guldas, H., Cemgil, T., Whiteley, N., & Heine, K. (2017). A practical introduction to butterfly and adaptive resampling in sequential monte carlo. In Y. Zhao (Ed.), *17th IFAC Symposium on System Identification SYSID 2015 – Beijing, China, 19–21 October 2015. IFAC-PapersOnLine* (Vol. 28, pp. 787–792). Amsterdam: Elsevier.

Hartley, H. O., & Rao, J. N. K. (1962). Sampling with unequal probabilities and without replacement. *The Annals of Mathematical Statistics, 33*(2), 350–374.

Higuchi, T. (1997). Monte Carlo filter using the genetic algorithm operators. *Journal of Statistical Computation and Simulation, 59*(1), 1–23.

Kitagawa, G. (1996). Monte Carlo filter and smoother for non-Gaussian nonlinear state space models. *Journal of Computational and Graphical Statistics, 5*(1), 1–25.

Künsch, H. R. (2005). Recursive Monte Carlo filters: Algorithms and theoretical analysis. *Annals of Statistics, 33*(5), 1983–2021.

Liu, J. S., & Chen, R. (1998). Sequential Monte Carlo methods for dynamic systems. *Journal of the American Statistical Association, 93*(443), 1032–1044.

Murray, L. M., Lee, A., & Jacob, P. E. (2016). Parallel resampling in the particle filter. *Journal of Computational and Graphical Statistics, 25*(3), 789–805.

Reich, S. (2013). A nonparametric ensemble transform method for Bayesian inference. *SIAM Journal on Scientific Computing, 35*(4), A2013–A2024.

Whitley, D. (1994). A genetic algorithm tutorial. *Statistics and Computing, 4*, 65–85.

Chapter 10
Particle Filtering

Summary We now have all the ingredients in place to describe particle filtering at a sufficient level of generality: on one hand, the chosen Feynman-Kac model defines the recursive quantities we wish to approximate; on the other hand, importance sampling and resampling gives us the right tools to actually compute these approximations.

In that spirit, we start this chapter by defining a completely generic particle filtering algorithm. Then we show that specific particle filtering algorithms such as the bootstrap filter, the guided filter, and the auxiliary particle filter, are special cases of this generic algorithm, obtained by choosing an appropriate Feynman-Kac model. For a given state-space model, these particle algorithms compute recursively approximations of the filtering distributions, the predictive distributions, and the likelihood of the data up to time t (for $t = 0 : T$).

10.1 The Generic PF for a Given Feynman-Kac Model

The box below gives the necessary ingredients to define our generic particle filter.

Input of Generic PF Algorithm

- A Feynman-Kac model $\{M_t, G_t\}$ such that:
  - the weight function G_t may be evaluated pointwise (for all t);
  - it is possible to simulate from $\mathbb{M}_0(\mathrm{d}x_0)$ and from $M_t(x_{t-1}, \mathrm{d}x_t)$ (for any x_{t-1} and t).
- The number of particles N.
- The choice of an unbiased resampling scheme (e.g. multinomial, residual, stratified, systematic), that is, an algorithm to generate variables $A_t^{1:N}$ in $1 : N$ such that $\mathbb{E}\left[\sum_{m=1}^N \mathbb{1}\{A_t^m = n\}\right] = NW_{t-1}^n$; see previous chapter for more details.

© Springer Nature Switzerland AG 2020

N. Chopin, O. Papaspiliopoulos, *An Introduction to Sequential Monte Carlo*, Springer Series in Statistics, https://doi.org/10.1007/978-3-030-47845-2_10

Once these inputs are provided, one may run the generic PF algorithm defined as Algorithm 10.1.

Algorithm 10.1: Generic PF algorithm

```
Operations involving index n must be performed
 for n = 1,...,N.
```

$X_0^n \sim \mathbb{M}_0(\mathrm{d}x_0)$

$w_0^n \leftarrow G_0(X_0^n)$

$W_0^n \leftarrow w_0^n / \sum_{m=1}^N w_0^m$

for $t = 1$ **to** T **do**

$\quad A_t^{1:N} \sim$ `resample`$(W_t^{1:N})$ ▷ `user-defined resampling scheme`

$\quad X_t^n \sim M_t(X_{t-1}^{A_t^n}, \mathrm{d}x_t)$

$\quad w_t^n \leftarrow G_t(X_{t-1}^{A_t^n}, X_t^n)$

$\quad W_t^n \leftarrow w_t^n / \sum_{m=1}^N w_t^m$

Basically, Algorithm 10.1 is the Monte Carlo implementation of the forward recursion seen in Sect. 5.2. In particular, at time 0 (first three lines), the algorithm performs an importance sampling step from $\mathbb{M}_0(\mathrm{d}x_0)$ to

$$\mathbb{Q}_0(\mathrm{d}x_0) = \frac{1}{L_0}\mathbb{M}_0(\mathrm{d}x_0)G_0(x_0)\,,$$

and generates

$$\mathbb{Q}_0^N(\mathrm{d}x_0) := \sum_{n=1}^N W_0^n \delta_{X_0^n}(\mathrm{d}x_0), \quad W_0^n = \frac{G_0(X_0^n)}{\sum_{m=1}^N G_0(X_0^m)}$$

as an importance sampling approximation of $\mathbb{Q}_0(\mathrm{d}x_0)$.

Then, to progress from time 0 to time 1, recall that one must first *extend* $\mathbb{Q}_0$:

$$\mathbb{Q}_0(\mathrm{d}x_{0:1}) = \mathbb{Q}_0(\mathrm{d}x_0)M_1(x_0, \mathrm{d}x_1) \tag{10.1}$$

and then apply the change of measure:

$$\mathbb{Q}_1(\mathrm{d}x_{0:1}) = \frac{1}{\ell_1}\mathbb{Q}_0(\mathrm{d}x_{0:1})G_1(x_0, x_1)\,. \tag{10.2}$$

For the extension step, if we plug in $\mathbb{Q}_0^N$ for $\mathbb{Q}_0$ in (10.1), we obtain

$$\mathbb{Q}_0^N(\mathrm{d}x_{0:1}) = \sum_{n=1}^{N} W_0^n \delta_{X_0^n}(\mathrm{d}x_0) M_1(X_0^n, \mathrm{d}x_1),$$

from which we may simulate from N times, so as to obtain pairs $(X_0^{A_1^n}, X_1^n)$; this is done in the first too lines inside the loop, (for $t = 1$). This is exactly the importance resampling strategy described in the previous chapter.

Finally, to apply change of measure (10.2), we simply perform importance sampling from $\mathbb{Q}_0^N(\mathrm{d}x_{0:1})$ to $\mathbb{Q}_1^N(\mathrm{d}x_{0:1})$, i.e. we reweight according to function G_1; see the third line inside the loop.

To progress to times 2, 3, and so on, we proceed exactly as above. The algorithm delivers the following approximations at each time t:

Output of Generic PF Algorithm

$$\frac{1}{N}\sum_{n=1}^{N} \delta_{X_t^n} \quad \text{approximates} \mathbb{Q}_{t-1}(\mathrm{d}x_t)$$

$$\mathbb{Q}_t^N(\mathrm{d}x_t) = \sum_{n=1}^{N} W_t^n \delta_{X_t^n} \quad \text{approximates } \mathbb{Q}_t(\mathrm{d}x_t)$$

$$\ell_t^N = \frac{1}{N}\sum_{n=1}^{N} w_t^n \quad \text{approximates } \ell_t$$

$$L_t^N = \prod_{s=0}^{t} \ell_s^N \quad \text{approximates } L_t = \prod_{s=0}^{t} \ell_s$$

Of course, in the two first cases, one may take the expectation with respect to some test function φ, to obtain:

$$\frac{1}{N}\sum_{n=1}^{N} \varphi(X_t^n) \quad \text{approximates } \mathbb{E}_{\mathbb{Q}_{t-1}}[\varphi(X_t)]$$

$$\mathbb{Q}_t^N(\varphi) = \sum_{n=1}^{N} W_t^n \varphi(X_t^n) \quad \text{approximates } \mathbb{Q}_t(\varphi)\,.$$

The word "approximates" is vague for the moment. We shall see in Chap. 11 that one way to formalise this is to establish that the approximation error goes to zero at the standard Monte Carlo rate (that is $N^{-1/2}$) as $N \to +\infty$. We can already remark that, following our basic description, iteration t performs a Monte Carlo approximation of a certain distribution, which depends itself on a Monte Carlo approximation computed at iteration $t-1$, and so on. In other words, the error at time t is the result of an accumulation of Monte Carlo errors at times 0 to t, and will be studied as such.

The complexity of Algorithm 10.1 is $\mathcal{O}(N)$ per time step, $\mathcal{O}(TN)$ overall. Recall in particular that the complexity of the resampling schemes seen in the previous chapter is always $\mathcal{O}(N)$.

10.2 Adaptive Resampling

In our generic particle filter, Algorithm 10.1, resampling is performed at every time step. In Chap. 9, we explained that we can always decide to resample or not, and that this decision amounts to a trade-off between an increase of variance in the past (because ancestors are resampled) and a decrease of variance in the present (because particles are then more diverse, provided kernel M_t forgets its past in some way).

What happens if we decide to *never* resample? This gives us sequential importance sampling; see Algorithm 10.2.

Algorithm 10.2: Sequential importance sampling (do not use!)

Operations involving index n must be performed for $n = 1, \ldots, N$.

$X_0^n \sim \mathbb{M}_0(\mathrm{d}x_0)$

$w_0^n \leftarrow G_0(X_0^n)$

$W_0^n \leftarrow w_0^n / \sum_{m=1}^N w_0^m$

for $t = 1$ **to** T **do**

  $X_t^n \sim M_t(X_{t-1}^n, \mathrm{d}x_t)$

  $w_t^n \leftarrow w_{t-1}^n G_t(X_{t-1}^n, X_t^n)$

  $W_t^n \leftarrow w_t^n / \sum_{m=1}^N w_t^m$

This algorithm is simpler to understand than our generic particle filter, but unfortunately it tends to perform poorly. Iterations 0 to t are equivalent to a single importance sampling step, from proposal $\mathbb{M}_t(\mathrm{d}x_{0:t})$ to target

$$\mathbb{Q}_t(\mathrm{d}x_{0:t}) \propto G_0(x_0) \prod_{s=1}^{t} G_s(x_{s-1}, x_s)\mathbb{M}_t(\mathrm{d}x_{0:t}) .$$

Since both $\mathbb{M}_t$ and $\mathbb{Q}_t$ are of increasing dimension, the variance of the weights should diverge quickly over time, because of the curse of dimensionality we discussed in Sect. 8.7. (See Sect. 10.5 for a numerical illustration.)

Thus, as discussed in Sect. 9.2, resampling at every time step seems to be the safe option, as this protects from weight degeneracy, at least as far as inference on current time (i.e. inference on $\mathbb{Q}_t(\mathrm{d}x_t)$) is concerned.

That said, one may contemplate resampling from time to time. Common practice in recent years has converged to the following recipe: trigger the resampling step whenever the variability of the weights is too large. In particular, the ESS (effective sample size) criterion seen in Sect. 8.6 is typically used; i.e.

$$\text{resample whenever } \mathrm{ESS}(W_{t-1}^{1:N}) < \mathrm{ESS}_{\min} \,;$$

a common choice is $\mathrm{ESS}_{\min} = N/2$. The performance of the algorithm is usually not very sensitive to $\mathrm{ESS}_{\min}$, see the numerical experiments of Sect. 10.5.

The caveat of course is that this recipe is completely ad hoc: the ESS criterion is already somehow arbitrary in the independent case, and is even more so in particle filtering, where successive steps introduce non-trivial dependencies between particles. That said, adaptive resampling (resampling only when the ESS is too low) may allow for reducing the CPU time of the algorithm; again see Sect. 10.5 for a numerical illustration.

The generic particle filter with adaptive resampling steps is described as Algorithm 10.3.

Input of Generic PF Algorithm (ESS-Based Adaptive Resampling)

- Same input as Algorithm 10.1
- a scalar $\mathrm{ESS}_{\min}$ (e.g. $\mathrm{ESS}_{\min} = \alpha N$, with $\alpha \in (0, 1)$)

Algorithm 10.3: Generic particle filter (ESS-based adaptive resampling)

```
Operations involving index n must be performed
 for n = 1, ..., N.
```

$X_0^n \sim \mathbb{M}_0(dx_0)$
$w_0^n \leftarrow G_0(X_0^n)$
$W_0^n \leftarrow w_0^n / \sum_{m=1}^N w_0^m$
for $t = 1$ **to** T **do**
  if $\text{ESS}(W_{t-1}^{1:N}) < \text{ESS}_{\min}$ **then**
    $A_t^{1:N} \leftarrow \texttt{resample}(W_{t-1}^{1:N})$
    $\widehat{w}_{t-1}^n \leftarrow 1$
  else
    $A_t^n \leftarrow n$
    $\widehat{w}_{t-1}^n \leftarrow w_{t-1}^n$
  $X_t^n \sim M_t(X_{t-1}^{A_t^n}, dx_t)$
  $w_t^n \leftarrow \widehat{w}_{t-1}^n G_t(X_{t-1}^{A_t^n}, X_t^n)$
  $W_t^n \leftarrow w_t^n / \sum_{m=1}^N w_t^m$

Adaptive resampling requires to adapt a little bit the expression of the estimates of ℓ_t and of the predictive distribution.

Output of PF Algorithm with Adaptive Resampling

Same as output of Algorithm 10.1 except

$$\ell_t^N = \begin{cases} \frac{1}{N} \sum_{n=1}^N w_t^n \text{ if resampling occurred at time } t \\ \frac{\sum_{n=1}^N w_t^n}{\sum_{n=1}^N w_{t-1}^n} \text{ otherwise} \end{cases} \tag{10.3}$$

(continued)

and the estimate of the predictive distribution is now

$$\frac{1}{\sum_{n=1}^{N} \widehat{w}_{t-1}^{n}} \sum_{n=1}^{N} \widehat{w}_{t-1}^{n} \delta_{X_t^n}(\mathrm{d}x_t) .$$

As before, the expression of ℓ_t^N is based on the identity $\ell_t = (\mathbb{Q}_{t-1}M_t)(G_t)$, and on how $\mathbb{Q}_{t-1}M_t$ is approximated by the algorithm: by the unweighted Monte Carlo sample $\{(X_{t-1}^{A_t^n}, X_t^n)\}_{n=1:N}$ when we resample; by the weighted Monte Carlo sample $\{(X_{t-1}^n, X_t^n)\}$, with weights W_{t-1}^n, when we do not. The same remarks apply to the predictive distribution.

Both previous algorithms may be recovered as special cases of Algorithm 10.3, by taking either $\mathrm{ESS}_{\min} = N$ (Algorithm 10.1, where resampling is performed at every step), or $\mathrm{ESS}_{\min} = 0$ (Algorithm 10.2, where resampling is never performed).

The specific particle algorithms discussed in the rest of the chapter will be written as special cases of this generic particle algorithm with adaptive resampling.

10.3 Application to State-Space Models: Filtering Algorithms

We now consider a particular state-space model, $\{(X_t, Y_t)\}$, with initial law $\mathbb{P}_0$, transition kernels $P_t(x_{t-1}, \mathrm{d}x_t)$, and observation densities $f_t(y_t|x_t)$, and we discuss how to choose a given Feynman-Kac representation to obtain particle algorithms that may perform sequential inference for such a model.

10.3.1 The Bootstrap Filter

We start with the most basic Feynman-Kac representation of our state-space model,

$$\mathbb{M}_0(\mathrm{d}x_0) = \mathbb{P}_0(\mathrm{d}x_0) , \qquad G_0(x_0) = f_0(y_0|x_0) , \tag{10.4a}$$

$$M_t(x_{t-1}, \mathrm{d}x_t) = P_t(x_{t-1}, \mathrm{d}x_t) , \qquad G_t(x_{t-1}, x_t) = f_t(y_t|x_t) , \tag{10.4b}$$

for $t \geq 1$. Then $\mathbb{Q}_{t-1}(\mathrm{d}x_t)$ matches the predictive distribution $\mathbb{P}_t(X_t \in \mathrm{d}x_t|Y_{0:t-1} = y_{0:t-1})$, $\mathbb{Q}_t(\mathrm{d}x_t)$ matches the filtering distribution $\mathbb{P}_t(X_t \in \mathrm{d}x_t|Y_{0:t} = y_{0:t})$, and L_t the likelihood $p(y_{0:t})$, as explained in Sect. 5.1.2.

This specific choice for (M_t, G_t) turns Algorithm 10.1 into what is usually called the *bootstrap* filter; see Algorithm 10.4.

Input of the Bootstrap Filter

- A state-space model such that:
  - one may simulate from $\mathbb{P}_0(\mathrm{d}x_0)$ and from $P_t(x_{t-1}, \mathrm{d}x_t)$ for each x_{t-1} and t;
  - one may compute function $x_t \to f_t(y_t|x_t)$ point-wise.
- Standard inputs of a PF: number of particles N, choice of a resampling scheme, ESS threshold $\mathrm{ESS}_{\min}$.

Algorithm 10.4: Bootstrap filter

Operations involving index n must be performed for $n = 1, \ldots, N$.

$X_0^n \sim \mathbb{P}_0(\mathrm{d}x_0)$
$w_0^n \leftarrow f_0(y_0|X_0^n)$
$W_0^n \leftarrow w_0^n / \sum_{m=1}^N w_0^m$
for $t = 1$ **to** T **do**
  if $\mathrm{ESS}(W_{t-1}^{1:N}) < \mathrm{ESS}_{\min}$ **then**
    $A_t^{1:N} \sim \texttt{resample}(W_{t-1}^{1:N})$
    $\widehat{w}_{t-1}^n \leftarrow 1$
  else
    $A_t^n \leftarrow n$
    $\widehat{w}_{t-1}^n \leftarrow w_{t-1}^n$
  $X_t^n \sim P_t(X_{t-1}^{A_t^n}, \mathrm{d}x_t)$
  $w_t^n \leftarrow \widehat{w}_{t-1}^n f_t(y_t|X_t^n)$
  $W_t^n \leftarrow w_t^n / \sum_{m=1}^N w_t^m$

This algorithm has an intuitive interpretation: at every time t, we generate simulations from the law of Markov chain $\{X_t\}$; then we weight (grade) these simulations according to how compatible they are with the datapoint Y_t, and resample if necessary. In practice, the following approximations are obtained:

Output of the Bootstrap Filter

$$\frac{1}{\sum_{n=1}^{N} \widehat{w}_{t-1}^n} \sum_{n=1}^{N} \widehat{w}_{t-1}^n \varphi(X_t^n) \quad \text{approximates } \mathbb{E}[\varphi(X_t)|Y_{0:t-1} = y_{0:t-1}]$$

$$\sum_{n=1}^{N} W_t^n \varphi(X_t^n) \quad \text{approximates } \mathbb{E}[\varphi(X_t)|Y_{0:t} = y_{0:t}]$$

$$\ell_t^N \text{(as defined in (10.3))} \quad \text{approximates } p_t(y_t|y_{0:t-1})$$

$$L_t^N = \prod_{s=0}^{t} \ell_s^N \quad \text{approximates } p(y_{0:t})$$

The advantages of the bootstrap filter is that it is very simple, and widely applicable: the only requirement is to be able to (a) simulate from $P_t(x_{t-1}, dx_t)$, and (b) to evaluate $f_t(y_t|x_t)$ pointwise. Indeed, we have discussed in Sect. 2.4 practical examples where the Markov transition of $\{X_t\}$ is defined through a complex simulation algorithm, but which is intractable in other respects, e.g., one may not be able to compute the probability density of $P_t(x_{t-1}, dx_t)$. This is a hindrance in many algorithms such as MCMC (see Chap. 15), but not for the bootstrap filter.

The main drawback of the bootstrap filter is that it samples particles X_t "blindly" from $P_t(x_{t-1}, dx_t)$, without guarantee that many of these simulated particles will be compatible with datapoint y_t (i.e. have non-negligible weights). This is especially problematic when the datapoints are very informative; i.e. the likelihood $x_t \to f_t(y_t|x_t)$ is a peaked function. In such a scenario, one may need to take N to be very large to obtain decent performance. The two next sections consider extensions of the bootstrap filter that address these issues.

10.3.2 The Guided Particle Filter

We have noticed in Sect. 5.1.2 that (10.4) is not the unique Feynman-Kac model such that $\mathbb{Q}_t$ coincides with the filtering distribution of the state-space model

$\{(X_t, Y_t)\}$: any (M_t, G_t) such that

$$G_t(x_{t-1}, x_t)M_t(x_{t-1}, \mathrm{d}x_t) = f_t(y_t|x_t)P_t(x_{t-1}, \mathrm{d}x_t) \tag{10.5}$$

has this property. We can exploit this property to produce more general, and hopefully more effective, particle filters. In particular, we can *guide* the particles to areas where the weight is expected to be high, instead of following the system dynamics as in the bootstrap filter. For a given choice of the kernel M_t, the potential function should be chosen as

$$G_t(x_{t-1}, x_t) = \frac{f_t(y_t|x_t)P_t(x_{t-1}, \mathrm{d}x_t)}{M_t(x_{t-1}, \mathrm{d}x_t)} \tag{10.6}$$

for (10.5) to hold. This requires that $M_t(x_{t-1}, \mathrm{d}x_t) \gg P_t(x_{t-1}, \mathrm{d}x_t)f_t(y_t|x_t)$ for any x_{t-1} (but not necessarily that $M_t(x_{t-1}, \mathrm{d}x_t) \gg P_t(x_{t-1}, \mathrm{d}x_t)$). In many practical cases, $M_t(x_{t-1}, \mathrm{d}x_t)$ and $P_t(x_{t-1}, \mathrm{d}x_t)$ admit probability densities $m_t(x_t|x_{t-1})$ and $p_t(x_t|x_{t-1})$, respectively, with respect to a common dominating measure $\mu(\mathrm{d}x_t)$. Then G_t simplifies to

$$G_t(x_{t-1}, x_t) = \frac{f_t(y_t|x_t)p_t(x_t|x_{t-1})}{m_t(x_t|x_{t-1})}\,. \tag{10.7}$$

One should bear in mind however that (10.6) is more general than (10.7), and that the guided particle filter does not require probability densities for M_t and P_t to exist; see below for an example.

In this specification we obtain the following correspondences, which can be checked by basic calculations.

Guided Feynman-Kac Models

For a given state-space model, we define a Feynman-Kac model with the following components:

- An initial distribution $\mathbb{M}_0$ and kernels M_t such that $M_t(x_{t-1}, \mathrm{d}x_t) \gg P_t(x_{t-1}, \mathrm{d}x_t)f_t(y_t|x_t)$ for any x_{t-1}.
- Potential functions

$$G_0(x_0) = \frac{f_0(y_0|x_0)\mathbb{P}_0(\mathrm{d}x_0)}{\mathbb{M}_0(\mathrm{d}x_0)} \tag{10.8}$$

$$G_t(x_{t-1}, x_t) = \frac{f_t(y_t|x_t)P_t(x_{t-1}, \mathrm{d}x_t)}{M_t(x_{t-1}, \mathrm{d}x_t)}\,. \tag{10.9}$$

(continued)

Then

- $\mathbb{Q}_t(\mathrm{d}x_{0:t}) = \mathbb{P}_t(X_{0:t} \in \mathrm{d}x_{0:t} | Y_{0:t} = y_{0:t})$
- $L_t = p_t(y_{0:t})$ and $\ell_t = p_t(y_t | y_{0:t-1})$
- If we assume further that $P_t(x_{t-1}, \mathrm{d}x_t) \gg M_t(x_{t-1}, \mathrm{d}x_t)$, then

$$\mathbb{Q}_{t-1}(\mathrm{d}x_{0:t}) \frac{P_t(x_{t-1}, \mathrm{d}x_t)}{M_t(x_{t-1}, \mathrm{d}x_t)} = \mathbb{P}_t(X_t \in \mathrm{d}x_{0:t} | Y_{0:t-1} = y_{0:t-1}) .$$

With these correspondences available, we can apply the basic PF to the associated Feynman-Kac model. This yields what we will call the guided PF.

Input of the Guided PF

- A state-space model (with components $\mathbb{P}_0(\mathrm{d}x_0)$, $P_t(x_{t-1}, \mathrm{d}x_t)$, and $f_t(y_t | x_t)$).
- An initial law $\mathbb{M}_0$ and a sequence of Markov kernels M_t such that (a) we are able to sample from $\mathbb{M}_0(\mathrm{d}x_0)$ and $M_t(x_{t-1}, \mathrm{d}x_t)$; and (b) the following Radon-Nikodym derivative s exist

$$\frac{f_t(y_t | x_t) P_t(x_{t-1}, \mathrm{d}x_t)}{M_t(x_{t-1}, \mathrm{d}x_t)}$$

and are computable up to a multiplicative constant with respect to x_t, x_{t-1}. In particular, if $P_t(x_{t-1}, \mathrm{d}x_t)$ and $M_t(x_{t-1}, \mathrm{d}x_t)$ admit probability densities $p_t(x_t | x_{t-1})$ and $m_t(x_t | x_{t-1})$, respectively, with respect to a common measure, the following ratio

$$\frac{f_t(y_t | x_t) p_t(x_t | x_{t-1})}{m_t(x_t | x_{t-1})}$$

must be computable up to a multiplicative constant with respect to x_t, x_{t-1}.
- Standards inputs of a PF: number of particles N, choice of a resampling scheme, ESS threshold.

Algorithm 10.5: Guided particle filter (GPF)

Operations involving index n must be performed for $n = 1, \ldots, N$.

$X_0^n \sim \mathbb{M}_0(dx_0)$

$w_0^n \leftarrow G_0(X_0^n)$ ▷ where $G_0(x_0) := \frac{\mathbb{P}_0(dx_0) f_0(y_0|x_0)}{\mathbb{M}_0(dx_0)}$

$W_0^n \leftarrow w_0^n / \sum_{m=1}^N w_0^m$

for $t = 1$ **to** T **do**

  if $\text{ESS}(W_{t-1}^{1:N}) < \text{ESS}_{\min}$ **then**

    $A_t^{1:N} \leftarrow \texttt{resample}(W_{t-1}^{1:N})$

    $\widehat{w}_{t-1}^n \leftarrow 1$

  else

    $A_{t-1}^n \leftarrow n$

    $\widehat{w}_{t-1}^n \leftarrow w_{t-1}^n$

  $X_t^n \sim M_t(X_{t-1}^{A_t^n}, dx_t)$

  $w_t^n \leftarrow \widehat{w}_{t-1}^n G_t(X_{t-1}^{A_t^n}, X_t^n)$ ▷ where $G_t(x_{t-1}, x_t) := \frac{f_t(y_t|x_t) P_t(x_{t-1}, dx_t)}{M_t(x_{t-1}, dx_t)}$

  $W_t^n \leftarrow w_t^n / \sum_{m=1}^N w_t^m$

The output can be used much like that of the bootstrap filter, with one important difference.

Output of the Guided PF

Same output as for the bootstrap filter:

$$\sum_{n=1}^N W_t^n \varphi(X_t^n) \quad \text{approximates } \mathbb{E}[\varphi(X_t)|Y_{0:t} = y_{0:t}]$$

$$\ell_t^N \text{ (as defined in (10.3))} \quad \text{approximates } p_t(y_t|y_{0:t-1})$$

$$L_t^N = \prod_{s=0}^t \ell_s^N \quad \text{approximates } p(y_{0:t})$$

(continued)

except for expectations of the predictive distributions $\mathbb{E}[\varphi(X_t)|Y_{0:t-1} = y_{0:t-1}]$ which are now approximated by:

$$\frac{1}{\sum_{n=1}^N \widehat{w}_{t-1}^n} \sum_{n=1}^N \widehat{w}_{t-1}^n \psi(X_{t-1}^{A_t^n}, X_t^n)\varphi(X_t^n) \qquad \psi(x_{t-1}, x_t) := \frac{P_t(x_{t-1}, \mathrm{d}x_t)}{M_t(x_{t-1}, \mathrm{d}x_t)}$$

under the extra assumption that $M_t(x_{t-1}, \mathrm{d}x_t) \gg P_t(x_{t-1}, \mathrm{d}x_t)$ for any $x_{t-1} \in \mathcal{X}$.

We observe that the condition under which the predictive may be approximated, $M_t(x_{t-1}, \mathrm{d}x_t) \gg P_t(x_{t-1}, \mathrm{d}x_t)$, is stronger than the condition we imposed to define properly the potential functions of the guided particle filter, $M_t(x_{t-1}, \mathrm{d}x_t) \gg P_t(x_{t-1}, \mathrm{d}x_t) f_t(y_t|x_t)$. The simple example below shows that indeed that stronger condition is not always met.

Example 10.1 Consider the state-space model such that $X_t|X_{t-1} = x_{t-1} \sim \mathcal{N}(x_{t-1}, 1)$, $Y_t = 2\mathbb{1}(X_t > 0) - 1$, and take $M_t(x_{t-1}, \mathrm{d}x_t)$ to be

$$M_t(x_{t-1}, \mathrm{d}x_t) = \frac{\exp\{-(x_t - x_{t-1})^2/2\}}{\sqrt{2\pi}} \frac{\mathbb{1}(y_t x_t > 0)}{\Phi(y_t x_{t-1})} \mathrm{d}x_t \qquad (10.10)$$

i.e. the Gaussian distribution $\mathcal{N}(x_{t-1}, 1)$ truncated to either $\mathbb{R}^+$ (if $y_t = 1$) or $\mathbb{R}^-$ (if $y_t = -1$). (Function Φ denotes here the cumulative distribution function of the $\mathcal{N}(0, 1)$ distribution.)

Since $M_t(x_{t-1}, \mathrm{d}x_t) \gg P_t(x_{t-1}, \mathrm{d}x_t) f_t(y_t|x_t)$, we can implement the corresponding guided filter. The potential function is then

$$G_t(x_{t-1}, x_t) = \Phi(y_t x_{t-1}).$$

However, we cannot approximate the predictive distribution within this particular algorithm, since $M_t(x_{t-1}, \mathrm{d}x_t)$ does not dominate $P_t(x_{t-1}, \mathrm{d}x_t)$. Simply put, our proposal kernel simulates particles that are restricted to either $\mathbb{R}^+$ (if $y_t = 1$) or $\mathbb{R}^-$ (if $y_t = -1$), while the support of the predictive distribution is the whole real line.

Of course, M_t plays the same role as a proposal distribution in standard importance sampling; which is why we call M_t a proposal kernel. Thus, in line with our discussion in Sect. 8.3, a natural question is: how to choose M_t in an optimal manner?

Theorem 10.1 (Local Optimality) *For a given state-space model, suppose that* (G_s, M_s) *have been chosen to satisfy* (10.5) *for* $s \leq t-1$. *Among all pairs* (M_t, G_t) *that satisfy* (10.5), *the Markov kernel*

$$M_t^{\mathrm{opt}}(x_{t-1}, \mathrm{d}x_t) := \frac{f_t(y_t|x_t)}{\int_{\mathcal{X}} f_t(y_t|x')\, P_t(x_{t-1}, \mathrm{d}x')} P_t(x_{t-1}, \mathrm{d}x_t)$$

minimises the marginal variance of both the weights w_t^n *and the incremental weights,* $G_t(X_{t-1}^{A_t^n}, X_t^n)$, *and also minimises the variance of estimates* ℓ_t^N *and* L_t^N.

Proof For a generic pair (G_t, M_t), and for $\mathcal{F}_{t-1} = \sigma(X_{t-1}^{1:N}, A_t^{1:N})$ we have, by the law of total variance, that

$$\begin{aligned}\mathrm{Var}\left[G_t(X_{t-1}^{A_t^n}, X_t^n)\right] &\geq \mathrm{Var}\left[\mathbb{E}\left[G_t(X_{t-1}^{A_t^n}, X_t^n)\middle|\mathcal{F}_{t-1}\right]\right] \\ &= \mathrm{Var}\left[\int_{\mathcal{X}} f(y_t|x')\, P_t(X_{t-1}^{A_t^n}, \mathrm{d}x')\right]\end{aligned}$$

where the lower bound does not depend on M_t. This lower bound is in fact the variance of the weights when $M_t = M_t^{\mathrm{opt}}$, since, in this particular case, $G_t(x_{t-1}, x_t) = \int_{\mathcal{X}} f_t(y_t|x')\, P_t(x_{t-1}, \mathrm{d}x')$. The variance of w_t^n, ℓ_t^N and L_t^N may be bounded exactly in the same way. □

One recognises M_t^{opt} as the distribution of X_t conditionally on $X_{t-1} = x_{t-1}$ and $Y_t = y_t$, which is of course the perfect compromise between the information on X_t brought by x_{t-1}, and the information brought by y_t. On the other hand, the bootstrap kernel $M_t = P_t$ simulates blindly particles without taking into account y_t.

Example 10.2 Consider a slightly more general version of the state-model defined in Example 10.1: $X_t|X_{t-1} = x_{t-1} \sim \mathcal{N}(x_{t-1}, 1)$, $Y_t = 2\mathbb{1}_{[0,\varepsilon]}(X_t) - 1$, for $\varepsilon > 0$. The optimal kernel is then a Gaussian distribution $\mathcal{N}(x_{t-1}, 1)$ truncated to either interval $[0, \varepsilon]$ (if $y_t = 1$) or $\mathbb{R} - [0, \varepsilon]$ (if $y_t = -1$). Proposal kernel (10.10) discussed in Example 10.1 corresponds to the special case $\varepsilon = +\infty$. For this model and for $y_t = 1$, the bootstrap filter simulates $X_t \sim \mathcal{N}(X_{t-1}, 1)$, then re-weight according to $G_t(x_{t-1}, x_t) = \mathbb{1}_{[0,\varepsilon]}(X_t)$. If ε is small, many particles should get zero weight (because they fall outside of $[0, \varepsilon]$), making their simulation a wasted effort. In contrast, the guided filter with optimal proposal simulates all its particles directly inside interval $[0, \varepsilon]$.

Example 10.3 Consider state-space models of the form: $X_t|X_{t-1} = x_{t-1} \sim \mathcal{N}(\mu(x_{t-1}), \sigma_X^2)$, $Y_t|X_t = x_t \sim \mathcal{N}(x_t, \sigma_y^2)$. It is easy to show that the optimal proposal in this case is $\mathcal{N}(\xi_{\text{opt}}(x_{t-1}), \sigma_{\text{opt}}^2)$ with

$$\xi_{\text{opt}}(x_{t-1}) = \frac{\mu(x_{t-1})/\sigma_X^2 + y_t/\sigma_Y^2}{1/\sigma_X^2 + 1/\sigma_Y^2}$$

$$\frac{1}{\sigma_{\text{opt}}^2} = \frac{1}{\sigma_X^2} + \frac{1}{\sigma_Y^2}.$$

When $\sigma_Y \ll \sigma_X$ (datapoints are very informative), we expect the same situation as when ε is small in the previous example: the guided filter (based on the optimal proposal) outperforms significantly the bootstrap filter, as the former generates particles in a small region around y_t, while the latter generates particles "all over the place": in particular, the optimal proposal has variance $\sigma_{\text{opt}}^2 \approx \sigma_Y^2 \ll \sigma_X^2$, while the bootstrap proposal has variance σ_X^2. Of course, if function μ is linear, there is no point in implementing particle filtering, as one could implement instead the Kalman filter (Chap. 7), an exact method. But if μ is non-linear, it is worth noting that the optimal proposal is tractable for this class of models; this remark extends to models where the variance of $X_t|X_{t-1}$ depends on X_{t-1}, and to multivariate models with the same structure (Exercise 10.4).

There are at least three criticisms to the relevance of this local optimality criterion. The first is that it is not really clear what we achieve by minimising the variance of the weights. That does not imply for instance that we minimise the variance of a given particle estimate $\mathbb{Q}_t^N(\varphi)$. However, in doing so we also minimise the variance of ℓ_t^N and L_t^N, quantities of practical interest.

The second criticism is that a specification that is optimal for one step of the filter, is not necessarily optimal when considering several steps ahead. Consider the following example.

Example 10.4 Recall the bearings-only tracking model discussed in Sect. 2.4.1, where one tracks a vehicle in $\mathbb{R}^2$, and observes

$$Y_t = \text{atan}\left(\frac{X_t(2)}{X_t(1)}\right) + V_t, \quad V_t \sim \mathcal{N}_1(0, \sigma_Y^2)$$

(continued)

Example 10.4 (continued)
where $(X_t(1), X_t(2))$ is the position of the vehicle, and Y_t is the polar angle (plus noise) of that vehicle. The state X_t is four-dimensional, and comprises both the position and the velocity of the vehicle:

$$X_t = \begin{pmatrix} I_2 & I_2 \\ 0_2 & I_2 \end{pmatrix} X_{t-1} + \begin{pmatrix} 0_2 & 0_2 \\ 0_2 & U_t \end{pmatrix}, \quad U_t \sim \mathcal{N}_2(0, \sigma_X^2 I_2).$$

In this particular case, the optimal proposal kernel reduces to the Markov transition of the model itself (as defined in the equation above). Indeed, the observation density $f_t(y_t|x_t)$ depends only on the position, $(X_t(1), X_t(2))$, but the position itself is a deterministic function of X_{t-1}; i.e. $X_t(k) = X_{t-1}(k)+X_{t-1}(k+2)$ for $k = 1, 2$. However, if $\sigma_Y \ll 1$, it may be worthwhile to use as a proposal some approximation of the distribution of X_t conditional on X_{t-1} and on (at least) Y_{t+1}; see Exercise 10.1.

Finally, and similarly to the optimality result in importance sampling, the result above is mostly qualitative, in that in most practical cases simulating from M_t^{opt} is difficult. One should instead construct a certain M_t which is both easy to sample from and as close as possible to M_t^{opt}, as illustrated by the following example.

Example 10.5 (Stochastic Volatility) We consider the basic stochastic volatility model defined in Sect. 2.4.3: $X_t = \mu + \rho(X_{t-1} - \mu) + \sigma U_t$, $U_t \sim \mathcal{N}(0, 1)$, $Y_t|X_t = x_t \sim \mathcal{N}(0, e^{x_t})$. In this example, the optimal kernel has probability density:

$$\begin{aligned} m_t^{\text{opt}}(x_t|x_{t-1}) &\propto f_t(y_t|x_t)p_t(x_t|x_{t-1}) \\ &\propto \exp\left[-\frac{1}{2\sigma^2}\{x_t - \mu - \rho(x_{t-1} - \mu)\}^2 - \frac{e^{-x_t}}{2}y_t^2 - \frac{x_t}{2}\right] \end{aligned}$$

and is not easy to sample from. A popular approach is to approximate m_t^{opt} by a Gaussian distribution, obtained by linearising the e^{-x_t} above: $e^{-x} = e^{-x_0}(1 + x_0 - x + \ldots)$, and take $x_0 = \mu^\star(x_{t-1}) := \mu + \rho(x_{t-1} - \mu)$, the mean of the Gaussian density $p_t(x_t|x_{t-1})$. Then simple calculations lead to a Gaussian distribution $\mathcal{N}\left(\xi(x_{t-1}), \sigma^2\right)$, with

$$\xi(x_{t-1}) = \mu^\star(x_{t-1}) + \frac{\sigma^2}{4}\left[y_t^2 e^{-\mu^\star(x_{t-1})} - 2\right].$$

The Taylor expansion trick above (see also Exercise 10.2) is a simple example of a more general strategy for building proposal kernels: derive some Gaussian approximation of $X_t|X_{t-1}$, derive some Gaussian approximation of function $x_t \to f_t(y_t|x_t)$, and combine these two approximations using Kalman filtering (Chap. 5). There are a variety of ways to construct such Gaussian approximations; see the third numerical example in Sect. 10.5 for how to use Newton-Raphson to approximate $f_t(y_t|x_t)$.

Note however that this strategy implicitly assumes that the optimal kernel is close to a Gaussian, and thus may not work well when it is not; in particular when it has tails that are heavier than a Gaussian. In such a case, the weights may have a large or even infinite variance.

10.3.3 The Auxiliary Particle Filter

The point of the guided PF was to generalise the bootstrap filter to an arbitrary proposal kernel (for simulating X_t given X_{t-1}). The point of the auxiliary PF is to generalise the guided PF to an arbitrary distribution for the ancestor variables in the resampling step.

Recall that our interpretation of an iteration t of the guided filter where resampling occurs is that of an importance sampling step, which generates variables $(X_{t-1}^{A_t^n}, X_t^n)$ from proposal

$$\sum_{n=1}^{N} W_{t-1}^n \delta_{X_{t-1}^n}(\mathrm{d}x_{t-1}) M_t(X_{t-1}^n, \mathrm{d}x_t)\,,$$

and assign weight $G_t(X_{t-1}^{A_t^n}, X_t^n)$ to these variables.

In the APF, we sample the A_t^n's from a different set of probabilities,

$$\widetilde{W}_t^n = \frac{W_t^n \eta_t(X_t^n)}{\sum_{m=1}^N W_t^m \eta_t(X_t^m)}$$

where $\eta_t : \mathcal{X} \to \mathbb{R}^+$ is a certain auxiliary function. In return, the pair (A_t^n, X_t^n) is assigned weight

$$\left(\frac{W_{t-1}^{A_t^n}}{\widetilde{W}_{t-1}^{A_t^n}}\right) \times G_t(X_{t-1}^{A_t^n}, X_t^n)\,.$$

In particular, the likelihood factor $p_t(y_t|y_{0:t-1}) = \ell_t = \mathbb{Q}_{t-1} M_t(G_t)$ may be estimated by

$$\ell_t^N = \frac{1}{N} \sum_{n=1}^{N} \frac{W_{t-1}^{A_t^n}}{\widetilde{W}_{t-1}^{A_t^n}} G_t(X_{t-1}^{A_t^n}, X_t^n)\,, \tag{10.11}$$

in which one recognises a normalised importance sampling estimator (Chap. 8); see Exercise 10.6 for an alternative (less useful) estimator. The standard case is recovered by taking $\eta_t(x_t) = 1$.

With these remarks, we can now describe the auxiliary PF algorithm.

Input of the Auxiliary PF
Same input as guided PF, and additionally:

- Auxiliary functions $\eta_t : \mathcal{X} \to \mathbb{R}^+$, for $t = 0, \dots, T$, which may be evaluated pointwise.

Algorithm 10.6: Auxiliary particle filter (APF)

`Operations involving index` n `must be performed for` $n = 1, \dots, N$.

$X_0^n \sim \mathbb{M}_0(dx_0)$

$w_0^n \leftarrow G_0(X_0^n)$ ▷ where $G_0(x_0) := \frac{\mathbb{P}_0(dx_0) f_0(y_0|x_0)}{\mathbb{M}_0(dx_0)}$

$W_0^n \leftarrow w_0^n / \sum_{m=1}^N w_0^m$

$\widetilde{w}_0^n \leftarrow w_0^n \eta_0(X_0^n)$

$\widetilde{W}_0^n \leftarrow \widetilde{w}_0^n / \sum_{m=1}^N \widetilde{w}_0^m$

for $t = 1$ **to** T **do**

  if $\mathrm{ESS}(W_{t-1}^{1:N}) < \mathrm{ESS}_{\min}$ **then**

    $A_t^{1:N} \leftarrow$ `resample`$(\widetilde{W}_{t-1}^{1:N})$

    $\widehat{w}_{t-1}^n \leftarrow W_{t-1}^{A_t^n} / \widetilde{W}_{t-1}^{A_t^n}$

  else

    $A_{t-1}^n \leftarrow n$

    $\widehat{w}_{t-1}^n \leftarrow w_{t-1}^n$

  $X_t^n \sim M_t(X_{t-1}^{A_t^n}, dx_t)$

  $w_t^n \leftarrow \widehat{w}_{t-1}^n G_t(X_{t-1}^{A_t^n}, X_t^n)$ ▷ where $G_t(x_{t-1}, x_t) := \frac{f_t(y_t|x_t) P_t(x_{t-1}, dx_t)}{M_t(x_{t-1}, dx_t)}$

  $W_t^n \leftarrow w_t^n / \sum_{m=1}^N w_t^m$

  $\widetilde{w}_t^n \leftarrow w_t^n \eta_t(X_t^n)$

  $\widetilde{W}_t^n \leftarrow \widetilde{w}_t^n / \sum_{m=1}^N \widetilde{w}_t^m$

The notations have been adjusted so that the output of the algorithm is denoted the same way as for the guided filter; i.e. $\sum_{n=1}^{N} W_t^n \varphi(X_t^n)$ approximates $\mathbb{E}[\varphi(X_t)|Y_{0:t} = y_{0:t}]$ and so on. Hence the box describing the output of the guided PF applies here as well.

Although not obvious at first sight, the auxiliary PF is yet another instance of our generic particle algorithm, for the Feynman-Kac model with initial distribution $\mathbb{M}_0$, Markov kernels M_t, and potential function:

$$G_t^{\text{apf}}(x_{t-1}, x_t) = G_t(x_{t-1}, x_t)\frac{\eta_t(x_t)}{\eta_{t-1}(x_{t-1})}.$$

Notice in particular that, at an iteration where resampling occurs, the resampling weights $\widetilde{W}_t^n$ are indeed proportional to $G_t^{\text{apf}}(X_{t-1}^{A_t^n}, X_t)$. The exact correspondence between Algorithm 10.6 and this Feynman-Kac formalism is worked out below.

Auxiliary Feynman-Kac Models

For a given state-space model, we define a Feynman-Kac model with the following components. (To distinguish this Feynman-Kac model from the guided Feynman-Kac model discussed so far, we add the mention 'apf' when needed.)

- An initial distribution $\mathbb{M}_0$ and kernels M_t such that $M_t(x_{t-1}, \mathrm{d}x_t) \gg P_t(x_{t-1}, \mathrm{d}x_t) f_t(y_t|x_t)$ for any x_{t-1};
- Potential functions such that

$$\begin{aligned} G_0^{\text{apf}}(x_0) &= \frac{f_0(y_0|x_0)\mathbb{P}_0(\mathrm{d}x_0)}{\mathbb{M}_0(\mathrm{d}x_0)} \times \eta_0(x_0), \\ G_t^{\text{apf}}(x_{t-1}, x_t) &= \frac{f_t(y_t|x_t)P_t(x_{t-1}, \mathrm{d}x_t)}{M_t(x_{t-1}, \mathrm{d}x_t)} \times \frac{\eta_t(x_t)}{\eta_{t-1}(x_{t-1})} \end{aligned} \tag{10.12}$$

where $\eta_t : \mathcal{X} \to \mathbb{R}^+$. Then

$$\mathbb{P}_t(X_{0:t} \in \mathrm{d}x_{0:t}|Y_{0:t} = y_{0:t}) = \mathbb{Q}_{t-1}^{\text{apf}}(\mathrm{d}x_{0:t})\frac{G_t^{\text{apf}}(x_{t-1}, x_t)/\eta_t(x_t)}{\mathbb{Q}_{t-1}^{\text{apf}}[G_t^{\text{apf}}/\eta_t]}$$

$$p(y_{0:t}) = L_{t-1}^{\text{apf}} \times \mathbb{Q}_{t-1}^{\text{apf}}[G_t^{\text{apf}}/\eta_t]$$

and, assuming that $M_t(x_{t-1}, \mathrm{d}x_t) \gg P_t(x_{t-1}, \mathrm{d}x_t)$,

$$\mathbb{P}_t(X_{0:t} \in \mathrm{d}x_{0:t}|Y_{0:t-1} = y_{0:t-1}) = \mathbb{Q}_{t-1}^{\text{apf}}(\mathrm{d}x_{0:t})\frac{P_t(x_{t-1}, \mathrm{d}x_t)}{M_t(x_{t-1}, \mathrm{d}x_t)}\frac{\{\eta_{t-1}(x_{t-1})\}^{-1}}{\mathbb{Q}_{t-1}^{\text{apf}}[\eta_{t-1}^{-1}]}.$$

On the other hand, we find it instructive to write explicitly how the auxiliary PF depends on the *guided* potential function G_t (as we did in Algorithm 10.6), and to define two sets of weights, w_t^n (the 'inferential' weights) and $\widetilde{w}_t^n$ (the auxiliary weights). This clarifies the purpose of the auxiliary functions η_t (i.e. to twist the resampling probabilities), and it also makes it possible to recover exactly the same expressions (as in the guided filter) for the output of the algorithm: the X_t^n with weights W_t^n approximate the filtering distribution, and so on. See therefore the expressions given in the previous section. We emphasise only the expression for ℓ_t^N, the estimate of $p_t(y_t|y_{0:t-1})$: from the generic expression (10.3), we indeed recover (10.11) when resampling occurs:

$$\ell_t^N = \frac{1}{N}\sum_{n=1}^N w_t^N = \frac{1}{N}\sum_{n=1}^N \frac{W_{t-1}^{A_t^n}}{\widetilde{W}_{t-1}^{A_t^n}} G_t(X_{t-1}^{A_t^n}, X_t^n) .$$

And when resampling does not occur, (10.3) reduces to

$$\ell_t^N = \sum_{n=1}^N W_{t-1}^n G_t(X_{t-1}^n, X_t^n)$$

which is consistent with the fact that $p(y_t|y_{0:t-1}) = \ell_t = \mathbb{Q}_{t-1}M_t(G_t)$.

We now discuss how to choose η_t. To keep things simple, we assume that resampling occurs at every iteration (i.e. $\mathrm{ESS}_{\min} = N$).

Theorem 10.2 (Local Optimality in Auxiliary PF) *Consider the set-up of Algorithm 10.6 (for a given state-space model, a given distribution $\mathbb{M}_0$, and given proposal kernels M_t), and assume that auxiliary functions η_s are fixed for $s < t-1$, and that $\mathrm{ESS}_{\min} = N$ (resampling occurs at every iteration). The auxiliary function η_{t-1} that minimises the marginal variance of the inferential weights w_t^n at iteration $t \geq 1$ is*

$$\eta_{t-1}^{\mathrm{opt}}(x_{t-1}) = \sqrt{M_t(x_{t-1}, G_t^2)} \quad \textit{where } M_t(x_{t-1}, G_t^2) := \int_{\mathcal{X}} M_t(x_{t-1}, \mathrm{d}x_t)\{G_t(x_{t-1}, x_t)\}^2 .$$

Under multinomial resampling, this auxiliary function also minimises the variance of estimate ℓ_t^N.

Proof Let $\mathcal{F}_{t-1} = \sigma(X_{t-1}^{1:N})$, and recall that

$$w_t^n = \frac{W_{t-1}^{A_t^n}}{\widetilde{W}_{t-1}^{A_t^n}} G_t(X_{t-1}^{A_t^n}, X_t^n), \quad A_t^n \sim \mathcal{M}(\widetilde{W}_{t-1}^{1:N}),$$

hence $\mathbb{E}[w_t^n|\mathcal{F}_{t-1}] = \sum_{n=1}^N W_{t-1}^n M_t(X_{t-1}^n, G_t)$ does not depend on η_{t-1} (and neither does $\mathbb{E}[w_t^n]$). In addition,

$$\mathbb{E}\left[(w_t^n)^2 \,\middle|\, \mathcal{F}_{t-1}\right] = \sum_{n=1}^N \frac{(W_{t-1}^n)^2}{\widetilde{W}_{t-1}^n} M_t(X_{t-1}^n, G_t^2)$$

which is minimised with respect to the vector of probabilities $\widetilde{W}_{t-1}^n$ by taking $\widetilde{W}_{t-1}^n \propto W_{t-1}^n \sqrt{M_t(X_{t-1}^n, G_t^2)}$ (as shown by computing the corresponding Lagrangian). Hence taking $\eta_{t-1}(x_{t-1}) = \sqrt{M_t(x_{t-1}, G_t^2)}$ minimises $\mathbb{E}[(w_t^n)^2]$ and therefore $\text{Var}[w_t^n]$.

Under multinomial resampling, the pairs (A_t^n, X_t^n) are i.i.d. conditional on $\mathcal{F}_{t-1}$, hence $\text{Var}[\ell_t^N|\mathcal{F}_{t-1}] = N^{-1}\text{Var}[w_t^n|\mathcal{F}_{t-1}]$, and the same calculations apply. □

Remark 10.1 If we choose M_t to be the locally optimal kernel defined in Theorem 10.1, $M_t^{\text{opt}}(x_{t-1}, \text{d}x_t) \propto P_t(x_{t-1}, \text{d}x_t) f_t(y_t|x_t)$, then $G_t(x_{t-1}, x_t) = \int_{\mathcal{X}} P_t(x_{t-1}, dx_t) f_t(y_t|x_t)$ (which depends only on x_{t-1}), and

$$\eta_{t-1}^{\text{opt}}(x_{t-1}) = \int_{\mathcal{X}} P_t(x_{t-1}, \text{d}x_t) f_t(y_t|x_t)$$

which is the density of y_t, conditional on $X_{t-1} = x_{t-1}$. In that case, the inferential weights w_t^n are constant (their variance is zero). This particular function is often referred to as the "perfectly adapted" auxiliary function in the literature, but the theorem above shows that this particular choice is optimal only when M_t matches M_t^{opt}.

We may repeat the same criticisms we made in the previous section regarding the relevance of this type of result. What is optimal at time t may be very sub-optimal for inference several steps ahead. In fact, Johansen and Doucet (2008) exhibit an example where the variance of the weights some steps ahead is smaller for the bootstrap filter than for the perfectly adapted APF.

In addition, this optimality result considers only the effect of η_t on the variance of the weights at iterations where resampling occurs. However, η_t has also an impact on the decision to resample or not, and taking both effects into account seems difficult.

But the main difficulty is that coming up with a good approximation of η_t^{opt} may be more tricky than expected, as the example below illustrates.

Example 10.6 (Stochastic Volatility, Continued) We return to the stochastic volatility model of Example 10.5. For this model, the perfectly adapted auxiliary function is:

$$\eta_{t-1}(x_{t-1}) = \frac{1}{2\pi\sigma}\int_{\mathcal{X}} \exp\left[-\frac{1}{2\sigma^2}\{x_t - \mu - \rho(x_{t-1}-\mu)\}^2 - \frac{e^{-x_t}}{2}y_t^2 - \frac{x_t}{2}\right] \mathrm{d}x_t \,.$$

It has been recommended in the literature to approximate this integral by linearising the exponential term, as in Example 10.5: $e^{-x_t} = e^{-\mu^\star(x_t)}(1 + \mu^\star(x_t) - x_t + \ldots)$ for $\mu^\star(x_t) = \mu + \rho(x_t - \mu)$. One then obtains:

$$\eta_{t-1}(x_{t-1}) = \exp\left[\frac{1}{2\sigma^2}\left\{\xi(x_{t-1})^2 - \mu^\star(x_{t-1})^2\right\} - \frac{y_t^2}{2}e^{-\mu^\star(x_{t-1})}\left\{1 + \mu^\star(x_{t-1})\right\}\right]$$

with $\xi(x_{t-1})$ defined as in Example 10.5.

Unfortunately, the approximation error strongly diverges when $\rho > 0$ and $x_{t-1} \to -\infty$, as Fig. 10.1 shows. There is a risk that at least one particle ends up in the region on the left of the plot, and gets an extremely large resampling weight, despite being in an unlikely region. Then all the other particles get a negligible weight and disappear in the subsequent resampling step. We return to this point in the numerical experiments section. See also Exercise 10.9, which establishes that this particular auxiliary function leads to an infinite variance for the weights.

Perhaps the main lesson to draw from the previous example is that constructing a good auxiliary weight function may be a perilous exercise even for experts.

More generally, in our experience it is not easy to come up with examples where an auxiliary PF improves much over the corresponding guided PF; see the numerical examples of next section. We offer the following explanation: in scenarios where datapoints are informative (either because the observation noise is small, or the dimension X_t is large), the bootstrap filter is quite inefficient, and it becomes essential to construct a proposal kernel that approximates the law of $X_t|X_{t-1}, Y_t$, so as to implement a guided PF. But once this is done, by construction the proposal kernel is such that the simulated X_t^n depends little on $X_{t-1}^{A_t^n}$ (again because Y_t is quite informative). Hence changing the resampling probabilities may not have much impact on the results.

10.4 Rao-Blackwellised (Marginalised) Particle Filters

A very different way to construct more efficient particle filters is to "Rao-Blackwellise"; that is, to marginalise out certain components of the state process and apply a particle filter to the lower-dimensional process. This approach applies

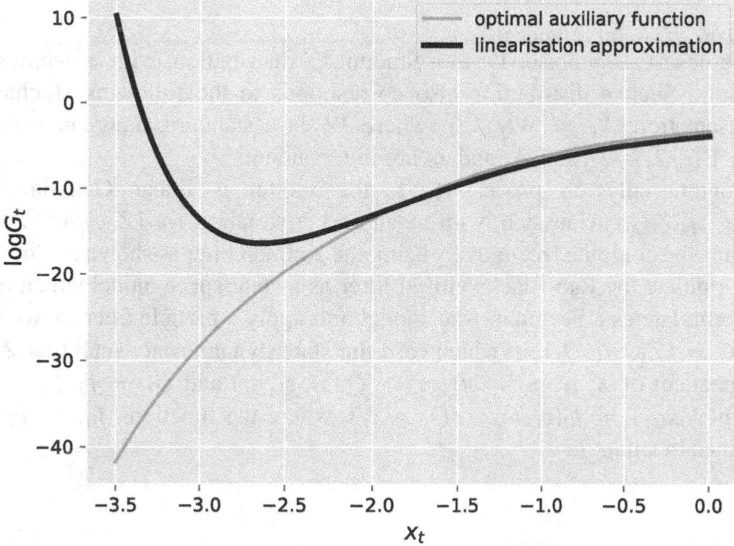

Fig. 10.1 Comparison of the log of the optimal auxiliary function η_t^{opt} (computed by numerical quadrature), and the log of the linearisation approximation for $y_t = 2.12$. Notice how the approximation diverges from the optimal function when $x_t \to -\infty$; see Sect. 10.5.2 for more details and comments

only to state-space models with a very specific structure. Specifically, consider a model where the observation process is $\{Y_t\}$, and the state process may be decomposed in two blocks; i.e. $\{X_t, \tilde{X}_t\}$. Furthermore, assume that the following conditional distributions are tractable in some way: $Y_t|Y_{0:t-1}, X_{0:t}$ and $X_t|X_{0:t-1}$. (These distributions typically *do not* simplify to e.g. $Y_t|X_t$ and $X_t|X_{t-1}$, as $\{X_t\}$ is not (marginally) Markov.)

In that case, we may implement a bootstrap filter for this "marginal" model, where the state process is the path $X_{0:t}$, and the potential function is the density of $Y_t|Y_{0:t-1}, X_{0:t}$. Often, this approach is applied to models which are partially linear and Gaussian; thus, computing the above conditional distributions amounts to implementing a Kalman Filter. The following examples illustrate this point.

Example 10.7 The general form of a linear Gaussian state-space model is

$$\tilde{X}_t = A\tilde{X}_{t-1} + U_t$$
$$Y_t = B\tilde{X}_t + V_t$$

where A, B are matrices, and U_t, V_t are (for now) Gaussian noises. A standard way to make this model more robust to outliers is to change the distribution

(continued)

Example 10.7 (continued)
of the observation noise, V_t, to a Student t's distribution, with κ degrees of freedom. Such a distribution also corresponds to the following stochastic representation: $V_t = W_t/\sqrt{Z_t}$, where W_t is a standard Gaussian variate, $Z_t \sim \Gamma(\kappa/2, \kappa/2)$, and W_t and Z_t are independent.

Conditionally on process $\{Z_t\}$, the model is linear Gaussian, so $Y_t|Y_{0:t-1}, Z_{0:t}$ is Gaussian, with mean and covariance, $(m_t(Z_{0:t}), Q_t(Z_{0:t}))$ that may be computed recursively using the Kalman filter as shown in Chap. 7. To formulate the Rao-Blackwellised filter as a state-space model which then we formulate as a Feynman-Kac model and apply a particle filter to, we can set $X_t = (Z_t, m_t, Q_t)$, in which case the state dynamics are such that Z_t is independent of X_{t-1}, $m_t = f_1(m_{t-1}, Q_{t-1}, z_t, y_t)$ and $Q_t = f_2(Q_{t-1}, z_t)$ and $p(y_t|x_{0:t}) = f_0(z_t, m_{t-1}, Q_{t-1}, y_t)$, where the functions f_0, f_1, f_2 are obtained in Chap. 7.

Example 10.8 Consider a navigation system for an airplane driven by an altimeter, and the corresponding state-space model, of the form:

$$X_t = AX_{t-1} + B\tilde{X}_{t-1} + U_t \tag{10.13}$$

$$\tilde{X}_t = \tilde{A}X_{t-1} + \tilde{B}X_{t-1} + \tilde{U}_t \tag{10.14}$$

$$Y_t = h(X_t) + V_t \tag{10.15}$$

where Y is the noisy height measurement, X_t is the "state" of the plane (velocity, roll, pitch, and similar characteristics), U_t and $\tilde{U}_t$ are Gaussian noises, and h a non-linear function. When the plane flies over dense forests, the altimeter may wrongly interpret the top of the trees as the actual ground. This is modelled by assigning a two-component Gaussian mixture distribution to V_t.

Contrary to the previous example, this model does not become linear and Gaussian when we condition on process $\{X_t\}$. However, we can compute $Y_t|X_{0:t}$ (which is simply $Y_t|X_t$, and is given by (10.15), see Exercise 10.10), and $X_t|X_{0:t-1}$. For the latter distribution, remark first that $(X_{0:t}, \tilde{X}_{0:t})$ is Gaussian, so clearly $X_t|X_{0:t-1}$ admits an exact expression. In fact, it may be computed recursively and efficiently by applying a Kalman filter to the following artificial model: $\{\tilde{X}_t\}$ is the state process, $\{X_t\}$ is the observation process, and thus computing $X_t|X_{0:t-1}$ amounts to computing the likelihood.

Rao-Blackwellised particle filters are popular in tracking and similar applications, where models of interest are often partially linear and Gaussian. Compared to a standard filter, they tend to be more CPU intensive, but, on the other hand, they tend to generate estimates with a much lower variance. See Exercise 10.11 for some intuition on this point.

10.5 Numerical Experiments

10.5.1 A Basic Linear Gaussian Example

We start with the basic linear Gaussian model discussed in Example 10.3 (see also Exercise 10.4):

$$\begin{aligned} X_0 &\sim \mathcal{N}(0, \sigma_X^2/(1-\rho^2)) \\ X_t|X_{t-1} = x_{t-1} &\sim \mathcal{N}(\rho x_{t-1}, \sigma_X^2) \\ Y_t|X_t = x_t &\sim \mathcal{N}(x_t, \sigma_Y^2)\,. \end{aligned}$$

Of course, this is a toy example, since we can use the Kalman filter to compute the filtering distributions exactly. However, we use this example to discuss to which extent the guided and auxiliary filters may improve on the bootstrap filter. In that spirit, we consider a scenario where datapoints are quite informative: we take $\rho = 0.9, \sigma_X = 1, \sigma_Y = 0.2$.

For this model, both the (locally) optimal proposal kernels and auxiliary functions are tractable, as noted in Example 10.3. Figure 10.2 compares the performance of the bootstrap filter, the guided filter (with M_t set to the optimal proposal), and the auxiliary particle filter (with M_t set as in the guided filter, and η_t set to the optimal auxiliary weight) when applied to simulated data ($T = 100$). These algorithms are run with $N = 1000$ and $\text{ESS}_{\text{min}} = 500$.

Since σ_Y is small in this case, we observe (as expected) that the guided and auxiliary PF outperform significantly the bootstrap filter. The top left plot of Fig. 10.2 shows that the ESS remains high for long periods of time for the guided and auxiliary PF; this means that the resampling step is triggered only from time to time (recall that $\text{ESS}_{\text{min}} = 500$). In contrast, the ESS is always below 300 for the bootstrap filter, and resampling is performed at every iteration.

The top right plot compares the square root of the MSE (mean square error) of the particle estimates of the filtering expectation obtained from the three algorithms. Again, the bootstrap filter is significantly outperformed. Same remark for log-likelihood estimates (box-plots in bottom left plot).

On the other hand, it is interesting to note that the auxiliary PF shows the same performance as the guided PF. One explanation that comes to mind is that, as we noted, the resampling step is performed only occasionally, whereas the point of the auxiliary PF is to increase the change of selecting a 'good' ancestor in the resampling step.

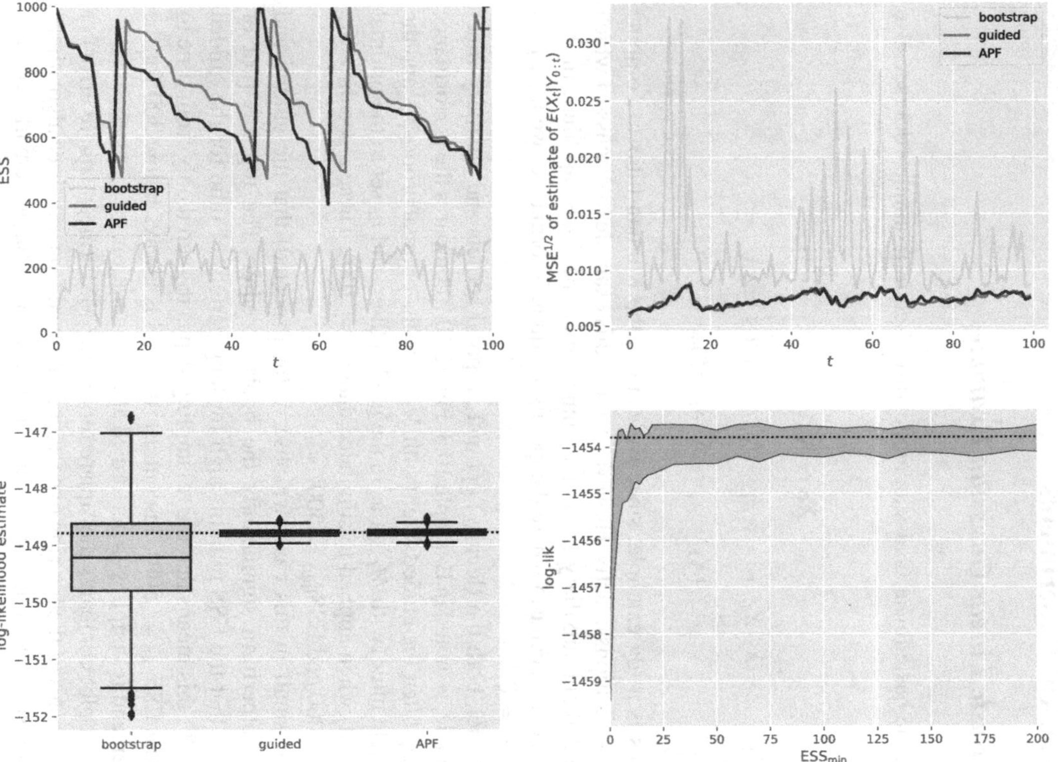

Fig. 10.2 Comparison of bootstrap, guided and auxiliary filter for a linear Gaussian model ($N = 1000$, $\mathrm{ESS}_{\min} = 500$, $T = 100$). Top left: ESS as a function of time from a single run; Top right: square root of MSE (over 1000 runs) of particle estimate of $\mathbb{E}[X_t|Y_{0:t} = y_{0:t}]$ as a function of time t; Bottom left: box-plots (over 1000 runs) of estimates of log-likelihood $\log p_T(y_{0:T})$; Bottom right: inter-quartile interval (over 200 runs) of log-likelihood estimates as a function of $\mathrm{ESS}_{\min}$ for the guided filter. For the last plot, $T = 1000$ was used instead. Dotted line represents the true value in both bottom plots

To investigate this point, we plot in the bottom right panel of Fig. 10.2 the inter-quartile interval (the interval between first and third quartiles) of the log-likelihood estimates L_T^N generated by 1000 runs of the guided filter, for different values of $\mathrm{ESS}_{\min}$. In this experiment, we took $N = 200$ and $T = 1000$. Remarkably, the performance of the algorithm seems insensitive to $\mathrm{ESS}_{\min}$: the inter-quartile range of these estimates seems roughly constant as soon as $\mathrm{ESS}_{\min} > 10$. It is only for very small values of $\mathrm{ESS}_{\min}$, that we see the performance drops quite significantly. Recall that $\mathrm{ESS}_{\min} \leq 1$ implies that resampling is never performed. In such a case, the algorithm generates estimates that are far off the true value. As we have already discussed, without resampling the algorithm collapses to importance sampling, which suffers from a curse of dimensionality.

There seems to be at least three lessons learnt from this simple experiment (which apply more generally in our experience): (a) the choice of $\mathrm{ESS}_{\min}$ should have little impact on the performance of particle algorithms; (b) the point of adaptive resampling (i.e. resampling when the ESS is below a certain threshold) is not so much variance reduction (we observe the same variability if we resample at every iteration), but CPU time reduction (since resampling is performed less often). On a standard computer and using our `particles` package, we observed up to 30% CPU reduction by taking a small $\mathrm{ESS}_{\min}$ in this particular example. (c) In informative data scenarios, the most dramatic improvement one may expect, relative to the bootstrap filter, is by constructing a good proposal kernel and implement a guided PF. On the other hand, the extra improvement brought by the APF is often smaller, and sometimes even negligible.

10.5.2 Stochastic Volatility

We now turn our attention to the stochastic volatility model and the guided and auxiliary PFs discussed in Examples 10.5 and 10.6. We consider the same data (100 times the compound daily returns on the U.S. dollar against the U.K. pound sterling from the first day of trading in 1997 and for the next 200 days of active trading) and parameter values ($\mu = -1.024$, $\sigma = 0.178$, $\rho = 0.9702$) as in Pitt and Shephard (1999). All the algorithms are run with $N = 10^3$, $\mathrm{ESS}_{\min} = N/2$, and systematic resampling.

Figure 10.3 shows the variability of estimate errors (estimate computed from the algorithm minus the "true value", where the latter was approximated by running a PF with a very large N) for filtering expectation $\mathbb{E}[X_t | Y_{0:t} = y_{0:t}]$. For the bootstrap and guided filters, we plot the inter-quartile intervals (over 50 runs). We see that the guided PF improves over the bootstrap filter only around time $t = 143$; this corresponds to a large outlier ($y_{143} = 2.17$, the largest observation in absolute value). Elsewhere, the intervals are nearly indistinguishable. Presumably, this is because the likelihood $f_t(y_t|x_t)$ is not very informative for this model (and the chosen parameter values), unless y_t is large.

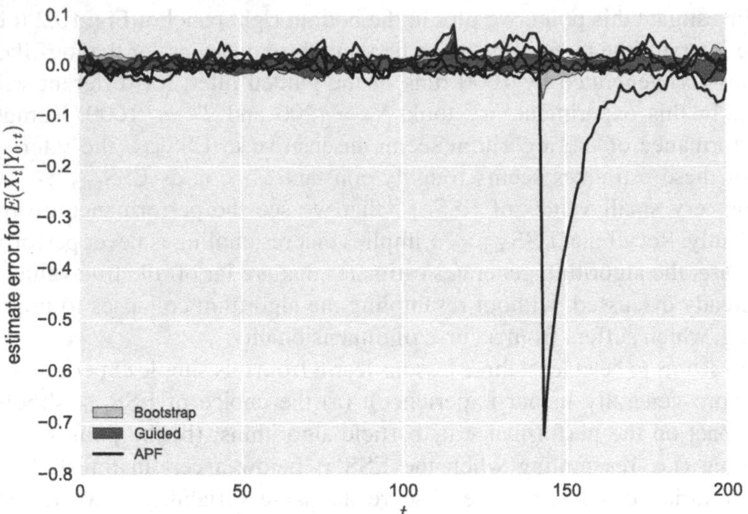

Fig. 10.3 Particle estimate errors as a function of t for filtering expectation $\mathbb{E}[X_t|Y_{0:t} = y_{0:t}]$: inter-quartile intervals (over 50 runs) for bootstrap and guided filters (light grey and dark grey regions), and errors from 5 runs of the APF (black lines); $N = 1000$ for all the runs

We also overlay the estimate error of 5 runs of the APF, with the auxiliary function chosen as discussed in Example 10.6. Strikingly, one run is completely "off-track" for some time after $t = 143$. And the problem gets *worse* when we increase N. For $N = 10^4$, more than one half of the runs completely deviates from the true value around time $t = 142$. (results not shown).

What seems to happen is that, by increasing N, we increase the probability of placing one particle in the region where the auxiliary weight diverges (i.e. on the left in Fig. 10.1); this seems enough to "derail" the corresponding auxiliary PF.

As we have noted in Example 10.6, this issue with the auxiliary function diverging when $x_t \to -\infty$ (and as a consequence the fact that the weights have infinite variance) should occur *at all times* t (unless $y_t = 0$). It just so happens that this phenomenon is easier to evidence at a time t where $|y_t|$ is large.

The bottom line is, of course, that one should not use the APF with the auxiliary weight chosen as in Example 10.6. It is not difficult to construct auxiliary weight functions that do not diverge, but we have been unable to find one that leads to a noticeable improvement over the guided PF for this class of models.

10.5.3 Neural Decoding

Finally, we consider a more challenging example, borrowed from Koyama et al. (2010), who considered the following neural decoding task: tracking the hand motion of a monkey from its neural activity (as measured by electrodes implanted in its motor cortex). The observation Y_t consists of the number of spikes generated by $d_y = 78$ neurons at time t; these numbers are independent, and Poisson-distributed:

$$Y_t(k)|X_t = x_t \sim \mathcal{P}\left(a_k + b_k' x_t\right)$$

where the state X_t is 6-dimensional, and consists of the position and velocity of the hand in space. In Koyama et al. (2010), process $\{X_t\}$ follows the three-dimensional variant of motion model (2.3). Instead, we assume that each of the three (position, velocity) pairs correspond to a discretised integrated Brownian motion, that is: independently, for $i = 1, \ldots, 3$, and some discretisation step $\delta > 0$,

$$\begin{pmatrix} X_t(i) \\ X_t(i+3) \end{pmatrix} | X_{t-1} = x_{t-1} \sim \mathcal{N}_2\left(\begin{pmatrix} 1 & \delta \\ 0 & 1 \end{pmatrix}\begin{pmatrix} X_{t-1}(i) \\ X_{t-1}(i+3) \end{pmatrix}, \sigma^2 \begin{pmatrix} \delta^3/3 & \delta^2/2 \\ \delta^2/2 & \delta \end{pmatrix}\right).$$

The advantage of this model (compared to a 3D version of (2.3)) is that X_t is not a deterministic function of X_{t-1}; hence this avoids the difficulty discussed in Example 10.4 regarding the implementation of a guided filter (see below).

Koyama et al. (2010) report poor performance for the bootstrap filter when applied to this type of model. This is confirmed by Fig. 10.4, which plots the ESS across time for $N = 10^4$ particles and 50 runs of the algorithm. Indeed, the ESS gets dangerously close to zero a few times, despite the fact that N is quite large. If run for a smaller value of N, the algorithm degenerates and generates estimates with a noticeable bias. (The dataset was simulated from the model, taking $T + 1 = 25$,

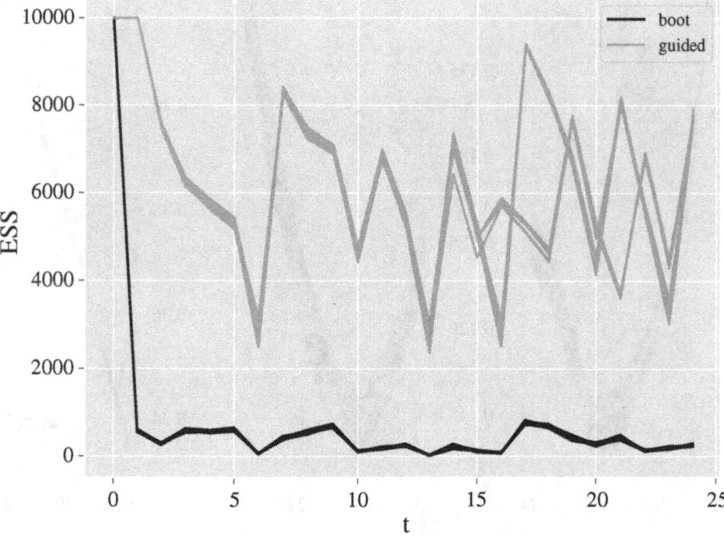

Fig. 10.4 Neural decoding example: ESS versus time for 50 runs of the following algorithms: the bootstrap filter, and a guided filter based on the proposal discussed in the text

$\sigma = 1$, $a_k \sim \mathcal{N}(2.5, 1)$, and the b_k's uniformly distributed on the 6-dimensional sphere; for simplicity $X_0(i) = 0$ for all i.)

To obtain better performance, we implement a guided filter. Contrary to the first example, the optimal proposal is not available in closed form; and, contrary to the second example, it does not seem easy to come up with some analytical approximation. Instead, we derive numerically a Gaussian approximation of the function $x_t \to f_t(y_t|x_t)$, for each time t, as follows: we use Newton-Raphson to find the maximum of $x_t \to \log f_t(y_t|x_t)$, $x^\star$, and the Hessian at the mode, $H^\star$. The corresponding Gaussian approximation has mean $x^\star$, and variance $-H^\star$. We observe empirically that this Gaussian approximation is quite accurate, which is hardly surprising, given that $f_t(y_t|x_t)$ is a product of 78 independent factors.

Since the state process is linear and Gaussian, it is easy to obtain a Gaussian approximation of the optimal proposal (the law of $X_t|X_{t-1}, Y_t$) by applying the filtering step of the Kalman filter (using as the predictive distribution, the law of $X_t|X_{t-1}$, as the observation, the mode $x^\star$, and as the observation noise, $-H^\star$).

Figure 10.4 also plots the ESS of the so-obtained guided filter. This guided filter seems to be much more stable than the bootstrap filter. This is confirmed by Fig. 10.5, which plots the inter-quartile ranges (over the 50 independent runs) for the particle estimates of the filtering expectation of the two algorithms (at selected times and for selected components, see caption for details).

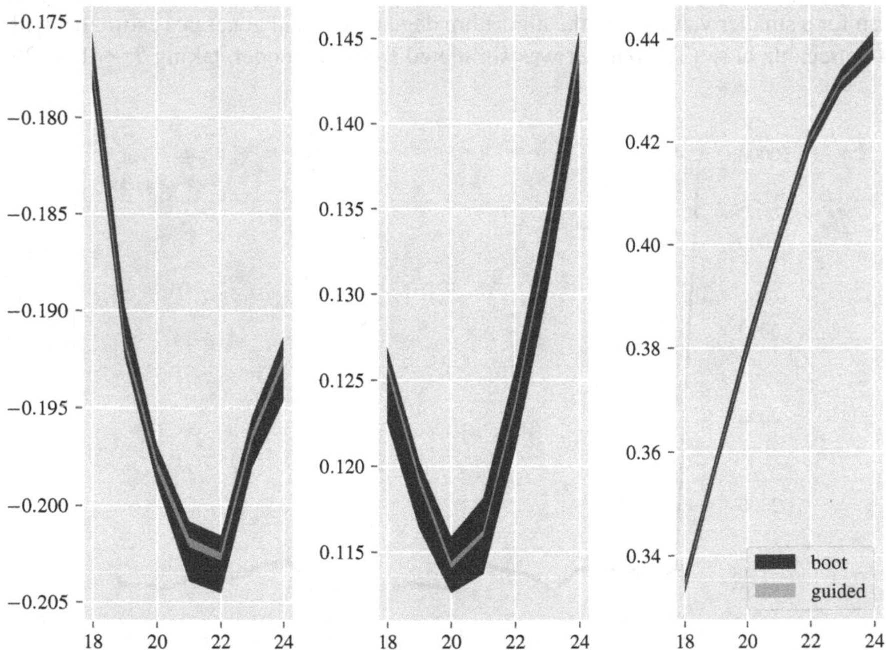

Fig. 10.5 Neural decoding example: inter-quartile range (over 50 runs) versus time, for the filtering expectation of $X_t(i)$, $i = 1, \ldots, 3$, at times $t = 18, \ldots, 24$, estimated by either the bootstrap filter or the guided filter discussed in the text

More generally, many non-trivial models are such that (a) the bootstrap filter performs poorly; but (b) deriving a guided filter with reasonable performance (using some form of e.g. Gaussian approximation) is reasonably straightforward.

For the record, we also implemented an auxiliary particle filter, where the auxiliary function was set to the same type of Gaussian approximation, this time applied to the density of Y_{t+1} given X_t; however, we did not notice any performance improvement relative to the guided filer, as in the previous examples.

Exercises

10.1 *For the bearings-only tracking model discussed in Example 10.4, derive the density of Y_{t+1} conditional on X_t. Construct a kernel proposal that approximates the law of $X_t|X_{t-1}, Y_{t+1}$, under the assumption that $\sigma_Y \ll 1$ (discuss), and derive the corresponding weight function G_t. As an intermediate step, you may establish that the position at time $t+1$, conditional on X_{t-1}, is a bivariate Gaussian vector, such that, (i) the angle relative to the origin is strongly determined by y_t when σ_Y is small; (i.e. one may consider $\mathcal{N}(y_{t+1}, \sigma_Y^2)$ as a proposal for that angle); (ii) conditional on that angle, one component of this bivariate vector is still Gaussian.*

10.2 *Adapt the linearisation technique discussed in Example 10.5 to derive a proposal kernel for the following model: $X_t|X_{t-1} = x_{t-1} \sim \mathcal{N}(\mu + \rho(x_{t-1} - \mu), \sigma^2)$, $Y_t|X_t = x_t \sim \mathcal{P}(e^{x_t})$.*

10.3 *In Example 10.5, use the convexity of the exponential function to show that the Gaussian proposal distribution dominates the optimal kernel (for a fixed x_{t-1}); i.e. $m_t(x_t|x_{t-1}) \geq C(x_{t-1}) m_t^{\mathrm{opt}}(x_t|x_{t-1})$, where $C(x_{t-1})$ is a constant that depends only on x_{t-1}. Use this remark to propose a rejection sampler to sample from the optimal kernel. Discuss the performance of this rejection algorithm when x_{t-1} goes to either $+\infty$ or $-\infty$. Could such a rejection sampler be used in practice in a guided particle filter?*

10.4 *Derive the optimal proposal for the following multivariate generalisation of the models considered in Example 10.3: $\mathcal{X} = \mathbb{R}^{d_x}$, $\mathcal{Y} = \mathbb{R}^{d_y}$, $X_t|X_{t-1} = x_{t-1} \sim \mathcal{N}_d(\mu(x_{t-1}), \Sigma_X(x_{t-1}))$ (both mean and variance depend on x_{t-1}), $Y_t|X_t = x_t \sim \mathcal{N}_d(Ax_t, \Sigma_y)$, where $\Sigma_X(x_{t-1})$ (resp. Σ_y) is a $d_x \times d_x$ (resp. $d_y \times d_y$) covariance matrix, and A is $d_y \times d_x$ matrix.*

Show that the optimal auxiliary function may also be derived for this class of model.

10.5 *Show that when the optimal proposal kernel is used in the guided PF, the weight function $G_t(x_{t-1}, x_t)$ depends only on x_{t-1}, and explain what is this particular function. It is sometimes said that auxiliary particle filtering amounts to exchange the order of the reweighting step and the resampling step. Explain this statement in the case where an optimal kernel is used.*

10.6 *Derive an alternative estimator for* (10.11)*, based on auto-normalised sampling. Note: a drawback of this alternative estimator is that it is biased, whereas unbiased estimators of the likelihood are required to use the PMCMC algorithms covered in Chap. 16.*

10.7 *Regarding the Feynman-Kac formalism of the auxiliary PF, e.g.* (10.12) *and around, show that an additional connection between the filtering distribution and* $\mathbb{Q}_t^{\mathrm{apf}}$ *is:*

$$\mathbb{Q}_t^{\mathrm{apf}}(\mathrm{d}x_{0:t}) \propto \mathbb{P}_t(X_{0:t} \in \mathrm{d}x_{0:t}|Y_{0:t} = y_{0:t})\eta_t(x_t)$$

where as usual the proportionality constant may be recovered by normalisation. Explain why this connection is less relevant for Algorithm 10.6. In particular, it imposes a certain condition on η_t*, whereas* η_t *has no impact on the output of Algorithm 10.6 used to approximate the filtering distribution.*

10.8 *In line with the previous exercise, explain why the choice of* η_T*, the auxiliary function at the final step of the algorithm, has no impact on any particle estimates that may be computed from the algorithm. Thus we may as well set* $\eta_T(x_T) = 1$ *to simplify the notations. Establish the following identity for the APF:*

$$L_T^N = \left\{\frac{1}{N}\sum_{n=1}^{N} G_0(X_0^n)\eta_0(X_0^n)\right\} \prod_{t=1}^{T} \left\{\frac{1}{N}\sum_{n=1}^{N} G_t(X_{t-1}^{A_t^n}, X_t^n)\frac{\eta_t(X_t^n)}{\eta_{t-1}(X_{t-1}^{A_t^n})}\right\}$$

where $L_T^N = \prod_{t=0}^{T} \ell_t^N$ *is the estimate of the likelihood* $p(y_{0:T})$*, and* ℓ_t^N *is defined in* (10.11)*. Find a more direct way to recover the above expression based on the Feynman-Kac formalism of the APF.*

10.9 *The aim of this exercise is to establish that the variance of the weights of the auxiliary PF discussed in Example 10.6 may be infinite at time 0. For this purpose, compute*

$$G_0(x_0) = \frac{f_0(y_0|x_0)p_0(x_0)}{m_0(x_0)} \times \eta_0(x_0)$$

where proposal density m_0 *is set as in Example 10.5, and auxiliary function* $\eta_0(x_0)$ *is set as in Example 10.6. Note however that, since* $X_0 \sim \mathcal{N}(\mu, \sigma_0^2)$*, with* $\sigma_0^2 = \sigma^2/(1-\rho^2)$*, one needs to replace* σ *by* σ_0 *and* $\mu^\star(x_{t-1})$ *by* μ *in the expression for the proposal density* m_0*.*

Show that $\log \eta_0$ *is dominated (when* $x_0 \to -\infty$*) by a positive term which is* $\mathcal{O}(e^{-2\rho x_0})$*. Show then that the other terms in* $\log G_0(x_0)$ *are all negligible relative to* $e^{-2\rho x_0}$*, if we assume* $\rho > 1/2$*. (In particular,* $\log f_0(y_0|x_0)$ *involves a negative term which is* $\mathcal{O}(e^{-x_0})$*.) Use the fact that particles* X_0^n *are simulated from* $\mathcal{N}(\mu, \sigma_0^2)$ *to conclude.*

10.10 *Draw the undirected graph that corresponds to model* (10.13) *to* (10.15) *(see Sect. 4.6 for how undirected graphs are defined and applied to Markov processes). Deduce from this graph that $Y_t|X_{0:t}$ depends only on X_t and that $X_t|X_{0:t-1}$, on the other hand, does not reduce to $X_t|X_{t-1}$.*

10.11 *In connection with Sect. 10.4, show that marginalisation always reduces the variance in the context of importance sampling (IS). That is, consider a target distribution and a proposal distribution, both for a couple of variables $(X, \tilde{X})$; show that the variance (resp. asymptotic variance) of a normalised (resp. auto-normalised) estimator of the expectation of a function of X is always lower than, or equal to, when we sample marginally X, and reweight according to the marginal change of measure, rather than when we sample from the joint proposal, and reweight according to the joint change of measure. The term Rao-Blackwellization in Statistics refers to the following inequality:* $\mathrm{Var}[\mathbb{E}(X|Y)] \leq \mathrm{Var}[X]$, *which may be deduced from:* $\mathrm{Var}(X) = \mathrm{Var}[\mathbb{E}(X|Y)] + \mathbb{E}[\mathrm{Var}(X|Y)]$. *Explain the connection between Rao-Blackwellization and marginalisation.*

10.12 *Consider a hidden Markov model (i.e. a state-space model such that $\mathcal{X} = 1, \ldots, K$, see Chap. 6) such that K is too large for the exact filter to be practical. (Recall from Chap. 6) that the complexity of this algorithm is $\mathcal{O}(TK^2)$.) What is the complexity (with respect to K and N) of the following algorithms for such a state-space model: the bootstrap filter, the optimal guided filter, and the optimal auxiliary particle filter? Discuss. Explain how you may reduce the complexity of the guided filter if you know a upper bound for the observation likelihood, $f_t(y_t|k)$.*

10.13 *In the neural decoding example, X_0 is Dirac-distributed (i.e. it is equal to a constant with probability one). Does the output of a particle filter depends on y_0 in such a case? Find two ways to justify your answer; one by looking at the operations performed by the algorithm; the other by using conditional independence properties.*

Python Corner

As we have stressed already, the operations performed by a particle filter are defined by the corresponding Feynman-Kac model: distribution $\mathbb{M}_0(\mathrm{d}x_0)$, defines how we sample particles at time 0, kernel $M_t(x_{t-1}, \mathrm{d}x_t)$ defines how we sample particles X_t^n given their ancestors, and so on.

In that spirit, here is how one initialises a particle filter in `particles`:

```
import particles

mypf = particles.SMC(fk=fk_boot_sv, N=100, resampling='systematic',
                     ESSrmin=0.5)
```

Object `fk_boot_sv` was defined in the Python corner of Chap. 5, and represents the Feynman-Kac model associated to the bootstrap filter of a particular stochastic volatility model. The meaning of the other arguments is pretty transparent: `N` is the number of particles, `resampling` is the resampling scheme (possible choices are `'multinomial'`, `'residual'`, `'stratified'` and `'systematic'`), and `ESSrmin` is $\text{ESS}_{\text{min}}/N$; i.e. we resample whenever the ESS is below N times `ESSrmin`. If set to `1.` or above (respectively `0.` or below), resampling is performed at every iteration (resp. is never performed). Default values for arguments `resampling` and `ESSrmin` are, respectively, `'systematic'` and `0.5`; hence the line above is equivalent to:

```
mypf = particles.SMC(fk=fk_boot_sv, N=100)
```

The snippet above only initialises the particle algorithm. To actually run it, one executes:

```
mypf.run()
```

Upon completion, object `mypf` will have attributes `X` (an array that contains the N final particles X_t^n), `W` (an array that contains the N final normalised weights W_T^n), `summaries.logLts` (a list of log-likelihood estimates obtained at every iteration) and other objects; see the documentation for more details.

How do we specify other particle algorithms such as guided particle filters? First, we need to enrich our state space model with proposal distributions. The class below defines a basic linear Gaussian state-space model, plus the locally optimal proposal at time 0 and at times $t \geq 1$.

```
from particles import state_space_models as ssms
from particles import distributions as dists

class LinearGauss(ssms.StateSpaceModel):
    default_params = {'sigmaY': .2, 'rho': 0.9, 'sigmaX': 1.}

    def PX0(self):  # X_0 ~ N(0, sigmaX^2)
        return dists.Normal(scale=self.sigmaX)

    def PX(self, t, xp):  # X_t | X_{t-1} ~ N(rho * X_{t-1}, sigmaX^2)
        return dists.Normal(loc=self.rho * xp, scale=self.sigmaX)

    def PY(self, t, xp, x):  # Y_t | X_t ~ N(X_t, sigmaY^2)
        return dists.Normal(loc=x, scale=self.sigmaY)

    def proposal0(self, data):
        sig2post = 1. / (1. / self.sigmaX**2 + 1. / self.sigmaY**2)
        mupost = sig2post * (data[0] / self.sigmaY**2)
        return dists.Normal(loc=mupost, scale=np.sqrt(sig2post))

    def proposal(self, t, xp, data):
        sig2post = 1. / (1. / self.sigmaX**2 + 1. / self.sigmaY**2)
        mupost = sig2post * (self.rho * xp / self.sigmaX**2
                             + data[t] / self.sigmaY**2)
        return dists.Normal(loc=mupost, scale=np.sqrt(sig2post))
```

Method `proposal` returns the distribution of $M_t(x_{t-1}, dx_t)$ for a given x_{t-1} (argument `xp`) and a given time t. Method `proposal0` returns simply $\mathbb{M}_0(dx_0)$.

To implement a guided filter, we use class `GuidedPF` which automatically sets a Feynman-Kac model such that M_t is set to the proposal defined above, and weight function G_t is set to (10.9).

```
ssm = ssms.LinearGauss()  # default parameters
x, y = ssm.simulate(100)  # simulate data
fk_guided = ssms.GuidedPF(ssm=ssm, data=y)
pf = particles.SMC(fk=fk_guided, N=1000)
pf.run()
```

Auxiliary particle filters may be implemented similarly (see the documentation of `particles` for more details).

Next chapter will discuss convergence properties of SMC estimates. In practice, however, it is difficult to assess the error of a given estimate from a single run. (To be fair, there exist methods to estimate the variance of a given estimate from a single run; see Sect. 19.3 in the final chapter.)

Package `particles` provides a function called `multiSMC` to run several particle filters, optionally in parallel. For instance, this:

```
results = particles.multiSMC(fk=fk_boot_sv, N=100, nruns=10, nprocs=0)
```

runs 10 particle filters, using all available cores: in particular if we have 10 cores or more, each particle filter will be run on a different core, and the running time of the line above will be the same as a single particle filter.

Parallel computation is a technical topic (in Python, and generally!). Standard Python code tends to run on a single core (because of a feature known as the GIL, the Global Interpreter Lock). Two standard Python libraries offer some level of multi-processing: the aptly named `multiprocessing`, and `concurrent` (in module futures). Function `multiSMC` relies on the former. These libraries allow to spawn several processes that may carry out independent tasks (in our cases, different runs of a particle filter). The way these processes are spawned is actually operating system-dependent. On Unix-based system, it relies on a mechanism called forking, which allows to quickly "clone" a parent process; this mechanism is not available on Windows, and as result multiprocessing is allegedly less efficient on this operating system.

Programs relying on random number generation pose an additional difficulty for multi-processing: since the processes generated by library `multiprocessing` are "clones" of the parent process, their random number generators are in the same state. Thus, by default, each process is going to return exactly the same result. To address this issue, function `multiSMC` creates distinct RNG seeds and passes them as an argument to each particle filter.

Finally, note that function `multiSMC` may be used to run parallel filters with different arguments; for instance, this:

```
results = particles.multiSMC(fk=fk_boot_sv, N=[100, 500], nruns=10,
                             resampling=['multinomial', 'residual',
                                         'systematic'])
```

runs 60 particle filters: 10 filters with $N = 100$ based on multinomial resampling, 10 more with $N = 500$, 10 with $N = 100$ and residual resampling, and so on; see the documentation for more details.

Bibliographical Notes

The term "bootstrap filter" comes from the seminal paper of Gordon et al. (1993), one of the very first papers on particle filtering; see also Stewart and McCarty (1992). The name refers to the bootstrap, a well-known statistical procedure due to Efron (1979) which also involves sampling with replacement. An early reference for optimality results such as Theorem 10.1 is Doucet et al. (2000). The APF methodology was developed by Pitt and Shephard (1999); Examples 10.5 and 10.6 are adapted from this paper. See also Carpenter et al. (1999). The name itself comes from the original description of the method in terms of sampling an auxiliary variable in the resampling step. In this section we followed instead Johansen and Doucet (2008) in describing the APF as just another instance of the general PF methodology.

For a general overview of Rao-Blackwellised particle filters, see Chen and Liu (2000), Andrieu and Doucet (2002) and Schön et al. (2005); Example 10.8 is taken from the last reference.

Section 19.1 in the final chapter revisits the problem of designing efficient proposal distributions for guided or auxiliary particle filters, when either $\mathcal{X}$ or $\mathcal{Y}$ are high-dimensional, and give references to recent attempts at addressing this problem.

Bibliography

Andrieu, C., & Doucet, A. (2002). Particle filtering for partially observed Gaussian state space models. *Journal of the Royal Statistical Society: Series B (Statistical Methodology), 64*(4), 827–836.

Carpenter, J., Clifford, P., & Fearnhead, P. (1999). Improved particle filter for nonlinear problems. *IEE Proceedings - Radar, Sonar Navigation, 146*(1), 2–7.

Chen, R., & Liu, J. S. (2000). Mixture Kalman filters. *Journal of the Royal Statistical Society: Series B (Statistical Methodology), 62*(3), 493–508.

Doucet, A., Godsill, S., & Andrieu, C. (2000). On sequential Monte Carlo sampling methods for Bayesian filtering. *Statistics and Computing, 10*(3), 197–208.

Efron, B. (1979). Bootstrap methods: Another look at the jackknife. *Annals of Statistics, 7*(1), 1–26.

Gordon, N. J., Salmond, D. J., & Smith, A. F. M. (1993). Novel approach to nonlinear/non-Gaussian Bayesian state estimation. *IEE Proceedings F (Radar and Signal Processing), 140*(2), 107–113.

Johansen, A. M., & Doucet, A. (2008). A note on auxiliary particle filters. *Statistics & Probability Letters, 78*(12), 1498–1504.

Koyama, S., Castellanos Pérez-Bolde, L., Shalizi, C. R., & Kass, R. E. (2010). Approximate methods for state-space models. *Journal of the American Statistical Association, 105*(489), 170–180. With supplementary material available on line.

Pitt, M. K., & Shephard, N. (1999). Filtering via simulation: Auxiliary particle filters. *Journal of the American Statistical Association, 94*(446), 590–599.

Schön, T., Gustafsson, F., & Nordlund, P.-J. (2005). Marginalized particle filters for mixed linear/nonlinear state-space models. *IEEE Transactions on Signal Processing, 53*(7), 2279–2289.

Stewart, L., & McCarty, P. (1992). Use of Bayesian belief networks to fuse continuous and discrete information for target recognition, tracking, and situation assessment. In *Signal processing, sensor fusion, and target recognition. International society for optics and photonics* (Vol. 1699, pp. 177–186). Orlando: SPIE.

Chapter 11
Convergence and Stability of Particle Filters

Summary This chapter is an introduction to the theoretical study of particle estimates: how they converge as $N \to +\infty$, and whether their error stays stable over time. The focus is on results that are easy to prove from first principles. References to more technical results are given in the bibliography at the end of the chapter.

A key idea is that the error of a particle estimate at time t may be decomposed into a sum of 'local' errors that correspond to the previous time steps $0, 1, \ldots, t$. In that spirit, most proofs will rely on an induction argument.

11.1 Preliminaries

We recall the proof of basic Probability limit theorems. Let $X_1, X_2, \ldots$ be a sequence of i.i.d. real-valued random variables, with mean μ and variance σ^2. Without loss of generality, take $\mu = 0$ (i.e. replace X_n by $X_n - \mu$ in the following statements).

Notation/Terminology The following symbols are used to denote convergence: $\stackrel{a.s.}{\to}$ (almost surely), $\stackrel{P}{\to}$ (in probability), $\Rightarrow$ (in distribution). Let $\mathcal{C}_b(\mathcal{X})$ denote the set of functions $\varphi : \mathcal{X} \to \mathbb{R}$ that are measurable and bounded, and let $\|\varphi\|_\infty$ denote the supremum norm; i.e. $\|\varphi\|_\infty = \sup_{x \in \mathcal{X}} |\varphi(x)|$.

We start with the central limit theorem:

$$\sqrt{N}\left(\frac{1}{N}\sum_{n=1}^{N} X_n\right) \Rightarrow \mathcal{N}\left(0, \sigma^2\right).$$

© Springer Nature Switzerland AG 2020
N. Chopin, O. Papaspiliopoulos, *An Introduction to Sequential Monte Carlo*, Springer Series in Statistics, https://doi.org/10.1007/978-3-030-47845-2_11

Convergence in distribution is equivalent to the pointwise convergence of the characteristic function (Lévy's theorem). One proves the latter as follows:

$$\begin{aligned}\mathbb{E}\left\{\exp\left(\frac{iu}{\sqrt{N}}\sum_{n=1}^{N}X_n\right)\right\} &= \left[\mathbb{E}\left\{\exp\left(\frac{iu}{\sqrt{N}}X_1\right)\right\}\right]^N \\ &= \left[\mathbb{E}\left\{1+\frac{iu}{\sqrt{N}}X_1-\frac{u^2}{2N}X_1^2+\ldots\right\}\right]^N \\ &= \left(1-\frac{u^2}{2N}\sigma^2+\ldots\right)^N \to \exp\left(-\frac{\sigma^2u^2}{2}\right) \qquad (11.1)\end{aligned}$$

and the limit is the characteristic function of a $\mathcal{N}(0,\sigma^2)$ variable. (See Exercise 11.1 for how to deal properly with the neglected term.)

Now consider the law of large numbers. The weak version

$$\frac{1}{N}\sum_{n=1}^{N}X_n \xrightarrow{P} 0$$

is trivial to establish if we are ready to assume that $\mathbb{E}[X_n^2]<\infty$: then

$$\mathbb{E}\left\{\left(\frac{1}{N}\sum_{n=1}^{N}X_n\right)^2\right\} = \mathrm{Var}\left(\frac{1}{N}\sum_{n=1}^{N}X_n\right) = \frac{\sigma^2}{N}$$

and by Chebyshev's inequality:

$$\mathbb{P}\left(\left|\frac{1}{N}\sum_{n=1}^{N}X_n\right|>\epsilon\right) \le \frac{\mathrm{Var}\left(N^{-1}\sum_{n=1}^{N}X_n\right)}{\epsilon^2} = \frac{\sigma^2}{\epsilon^2 N} \to 0\,.$$

In other words, mean square error (MSE) convergence implies convergence in probability. One may actually show (through basic large deviations techniques) that the probability above converges exponentially fast.

The strong law of large numbers states that

$$\frac{1}{N}\sum_{n=1}^{N}X_n \xrightarrow{a.s.} 0\,.$$

Its proof is somewhat technical (see e.g. p. 301 of Billingsley 2012), but a simpler version may be derived if we assume that $\mathbb{E}\left[X_n^4\right]<\infty$. In that case

$$\mathbb{E}\left\{\left(\sum_{n=1}^{N}X_n\right)^4\right\} = N\mathbb{E}\left(X_1^4\right) + 3N(N-1)\left\{\mathbb{E}(X_1^2)\right\}^2 \le cN^2$$

for some $c > 0$, hence, by Markov's inequality:

$$\mathbb{P}\left(\left|\frac{1}{N}\sum_{n=1}^{N} X_n\right| > \epsilon\right) \leq \frac{1}{N^4\epsilon^4}\mathbb{E}\left\{\left(\sum_{n=1}^{N} X_n\right)^4\right\} \leq \frac{c}{\epsilon^4}N^{-2}$$

which implies that

$$\sum_{N=1}^{\infty} \mathbb{P}\left(\left|\frac{1}{N}\sum_{n=1}^{N} X_n\right| > \epsilon\right) < \infty .$$

This means that, by Borel–Cantelli lemma (Exercise 11.2), with probability one, the events $\left|N^{-1}\sum_{n=1}^{N} X_n\right| > \epsilon$ occur only for a finite number of N. Since this holds for any $\epsilon > 0$, this implies that (again see Exercise 11.2)

$$\frac{1}{N}\sum_{n=1}^{N} X_n \overset{a.s.}{\to} 0 .$$

We cannot apply directly these limit theorems to particles generated by an SMC algorithm: first, particles are not independent; and second they do not form a sequence, strictly speaking. That is, the distribution of a single particle X_t^n depends on N. (To see this, consider the conditional distribution of the indices $A_t^{1:N}$, given the particles at time $t-1$.) Thus, rigorously speaking, particles should be denoted $X_t^{n,N}$, as they form what is known in Probability as a triangular array; see Fig. 11.1.

However, we will be able to obtain similar results for particle estimates by adapting in some way the above calculations.

$$\begin{array}{cccccc}
X_t^{1,1} & & & & & \\
X_t^{1,2} & X_t^{2,2} & & & & \\
X_t^{1,3} & X_t^{2,3} & X_t^{3,3} & & & \\
\dots & \dots & \dots & \dots & & \\
X_t^{1,N} & X_t^{2,N} & X_t^{3,N} & \dots & X_t^{N,N} & \\
\dots & \dots & \dots & \dots & \dots & \dots
\end{array}$$

Fig. 11.1 Representation of the triangular array formed by the particles: each row corresponds to a fixed value of N. The cumbersome notation $X_t^{n,N}$ is not used anywhere else in the book, but it is worth keeping in mind that the joint distribution of particles $X_t^1, X_t^2, \dots$ does depend on N

11.2 Convergence of Particle Estimates

11.2.1 Set-Up

We are interested in establishing that the particles generated by an SMC algorithm are such that

$$\frac{1}{N}\sum_{n=1}^{N}\varphi(X_t^n) \to \mathbb{Q}_{t-1}M_t(\varphi)$$

$$\sum_{n=1}^{N}W_t^n\varphi(X_t^n) \to \mathbb{Q}_t(\varphi)$$

in some sense, as $N \to +\infty$. The general strategy to establish such results is to decompose the error at time t as a sum of contributions from each step (sampling, reweighting, resampling) of the current and previous time steps. Thus, most proofs will work by induction.

For the sake of simplicity, we consider the settings of Algorithm 10.1 (resampling occurs at every step), and assume multinomial resampling is used. The only assumption we make on the considered Feynman-Kac model is that the potential functions G_t are upper bounded. This assumption is often met in practice (as discussed later).

Notation/Terminology For the sake of space, our calculations are restricted to the case where potential functions G_t depend on argument x_t only; i.e. $G_t(x_t)$ rather than $G_t(x_{t-1}, x_t)$. Since we make no assumption at this stage on the Markov kernels M_t, this does not entail any loss of generality: i.e. we may redefine the considered Feynman-Kac model by replacing process $\{X_t\}$ by process $\{(X_t, X_{t-1})\}$, which is also Markov. Alternatively, the reader may want to repeat the forthcoming derivations in the general case (Exercise 11.3).

11.2.2 MSE Convergence

We consider first MSE (L^2) convergence, as this mode of convergence is easier to study.

As said above, we focus on multinomial resampling. This has the advantage of making the particles X_t^n i.i.d., conditional on $\mathcal{F}_{t-1} := \sigma(X_{t-1}^{1:N})$; their conditional distribution is:

$$\sum_{n=1}^{N} W_{t-1}^n M_t(X_{t-1}^n, \mathrm{d}x_t) \,.$$

The following lemma bounds the Monte Carlo error at time t, i.e. the error related to sampling N particles from this conditional distribution; recall that $M_t\varphi$ stands for the function $x_{t-1} \to \int_{\mathcal{X}} M_t(x_{t-1}, \mathrm{d}x_t)\varphi(x_t)$.

Lemma 11.1 *For all $\varphi \in \mathcal{C}_b(\mathcal{X})$,*

$$\mathbb{E}\left[\left\{\frac{1}{N}\sum_{n=1}^{N}\varphi(X_t^n) - \sum_{n=1}^{N} W_{t-1}^n (M_t\varphi)(X_{t-1}^n)\right\}^2\right] \leq \frac{\|\varphi\|_\infty^2}{N} \,.$$

Proof Since $\mathbb{E}\left[\varphi(X_t^n)|\mathcal{F}_{t-1}\right] = \sum_{n=1}^{N} W_{t-1}^n (M_t\varphi)(X_{t-1}^n)$, one has:

$$\begin{aligned}
\mathbb{E}\left[\left\{\frac{1}{N}\sum_{n=1}^{N}\varphi(X_t^n) - \sum_{n=1}^{N} W_{t-1}^n (M_t\varphi)(X_{t-1}^n)\right\}^2 \middle| \mathcal{F}_{t-1}\right] &= \frac{1}{N}\mathrm{Var}\left\{\varphi(X_t^1)|\mathcal{F}_{t-1}\right\} \\
&\leq \frac{1}{N}\mathbb{E}\left\{\varphi^2(X_t^1)|\mathcal{F}_{t-1}\right\} \\
&\leq \frac{\|\varphi\|_\infty^2}{N}
\end{aligned}$$

and one concludes by applying the tower property of conditional expectation. □

The next lemma bounds the error due to the normalisation of the weights:

Lemma 11.2 *Provided G_t is upper bounded, for all $\varphi \in \mathcal{C}_b(\mathcal{X})$,*

$$\mathbb{E}\left[\left\{\sum_{n=1}^{N} W_t^n\varphi(X_t^n) - \frac{1}{N}\sum_{n=1}^{N}\bar{G}_t(X_t^n)\varphi(X_t^n)\right\}^2\right] \leq \|\varphi\|_\infty^2\mathbb{E}\left[\left\{\frac{1}{N}\sum_{n=1}^{N}\bar{G}_t(X_t^n) - 1\right\}^2\right]$$

where $\bar{G}_t := G_t/\ell_t$. (Recall that $\ell_t = L_t/L_{t-1} = \mathbb{Q}_{t-1}M_t(G_t)$.)

Proof Since

$$\sum_{n=1}^{N} W_t^n\varphi(X_t^n) = \frac{\sum_{n=1}^{N}\bar{G}_t(X_t^n)\varphi(X_t^n)}{\sum_{n=1}^{N}\bar{G}_t(X_t^n)}$$

one has

$$\sum_{n=1}^{N} W_t^n \varphi(X_t^n) - \frac{1}{N}\sum_{n=1}^{N} \bar{G}_t(X_t^n)\varphi(X_t^n) = \left\{\sum_{n=1}^{N} W_t^n \varphi(X_t^n)\right\}\left\{1 - \frac{1}{N}\sum_{n=1}^{N} \bar{G}_t(X_t^n)\right\}$$

and the first factor is bounded by $\|\varphi\|_\infty$. □

We may now combine these two lemmas in order to bound the MSE of particle estimates.

Proposition 11.3 *For Algorithm 10.1 (using the multinomial resampling scheme), provided potential functions G_t are all upper bounded, there exist constants c_t and c_t' such that, for all $\varphi \in \mathcal{C}_b(\mathcal{X})$,*

$$\mathbb{E}\left[\left\{\frac{1}{N}\sum_{n=1}^{N} \varphi(X_t^n) - \mathbb{Q}_{t-1}M_t(\varphi)\right\}^2\right] \leq c_t \frac{\|\varphi\|_\infty^2}{N} \tag{11.2}$$

(replacing $\mathbb{Q}_{t-1}M_t$ by $\mathbb{M}_0$ for $t = 0$) and

$$\mathbb{E}\left[\left\{\sum_{n=1}^{N} W_t^n \varphi(X_t^n) - \mathbb{Q}_t(\varphi)\right\}^2\right] \leq c_t' \frac{\|\varphi\|_\infty^2}{N}. \tag{11.3}$$

Proof By induction: (11.2) holds at time $t = 0$ with $c_0 = 1$. Assume (11.2) holds at time $t \geq 0$. One has:

$$\sum_{n=1}^{N} W_t^n \varphi(X_t^n) - \mathbb{Q}_t(\varphi) = \left(\sum_{n=1}^{N} W_t^n \varphi(X_t^n) - \frac{1}{N}\sum_{n=1}^{N} \bar{G}_t(X_t^n)\varphi(X_t^n)\right) + \left(\frac{1}{N}\sum_{n=1}^{N} \bar{G}_t(X_t^n)\varphi(X_t^n) - \mathbb{Q}_t(\varphi)\right).$$

The MSE of the second term may be bounded by applying (11.2) to function $\bar{G}_t \times \varphi$. The MSE of the first term may be bounded by applying Lemma 11.2, then (11.2) to function $\bar{G}_t$ (since $\mathbb{Q}_{t-1}M_t(\bar{G}_t) = 1$). Using the fact that $\mathbb{E}\left\{(X+Y)^2\right\} \leq 2\left\{\mathbb{E}X^2 + \mathbb{E}Y^2\right\}$, we obtain (11.3) with $c_t' = 4c_t\|\bar{G}_t\|_\infty^2$.

We show that (11.3) at time $t-1$ implies (11.2) at time t using similar calculations:

$$\frac{1}{N}\sum_{n=1}^{N} \varphi(X_t^n) - \mathbb{Q}_{t-1}M_t(\varphi) = \left(\frac{1}{N}\sum_{n=1}^{N} \varphi(X_t^n) - \sum_{n=1}^{N} W_{t-1}^n (M_t\varphi)(X_{t-1}^n)\right) + \left(\sum_{n=1}^{N} W_{t-1}^n (M_t\varphi)(X_{t-1}^n) - \mathbb{Q}_{t-1}M_t(\varphi)\right).$$

The MSE of these two terms are bounded respectively by Lemma 11.1 and by (11.3) (at time $t-1$, replacing φ by $M_t\varphi$). Which implies that (11.2) holds with $c_t = 2\left(1+c'_{t-1}\right)$. □

11.2.3 Almost Sure Convergence

We can deduce immediately that, under the same assumptions, particle estimates converges in probability (using Chebyshev's inequality). To establish almost sure convergence, we need to adapt slightly the proof of the law of large numbers given in Sect. 11.1. In doing so, we are able to state the convergence of $\sum_{n=1}^N W_t^n \varphi(X_t^n)$ for a larger class of functions; that is, functions such that the product $G_t \times \varphi$ is bounded.

Proposition 11.4 *Under the same settings as Proposition 11.3, for $t \geq 0$ and $\varphi \in \mathcal{C}_b(\mathcal{X})$*

$$\frac{1}{N}\sum_{n=1}^N \varphi(X_t^n) \overset{a.s.}{\to} \mathbb{Q}_{t-1}M_t(\varphi) \tag{11.4}$$

(where $\mathbb{Q}_{t-1}M_t$ stands for $\mathbb{M}_0$ when $t=0$), and for φ such that $\varphi \times G_t \in \mathcal{C}_b(\mathcal{X})$

$$\sum_{n=1}^N W_t^n \varphi(X_t^n) \overset{a.s.}{\to} \mathbb{Q}_t(\varphi)\,. \tag{11.5}$$

Proof Again we work by induction. At time 0, $N^{-1}\sum_{n=1}^N \varphi(X_0^n) \overset{a.s.}{\to} \mathbb{M}_0(\varphi)$ by the strong law of large numbers. If (11.4) holds for any $\varphi \in \mathcal{C}_b(\mathcal{X})$ at time $t \geq 0$, then

$$\sum_{n=1}^N W_t^n \varphi(X_t^n) = \frac{N^{-1}\sum_{n=1}^N \bar{G}_t(X_t^n)\varphi(X_t^n)}{N^{-1}\sum_{n=1}^N \bar{G}_t(X_t^n)}\,.$$

Both the numerator and the denominator converge almost surely (respectively to $\mathbb{Q}_{t-1}M_t(\bar{G}_t \times \varphi) = \mathbb{Q}_t(\varphi)$, and to $\mathbb{Q}_{t-1}M_t(\bar{G}_t) = 1$); in particular note that the numerator converges as soon as $G_t \times \varphi$ is bounded. (The class of such functions is larger than the class of bounded functions, since G_t itself is bounded.)

Now assume that (11.5) holds for any $\varphi \in \mathcal{C}_b(\mathcal{X})$ at time $t-1 \geq 0$, and let $Z_t^n = \varphi(X_t^n) - \sum_{n=1}^N W_{t-1}^n (M_t\varphi)(X_{t-1}^n)$. Conditional on $\mathcal{F}_{t-1} = \sigma(X_{t-1}^{1:N})$, the

Z_t^n's are i.i.d., with zero mean and

$$\mathbb{E}\left[\left(\sum_{n=1}^{N} Z_t^n\right)^4 \middle| \mathcal{F}_{t-1}\right] = N\mathbb{E}\left[\left(Z_t^1\right)^4 \middle| \mathcal{F}_{t-1}\right] + 3N(N-1)\left\{\mathbb{E}\left[\left(Z_t^1\right)^2 \middle| \mathcal{F}_{t-1}\right]\right\}^2$$
$$\leq cN^2$$

for some constant $c > 0$. Applying the tower property, we obtain the same bound for the unconditional expectation. We then proceed exactly as in the proof of the strong law of large numbers in Sect. 11.1, and obtain that

$$\frac{1}{N}\sum_{n=1}^{N}\varphi(X_t^n) - \sum_{n=1}^{N} W_{t-1}^n (M_t\varphi)(X_{t-1}^n) = \frac{1}{N}\sum_{n=1}^{N} Z_t^n \overset{a.s.}{\to} 0 .$$

Since $\|M_t\varphi\|_\infty \leq \|\varphi\|_\infty$, by (11.5) $\sum_{n=1}^N W_{t-1}^n (M_t\varphi)(X_{t-1}^n) \overset{a.s.}{\to} \mathbb{Q}_{t-1}M_t(\varphi)$. Thus, (11.4) holds. □

11.2.4 Discussion

The bottom line of this section is that convergence (as $N \to +\infty$) of particle estimates is easy to establish, and holds under reasonable hypotheses (i.e. G_t is upper bounded). However, Proposition 11.3 has two limitations. First, it applies only to bounded functions. Note however that we have seen that almost sure convergence actually holds for a larger class of functions.

Example 11.1 Consider the bootstrap filter associated to a state-space model such that the emission kernel is defined by equation $Y_t = F(X_t) + V_t$, $V_t \sim \mathcal{N}(0, \sigma^2)$. Then it is easy to see that

$$G_t(x_t) = (2\pi\sigma^2)^{-1/2}\exp\left\{-\left(y_t - F(x_t)^2\right)/2\sigma^2\right\} \leq (2\pi\sigma^2)^{-1/2} .$$

In addition, the class of functions such that $G_t \times \varphi$ is bounded (and therefore such that $\sum_{n=1}^N W_t^n\varphi(X_t^n)$ converges almost surely) includes, in particular, all polynomial functions.

Second, the constants c_t and c_t' that appear in Proposition 11.3 typically grow exponentially with time. It is possible to establish that the error of particle estimates remains bounded, however this requires stronger assumptions, in particular on the Markov kernels; see Sect. 11.4.

11.3 Central Limit Theorems

11.3.1 Objective

We now consider central limit theorems of the form:

$$\sqrt{N}\left(\frac{1}{N}\sum_{n=1}^{N}\varphi(X_t^n) - \mathbb{Q}_{t-1}M_t(\varphi)\right) \Rightarrow \mathcal{N}\left(0, \widetilde{\mathcal{V}}_t(\varphi)\right) \tag{11.6}$$

$$\sqrt{N}\left(\sum_{n=1}^{N}W_t^n\varphi(X_t^n) - \mathbb{Q}_t(\varphi)\right) \Rightarrow \mathcal{N}\left(0, \mathcal{V}_t(\varphi)\right) \tag{11.7}$$

which basically say that the error of a particle estimate is (asymptotically) distributed according to a Gaussian distribution with scale $\mathcal{O}(N^{-1/2})$. The asymptotic variances may be defined recursively as follows: $\widetilde{\mathcal{V}}_0(\varphi) := \mathrm{Var}_{\mathbb{M}_0}(\varphi)$,

$$\widetilde{\mathcal{V}}_t(\varphi) := \mathcal{V}_{t-1}(M_t\varphi) + \mathrm{Var}_{\mathbb{Q}_{t-1}M_t}(\varphi), \quad \text{for } t \geq 1 \tag{11.8}$$

$$\mathcal{V}_t(\varphi) := \widetilde{\mathcal{V}}_t\left\{\bar{G}_t\left(\varphi - \mathbb{Q}_t\varphi\right)\right\}, \quad \text{for } t \geq 0 \tag{11.9}$$

where $\mathrm{Var}_{\mathbb{Q}_{t-1}M_t}(\varphi) = \mathbb{Q}_{t-1}M_t\left[\{\varphi(X_t) - \mathbb{Q}_{t-1}M_t(\varphi)\}^2\right]$, and (as before) $\bar{G}_t = G_t/\ell_t$.

Before stating these results rigorously, we provide some intuition regarding these recursive formulae. Expression (11.9) is reminiscent of the asymptotic variance of auto-normalised importance sampling (8.4). In fact, it may be derived in exactly the same way:

$$\sqrt{N}\left\{\sum_{n=1}^{N}W_t^n\varphi(X_t^n) - \mathbb{Q}_t(\varphi)\right\} = \frac{N^{-1/2}\left\{\sum_{n=1}^{N}\bar{G}_t(X_t^n)\bar{\varphi}(X_t^n)\right\}}{N^{-1}\sum_{n=1}^{N}\bar{G}_t(X_t^n)}$$

where $\bar{\varphi} = \varphi - \mathbb{Q}_t(\varphi)$, so if (11.6) holds for function $\bar{G}_t \times \bar{\varphi}$, and if the denominator converges in probability to the desired limit (which is one, as $\mathbb{Q}_{t-1}M_t(\bar{G}_t) = 1$), then by Slutsky's theorem, the left hand side converges to a Gaussian distribution with variance $\mathcal{V}_t(\varphi) = \widetilde{\mathcal{V}}_t\left\{\bar{G}_t\left(\varphi - \mathbb{Q}_t\varphi\right)\right\}$.

Notation/Terminology As in the previous section, we do all of our derivations for potential functions that depend on x_t only; i.e. $G_t(x_t)$ instead of $G_t(x_{t-1}, x_t)$. We will not make any assumption on the Markov kernels M_t so again this does not incur any loss in generality.

Expression (11.8) may be viewed as the asymptotic version of the law of total variance:

$$\begin{aligned}\mathrm{Var}\left[\frac{1}{N}\sum_{n=1}^{N}\varphi(X_t^n)\right] =&\mathrm{Var}\left[\mathbb{E}\left\{\frac{1}{N}\sum_{n=1}^{N}\varphi(X_t^n)\middle|\mathcal{F}_{t-1}\right\}\right]\\ &+\mathbb{E}\left[\mathrm{Var}\left\{\frac{1}{N}\sum_{n=1}^{N}\varphi(X_t^n)\middle|\mathcal{F}_{t-1}\right\}\right]\\ =&\mathrm{Var}\left[\sum_{n=1}^{N}W_{t-1}^n\left(M_t\varphi\right)\left(X_{t-1}^n\right)\right]\\ &+\frac{1}{N}\mathbb{E}\left[\mathrm{Var}\left\{\varphi(X_t^1)\middle|\mathcal{F}_{t-1}\right\}\right]\end{aligned}$$

where $\mathcal{F}_{t-1} := \sigma(X_{t-1}^{1:N})$ and we have used again the fact that, conditional on $\mathcal{F}_{t-1}$, the X_t^n's are i.i.d., with distribution $\sum_{n=1}^{N} W_{t-1}^n M_t(X_{t-1}^n, \mathrm{d}x_t)$. Since that distribution should converge to $\mathbb{Q}_{t-1}M_t$, we expect the second term above (times N) to converge to $\mathrm{Var}_{\mathbb{Q}_{t-1}M_t}(\varphi)$, the second term in (11.8). As said above, these remarks are only meant to provide some intuition; the next section gives a formal proof of these expressions.

11.3.2 Formal Statement and Proof

Proposition 11.2 *Under the same settings as Proposition 11.3, at any $t \geq 1$, and for any $\varphi \in \mathcal{C}_b(\mathcal{X})$,*

$$\sqrt{N}\left(\frac{1}{N}\sum_{n=1}^{N}\varphi(X_t^n) - \mathbb{Q}_{t-1}M_t(\varphi)\right) \Rightarrow \mathcal{N}\left(0, \widetilde{\mathcal{V}}_t(\varphi)\right) \tag{11.10}$$

and at any $t \geq 0$, and for any $\varphi : \mathcal{X} \to \mathbb{R}$ such that $\varphi \times G_t \in \mathcal{C}_b(\mathcal{X})$,

$$\sqrt{N}\left(\sum_{n=1}^{N} W_t^n\varphi(X_t^n) - \mathbb{Q}_t(\varphi)\right) \Rightarrow \mathcal{N}\left(0, \mathcal{V}_t(\varphi)\right) \tag{11.11}$$

where the asymptotic variances are defined recursively by (11.8) and (11.9), and are finite.

Proof Again, the proof relies on induction: at time 0, (11.10) holds (replacing $\mathbb{Q}_{t-1}M_t$ by $\mathbb{M}_0$) by the standard CLT. We have already explained in the previous section how (11.11) may be deduced from (11.10) at any time $t \geq 0$ (by applying

Slutsky's theorem). Note in particular this holds whenever (11.10) holds for function $\bar{G}_t \times \varphi$, i.e. whenever this function is bounded.

The difficult part is to prove that, if (11.11) holds at time $t-1 \geq 0$ for any bounded function, then (11.10) holds at time t. Let $\mathcal{F}_{t-1} = \sigma(X_{t-1}^{1:N})$,

$$\Delta_1 := \sqrt{N}\left(\frac{1}{N}\sum_{n=1}^{N}\varphi(X_t^n) - \sum_{n=1}^{N} W_{t-1}^n (M_t\varphi)(X_{t-1}^n)\right)$$

$$\Delta_2 := \sqrt{N}\left(\sum_{n=1}^{N} W_{t-1}^n (M_t\varphi)(X_{t-1}^n) - \mathbb{Q}_{t-1}M_t(\varphi)\right)$$

then the characteristic function of the left hand side of (11.10) equals

$$\mathbb{E}\left[\exp\{iu(\Delta_1+\Delta_2)\}\right] = \mathbb{E}\left[\exp\{iu\Delta_2\}\,\mathbb{E}\left[\exp\{iu\Delta_1\}\,|\,\mathcal{F}_{t-1}\right]\right]. \tag{11.12}$$

Recall that variables $Z_t^n = \varphi(X_t^n) - \sum_{n=1}^{N} W_{t-1}^n (M_t\varphi)(X_{t-1}^n)$ are i.i.d. conditional on $\mathcal{F}_{t-1}$, hence (repeating the same calculations as in Sect. 11.1):

$$\begin{aligned}\mathbb{E}\left[\exp\{iu\Delta_1\}\,|\,\mathcal{F}_{t-1}\right] &= \mathbb{E}\left[\exp\left\{iuN^{-1/2}\sum_{n=1}^{N} Z_t^n\right\}\middle|\,\mathcal{F}_{t-1}\right]\\ &= \left(\mathbb{E}\left[\exp\left\{iuN^{-1/2}Z_t^1\right\}\middle|\,\mathcal{F}_{t-1}\right]\right)^N.\end{aligned}$$

We wish to prove that this quantity converges in probability to $\exp(-\sigma^2u^2/2)$, where $\sigma^2 := \mathrm{Var}_{\mathbb{Q}_{t-1}M_t}(\varphi)$. Using inequality $|u^N - v^N| \leq N|u-v|$ for complex numbers such that $|u|, |v| \leq 1$ (Exercise 11.1), one has

$$\begin{aligned}&\left|\mathbb{E}\left[\exp\{iu\Delta_1\}\,|\,\mathcal{F}_{t-1}\right] - \exp\left(-\frac{\sigma^2u^2}{2}\right)\right|\\ &\qquad\leq N\left|\mathbb{E}\left[\exp\left\{iuN^{-1/2}Z_t^1\right\}\middle|\,\mathcal{F}_{t-1}\right] - \exp\left(-\frac{\sigma^2u^2}{2N}\right)\right|.\end{aligned} \tag{11.13}$$

Furthermore, since $\left|\exp(ix) - 1 - ix + x^2/2\right| \leq |x|^3/6$ (Exercise 11.1), $\mathbb{E}[Z_t^n|\mathcal{F}_{t-1}] = 0$,

$$\mathbb{E}[(Z_t^n)^2\,|\mathcal{F}_{t-1}] = \sigma_N^2 := \sum_{n=1}^{N} W_{t-1}^n (M_t\varphi^2)(X_{t-1}^n) - \left(\sum_{n=1}^{N} W_{t-1}^n (M_t\varphi)(X_{t-1}^n)\right)^2,$$

and $|Z_t^n| \leq 2\|\varphi\|_\infty$, one has

$$\left|\mathbb{E}\left[\exp\left\{iuN^{-1/2}Z_t^1\right\}\Big|\mathcal{F}_{t-1}\right] - 1 + \frac{u^2}{2N}\sigma_N^2\right| \leq \frac{8\,|u|^3}{6N^{3/2}}\|\varphi\|_\infty^3$$

which implies that

$$N\left[\mathbb{E}\left[\exp\left\{iuN^{-1/2}Z_t^1\right\}\Big|\mathcal{F}_{t-1}\right] - 1 + \frac{u^2}{2N}\sigma_N^2\right] \stackrel{a.s.}{\rightarrow} 0\,.$$

Since we assumed that (11.11) holds at time $t-1 \geq 0$, one has $\sigma_N^2 \stackrel{P}{\rightarrow} \sigma^2$, and clearly $N\left\{1 - \exp\left(-\sigma^2u^2/2N\right)\right\} \rightarrow \sigma^2u^2/2$, hence

$$N\left[\mathbb{E}\left[\exp\left\{iuN^{-1/2}Z_t^1\right\}\Big|\mathcal{F}_{t-1}\right] - \exp\left(-\frac{\sigma^2u^2}{2N}\right)\right] \stackrel{P}{\rightarrow} 0.$$

This implies, from (11.13), that indeed

$$\mathbb{E}\left[\exp\{iu\Delta_1\}\,|\,\mathcal{F}_{t-1}\right] \stackrel{P}{\rightarrow} \exp\left(-\frac{\sigma^2u^2}{2}\right).$$

We now return to (11.12): from (11.11) at time $t-1 \geq 0$, $\Delta_2 \Rightarrow Y$, where $Y \sim N\left(0, \mathcal{V}_{t-1}(M_t\varphi)\right)$, hence, by Slutsky's theorem:

$$\exp\{iu\Delta_2\}\,\mathbb{E}\left[\exp\{iu\Delta_1\}\,|\,\mathcal{F}_{t-1}\right] \Rightarrow \exp\left\{iuY - \frac{\sigma^2u^2}{2}\right\}$$

and we may take the expectation on both sides (dominated convergence theorem) to obtain the desired result. □

11.3.3 Discussion

Propositions 11.3 and 11.2 essentially deliver the same message: the error of particle estimates converges at the standard Monte Carlo rate, $\mathcal{O}_P(N^{-1/2})$, where $\mathcal{O}_P$ is the "big O in Probability" notation. Proposition 11.3 is the stronger result of the two, since it is non-asymptotic. On the other hand, Proposition 11.2 holds for a larger class of functions (all functions φ such that $G_t \times \varphi$ is bounded; in fact CLTs for much larger classes of functions have been stated in the literature, see the bibliography). Another advantage of Proposition 11.2 is that it gives us explicit expressions for the asymptotic variances, which we use to study the stability of particle estimates in next section.

11.4 Stability of Particle Algorithms

We have seen that particle estimates computed at a *fixed* time t converge as $N \to +\infty$. We now study how the error of a particle estimate may evolve over time.

For instance, it is easy to establish, from (11.8) and (11.9), that the asymptotic variance $\mathcal{V}_t(\varphi)$ may be rewritten as:

$$\mathcal{V}_t(\varphi) = \sum_{s=0}^{t} (\mathbb{Q}_{s-1}M_s)\left[\left\{\bar{G}_s R_{s+1:t}(\varphi - \mathbb{Q}_t\varphi)\right\}^2\right] \tag{11.14}$$

where $R_t(\varphi) = M_t(\bar{G}_t \times \varphi)$, and $R_{s+1:t}(\varphi) = R_{s+1} \circ \ldots \circ R_t(\varphi)$. These operators are functionals: $R_t(\varphi)$ is the function $x_{t-1} \to \int_{\mathcal{X}} M_{t-1}(x_{t-1}, \mathrm{d}x_t)\bar{G}_t(x_t)\varphi(x_t)$.

Formula (11.14) formalises the intuition we have already used in the previous proofs: the error at time t is an accumulation of sampling errors from the previous time steps. To prevent $\mathcal{V}_t(\varphi)$ from blowing up, we need to make sure that the impact of past time steps vanishes in some sense; that is, that the Markov kernels M_t forget the past in some way. Without such assumptions, it is easy to come up with examples where $\mathcal{V}_t(\varphi)$ diverges at a geometric (or higher) rate; see Exercise 11.4.

Next section discusses properties on Markov kernels M_t that formalise this notion of forgetting the past.

11.4.1 Strongly Mixing Markov Kernels

We first define the total variation distance between two probability measures $\mathbb{P}$, $\mathbb{Q}$ on $(\mathcal{X}, \mathcal{B}(\mathcal{X}))$, as

$$\|\mathbb{P} - \mathbb{Q}\|_{\mathrm{TV}} := \sup_{A \in \mathcal{B}(\mathcal{X})} |\mathbb{P}(A) - \mathbb{Q}(A)| .$$

Equivalent definitions are:

$$\|\mathbb{P} - \mathbb{Q}\|_{\mathrm{TV}} := \sup_{\Delta\varphi \leq 1} |\mathbb{P}(\varphi) - \mathbb{Q}(\varphi)|$$

where $\Delta\varphi$ is the variation of function $\varphi : \mathcal{X} \to \mathbb{R}$,

$$\Delta\varphi := \sup_{x,x'} \left|\varphi(x) - \varphi(x')\right| ;$$

and, provided $\mathbb{P}$, $\mathbb{Q}$ admit densities p, q with respect to a dominating measure $\mathrm{d}x$,

$$\|\mathbb{P} - \mathbb{Q}\|_{\mathrm{TV}} := \frac{1}{2} \int_{\mathcal{X}} |p(x) - q(x)| \, \mathrm{d}x ;$$

that is half the L^1 distance between p and q. For a proof of the equivalence of these three definitions, see Exercise 11.6. (As the notation suggests, the total variation distance is actually a norm on the space of signed measures. However, treating it as a distance is sufficient for our purposes.)

The total variation distance is not always very convenient to deal with (Exercise 11.5), but, in the context of Markov kernels, it provides a nice way to quantify how quickly they forget the past.

Definition 11.6 The contraction coefficient of a Markov kernel M_t on $(\mathcal{X}, \mathcal{B}(\mathcal{X}))$ is the quantity $\rho_M \in [0, 1]$ defined as

$$\rho_M := \sup_{x_{t-1}, x'_{t-1}} \|M_t(x_{t-1}, \mathrm{d}x_t) - M_t(x'_{t-1}, \mathrm{d}x_t)\|_{\mathrm{TV}} .$$

A Markov kernel M_t is said to be strongly mixing if $\rho_M < 1$.

The contraction coefficient measures to which extent the Markov kernel variation of test functions:

Lemma 11.7 *For a Markov kernel M_t on $(\mathcal{X}, \mathcal{B}(\mathcal{X}))$ with contraction coefficient ρ_M, and $\phi \in \mathcal{C}_b(\mathcal{X})$, one has:*

$$\Delta\left(M_t(\varphi)\right) \leq \rho_M \Delta\varphi .$$

The proof is left as an Exercise.

The following lemma gives a sufficient condition to ensure that a Markov kernel is strongly mixing.

Proposition 11.8 *If a Markov kernel M_t admits a probability density $m_t(x_t|x_{t-1})$ (with respect to a fixed dominating measure) such that*

$$\frac{m_t(x_t|x_{t-1})}{m_t(x_t|x'_{t-1})} \leq c_M \tag{11.15}$$

where $c_M \geq 1$, then it is strongly mixing, with contraction coefficient $\rho_M \leq 1 - c_M^{-1}$.

Proof For any $A \in \mathcal{B}(\mathcal{X})$, $x_{t-1}, x'_{t-1} \in \mathcal{X}$,

$$\begin{aligned}\int_A \left\{m_t(x_t|x_{t-1}) - m_t(x_t|x'_{t-1})\right\} \mathrm{d}x_t &\leq (1 - c_M^{-1}) \int_A m_t(x_t|x_{t-1})\, \mathrm{d}x_t \\ &\leq (1 - c_M^{-1})\end{aligned}$$

hence $\|M_t(x_{t-1}, \mathrm{d}x) - M_t(x'_{t-1}, \mathrm{d}x)\|_{\mathrm{TV}} \leq 1 - c_M^{-1}$. □

Of course, condition (11.15) is restrictive; it is rarely met in realistic models. However, it greatly simplifies the analysis. Thus we assume from now on that the Markov kernels of the considered Feynman-Kac model fulfil this condition.

Assumption (M) *The considered Feynman-Kac model is such that all the Markov kernels M_t (for $t = 1, \ldots, T$) admit a probability density m_t such that*

$$\frac{m_t(x_t|x_{t-1})}{m_t(x_t|x'_{t-1})} \leq c_M$$

for some $c_M \geq 1$.

We have seen in Chap. 5 that, under distribution $\mathbb{Q}_t$, $\{X_t\}$ remains a Markov process. The following proposition shows that this Markov process is also strongly mixing under Assumption (M), and an extra assumption on the potential functions.

Assumption (G) *The potential functions G_t of the considered model are uniformly bounded: for some constants c_l, c_u and all times $t \geq 0$*

$$0 < c_l \leq G_t(x_{t-1}, x_t) \leq c_u$$

(where $G_t(x_{t-1}, x_t)$ must be replaced by $G_0(x_0)$ for $t = 0$).

Notation/Terminology Since we are now making assumptions on the kernels M_t, we can no longer afford to take, without loss of generality, G_t to be a function of x_t only (see the notation/terminology boxes in the two previous sections). Thus, we revert to the standard notation, $G_t(x_{t-1}, x_t)$.

The fact that G_t is lower bounded is a very strong condition, which is almost never met when $\mathcal{X}$ is not compact.

Proposition 11.9 *Under Assumptions (M) and (G), the Markov process defined by distribution $\mathbb{Q}_t$ is strongly mixing with contraction coefficient smaller than or equal to $\rho_Q := 1 - 1/c_M^2 c_G$, where $c_G = c_u/c_l$.*

Proof From (5.13), the density of $Q_{s|t}$ (transition density of the conditioned Markov process) is

$$q_{s|t}(x_s|x_{s-1}) = \frac{H_{s:t}(x_s)}{H_{s-1:t}(x_{s-1})} G_s(x_{s-1}, x_s) m_s(x_s|x_{s-1})$$

where

$$\begin{aligned} H_{s:t}(x_s) &= \int_{\mathcal{X}} G_{s+1}(x_s, x_{s+1}) H_{s+1:t}(x_{s+1}) m_{s+1}(x_{s+1}|x_s)\, \mathrm{d}x_{s+1} \\ &\leq c_M c_G H_{s:t}(x'_s) \end{aligned}$$

which implies that

$$\frac{q_{s|t}(x_s|x_{s-1})}{q_{s|t}(x_s|x'_{s-1})} = \frac{H_{s-1:t}(x'_{s-1})\,m_s(x_s|x_{s-1})}{H_{s-1:t}(x_{s-1})\,m_s(x_s|x'_{s-1})} \le c_M^2 c_G$$

and one concludes by applying Proposition 11.8. □

11.4.2 Stability of Asymptotic Variances

Before establishing the stability of asymptotic variance $\mathcal{V}_t(\varphi)$, we state a few technical lemmas.

Lemma 11.10 *Let $\varphi, \psi \in \mathcal{C}_b(\mathcal{X})$ such that $\psi \ge 0$, then*

$$\Delta(\varphi\psi) \le \|\psi\|_\infty \Delta\varphi\,.$$

The proof is straightforward.

Lemma 11.11 *Let $\varphi \in \mathcal{C}_b(\mathcal{X})$ a function such that $\mathbb{P}(\varphi) = 1$ for a certain probability distribution $\mathbb{P}$. Then*

$$\|\varphi\|_\infty \le 1 + \Delta\varphi\,.$$

Proof From the definition of $\Delta\varphi$, one has for any $x, x' \in \mathcal{X}$,

$$\varphi(x') - \Delta\varphi \le \varphi(x) \le \varphi(x') + \Delta\varphi$$

then take the expectation with respect to $X' \sim \mathbb{P}$ and observe that $|1-\Delta\varphi| \le 1+\Delta\varphi$ to conclude. □

The next lemma is the main technical result of this section. It bounds the contraction of the operators $R_{s+1:t}$ that appear in decomposition (11.14).

Lemma 11.12 *Under Assumptions (M) and (G), for any $\varphi \in \mathcal{C}_b(\mathcal{X})$, $t \ge 1$ and $s \le t-1$*

$$\Delta R_{s+1:t}(\varphi) \le \prod_{i=1}^{t-s}\left(1 + \rho_M \rho_Q^{i-1} c_G\right)\rho_Q^{t-s}\Delta\varphi\,. \tag{11.16}$$

Proof From (5.13), one deduces that

$$R_{s+1:t}(\varphi) = \bar{H}_{s:t} \times Q_{s+1:t|t}(\varphi)$$

where $Q_{s+1:t|t} = Q_{s+1|t} \cdots Q_{t|t}$ (i.e. the Markov kernel that gives the distribution of X_t, conditional on X_s, under the distribution $\mathbb{Q}_t$), and $\bar{H}_{s:t}$ is defined in the same way as the cost-to-go function $H_{s:t}$, except each function G_s is replaced by $\bar{G}_s$: $\bar{H}_{t:t}(x_t) = 1$ and, recursively,

$$\bar{H}_{s:t}(x_s) = \int_{\mathcal{X}} M_{s+1}(x_s, \mathrm{d}x_{s+1})\bar{G}_{s+1}(x_{s+1})\bar{H}_{s+1:t}(x_{s+1}) .$$

Thus, by Lemma 11.10 and Proposition 11.9,

$$\Delta R_{s+1:t}(\varphi) \leq \|\bar{H}_{s:t}\|_\infty \rho_Q^{t-s} \Delta\varphi$$

and what remains to do is to prove that

$$\|\bar{H}_{s:t}\|_\infty \leq \prod_{i=1}^{t-s} \left(1 + \rho_M \rho_Q^{i-1} c_G\right) .$$

We consider the case $s = t - 2$, other cases may be treated similarly. Since

$$Q_{t-1|t-1}(x_{t-2}, \mathrm{d}x_{t-1}) = M_{t-1}(x_{t-2}, \mathrm{d}x_{t-1}) \frac{\bar{G}_{t-1}(x_{t-1})}{M_{t-1}(\bar{G}_{t-1})(x_{t-2})}$$

one has:

$$\begin{aligned}\bar{H}_{t-2:t} &= M_{t-1}(\bar{G}_{t-1}\bar{H}_{t-1:t}) \\ &= M_{t-1}(\bar{G}_{t-1}) \times Q_{t-1|t-1}(\bar{H}_{t-1:t}) \\ &= M_{t-1}(\bar{G}_{t-1}) \times \left(Q_{t-1|t-1}M_t\right)(\bar{G}_t)\end{aligned}$$

and thus by Lemmas 11.11 and 11.7 (since $\mathbb{Q}_{t-2}M_{t-1}(\bar{G}_{t-1}) = 1$, and the expectation of $\left(Q_{t-1|t-1}M_t\right)(\bar{G}_t)$ with respect to $\mathbb{Q}_{t-1}(\mathrm{d}x_{0:t-2})$ is also one):

$$\|\bar{H}_{t-2:t}\|_\infty \leq (1 + \rho_M c_G)\left(1 + \rho_Q \rho_M c_G\right) . \qquad \square$$

Now that these technicalities are out of the way, the desired result is easy to establish.

Proposition 11.13 *Under Assumptions (G) and (M), for any* $\varphi \in \mathcal{C}_b(\mathcal{X})$, *the asymptotic variance* $\mathcal{V}_t(\varphi)$ *defined in Proposition (11.2) is bounded uniformly in time.*

Proof Note first that $R_{s+1:t}(\varphi - \mathbb{Q}_t\varphi)$ has expectation zero with respect to $\mathbb{Q}_s$; hence this function takes negative and positive values, and

$$\|R_{s+1:t}(\varphi - \mathbb{Q}_t\varphi)\|_\infty \leq \Delta R_{s+1:t}(\varphi - \mathbb{Q}_t\varphi) .$$

Therefore, applying (11.16)–(11.14) leads to

$$\begin{aligned}
\mathcal{V}_t(\varphi) &\leq c_G^2 \, (\Delta\varphi)^2 \sum_{s=0}^{t} \prod_{i=1}^{t-s} \left(1 + \rho_M \rho_Q^{i-1} c_G\right)^2 \rho_Q^{2(t-s)} \\
&= c_G^2 \, (\Delta\varphi)^2 \sum_{s=0}^{t} \exp\left(2 \sum_{i=1}^{t-s} \log(1 + \rho_M \rho_Q^{i-1} c_G)\right) \rho_Q^{2(t-s)} \\
&\leq c_G^2 \, (\Delta\varphi)^2 \sum_{s=0}^{t} \exp\left(2 \rho_M c_G \sum_{i=1}^{t-s} \rho_Q^{i-1}\right) \rho_Q^{2(t-s)} \\
&\leq c_G^2 \, (\Delta\varphi)^2 \exp\left(\frac{2\rho_M c_G}{1-\rho_Q}\right) \sum_{s=0}^{t} \rho_Q^{2s} \\
&\leq c_G^2 \, (\Delta\varphi)^2 \exp\left(\frac{2\rho_M c_G}{1-\rho_Q}\right) \times \frac{1}{1-\rho_Q^2}
\end{aligned}$$

since $\log(1+x) \leq x$ (third line). □

11.4.3 Discussion

The result above makes sense only in situations where the target distributions, $\mathbb{Q}_t(\mathrm{d}x_t)$, stay stable over time. This is often the case in filtering problems, but not necessarily in other cases. For instance, in Chap. 17, we will derive SMC algorithms for sequential parameter estimation; there $\mathbb{Q}_t(\mathrm{d}x_t)$ will stand for the posterior distribution of some fixed parameter θ, and will concentrate in a smaller and smaller region. In such a scenario, we would like $\mathcal{V}_t(\varphi)$ to get smaller and smaller as well, in a way that ensures that $\mathcal{V}_t(\varphi)$ remains small relative to the variance of $\mathbb{Q}_t(\mathrm{d}x_t)$. We return to this point in Chap. 17.

Exercises

11.1 *Show that* $\left|e^{iu} - 1 - iu + u^2/2\right| \leq \min\left(u^2, |u|^3/6\right)$ *for any* $u \in \mathbb{R}$. *(Hint: use Taylor's formula with an integral remainder.) Then show that* $\left|a^N - b^N\right| \leq N\,|a-b|$ *when* $|a|, |b| \leq 1$. *Use both inequalities to prove the convergence stated in (11.1).*

11.2 *Consider a sequence of events* A_n *such that* $\sum_{n=1}^{\infty} \mathbb{P}(A_n) < \infty$. *Explain what*

$$B = \cap_{p=1}^{\infty} \cup_{n \geq p} A_n$$

represents (in connection with the phrase "occurs infinitely often" at the end of Sect. 11.1). Show that $\mathbb{P}(B) \leq \sum_{n\geq p} \mathbb{P}(A_n)$ *for any* p*, and deduce that* $\mathbb{P}(B) = 0$*. This result is known as Borel-Cantelli lemma. Now take* $A_n = \{|S_n| \geq \epsilon\}$ *where* S_n *is a sequence of real-valued random variables; show that the event* $\{S_n \to 0\}$ *is included in* B^c *(the complement of* B*). Deduce the following property (which is used implicitly at the end of Sect. 11.1): if, for any* $\epsilon > 0$, $\sum_{n=1}^{\infty} \mathbb{P}(|S_n| \geq \epsilon) < \infty$*, then the event* $\{S_n \to 0\}$ *has probability one (i.e.* $S_n \to 0$ *almost surely). (You may consider an intersection over values* $\epsilon = 1/q$ *for* $q = 1, 2, \ldots$*)*

11.3 *Repeat the derivations of Sect. 11.2 in the general case where* G_t *depends on both* x_{t-1} *and* x_t*. To that purpose, show first that the pairs* $(X_{t-1}^{A_t^n}, X_t^n)$ *are i.i.d. variables conditional on* $X_{t-1}^{1:N}$*.*

11.4 *Consider a Feynman-Kac model such that* $M_t(x_{t-1}, \mathrm{d}x_t) = \delta_{x_{t-1}}$*, and* $G_t(x_t) = \mathbb{1}_{A_t}(x_t)$*, where* $A_t \subset A_{t-1} \subset \ldots \subset A_0$*. Assume first that* $\mathbb{M}_0(A_t) = 2^{-(t+1)}$*. Show that the first term of* (11.14) *diverges at a geometric rate. Adapt this example to obtain any desired rate of divergence.*

11.5 *Let* $X_1, X_2, \ldots$ *be a sequence of i.i.d. variables, such that* $\mathbb{P}(X_1 = 1) = \mathbb{P}(X_1 = -1) = 1/2$*. Show that* $N^{-1/2}\sum_{n=1}^{N} X_n \Rightarrow \mathcal{N}(0, 1/4)$*, and that the total variation distance between the distribution of* $N^{-1/2}\sum_{n=1}^{N} X_n$ *and* $\mathcal{N}(0, 1/4)$ *does not converge to zero.*

11.6 *Show that the three definitions of the total variation distance given in Sect. 11.4.1 are indeed equivalent. (Hint: for the third definition, consider the set* $A = \{x : p(x) \geq q(x)\}$*.)*

11.7 *Deduce from* (11.8) *the following CLT:*

$$\sqrt{N}\left(\frac{1}{N}\sum_{n=1}^{N} \varphi(X_t^{A_t^n}) - \mathbb{Q}_t(\varphi)\right) \Rightarrow \mathcal{N}\left(0, \hat{\mathcal{V}}_t(\varphi)\right).$$

for a particle estimate computed immediately after resampling. (Hint: to do so, consider a particular Markov kernel M_t*.) Then, rewrite the recursive formulae in terms of the three asymptotic variances,* $\tilde{\mathcal{V}}_t(\varphi)$*,* $\mathcal{V}_t(\varphi)$ *and* $\hat{\mathcal{V}}(\varphi)$*, and identify the term that corresponds to the extra variance due to the resampling step. Derive the expression of this term when residual resampling is used instead of multinomial resampling.*

11.8 *Assume that the bias of a given SMC estimate is* $cN^{-1} + \mathcal{O}(N^{-2})$*. Explain how you may combine SMC estimates obtained from two SMC runs, with different values for* N*, to obtain a new estimate with bias* $\mathcal{O}(N^{-2})$*. Discuss the impact on the variance of the so-obtained estimate (as a function of the two values of* N*).*

Python Corner

We recommended in the previous Python corner to run several times SMC algorithms in order to assess the variability of their output. Rigorously speaking, this approach gives us a way to assess the *variance* of SMC estimates, but not their *bias*. One result we did not state in this chapter is that the bias of a SMC estimate is (typically) $\mathcal{O}(N^{-1})$; thus its contribution to the mean square error is negligible relative to the variance (since MSE is $\mathcal{O}(N^{-1})$, and MSE equals variance plus bias squared). Hence, at first order, visualisations such as e.g. box-plots (of estimates obtained from independent runs based on the same N) remain a reasonable assessment of the error of SMC estimates.

In the same spirit, when we run k times a given SMC algorithm, we may report the average over these k runs of a given SMC estimate as our "best" estimate; its variance will be reduced by a k^{-1} factor (relative to the variance of each run), while the bias should be the same. To reduce the bias, one might try to combine the results of SMC runs with different values for N (Exercise 11.8).

Bibliographical Notes

Proposition 11.3 which bounds the MSE of particle estimates is adapted from Crisan and Doucet (2002). Many more such results (e.g. in terms of L^p or other norms) may be found in the monograph of Del Moral (2004, especially Chapter 7). See also e.g. Künsch (2001), Künsch (2005) for results based on the L^1 norm and Le Gland and Oudjane (2004) for results based on the Hilbert metric.

Proposition 11.2 is a simplified version of the central limit theorems of Chopin (2004) and Künsch (2005). The first central limit theorem for particle estimates was established by Del Moral and Guionnet (1999); see also Douc and Moulines (2008).

In the context of state-space models, the term 'stability' usually refers to the stability of the filter; that is, how the filtering distribution at time t depends on the initial distribution $\mathbb{P}_0$. It turns out that the stability of state-space models and the stability of particle filters are strongly related; in fact both notions are usually studied simultaneously. The classical approach (again see the book of Del Moral 2004 and references therein, e.g. Del Moral and Guionnet 2001) is to derive time uniform bounds for the error under Assumptions (M) and (G).

These assumptions are too strong for most practical problems (in particular when $\mathcal{X}$ is not compact). Stability results under more general conditions have been established by e.g. Oudjane and Rubenthaler (2005), van Handel (2009), and Douc et al. (2014). In particular, Oudjane and Rubenthaler (2005) introduced an innovative truncation approach, where the true filter is compared to a filter where G_t is set to zero outside a compact; the truncated filter has then the same stability properties of a filter defined on a compact state-space.

We have considered only filtering estimates in this section (or more generally estimates computed from the output a single time step). Cérou et al. (2011), Whiteley (2013) derive bounds for the MSE of $L_t^N = \prod_{s=0}^t \ell_t^N$, the estimate of the normalising constant; this quantity will play an important role in Chap. 16. For smoothing estimates (which are discussed in Chap. 12), see Douc et al. (2011).

In line with the literature, we stated all our results under multinomial resampling. Obtaining similar results under alternative resampling schemes turns out to be quite challenging. The CLT of Chopin (2004) holds for residual resampling as well (see comments in Exercise 11.7). The asymptotic variance is shown to be smaller under residual resampling than under multinomial resampling. Gerber et al. (2019) give sufficient and necessary conditions for a resampling scheme to be consistent; one of these conditions is the numbers of children that originate from each particle are negatively associated (i.e. negatively correlated under monotonous transformation). Interestingly, systematic resampling does not fulfil this condition. (Thus, systematic resampling is not consistent, under the notion of consistency considered in that paper, but its consistency in a weaker sense remains an open problem.) As already mentioned in Chap. 9, Gerber et al. (2019) propose a resampling scheme closely related to systematic resampling (i.e. the number of children varies at most by one) which fulfil this condition, and therefore is ensured to be consistent.

Bibliography

Billingsley, P. (2012). *Probability and measure (anniversary edition). Wiley series in probability and statistics*. Hoboken, NJ: Wiley.

Cérou, F., Del Moral, P., & Guyader, A. (2011). A nonasymptotic theorem for unnormalized Feynman-Kac particle models. *Annales de l'Institut Henri Poincaré - Probabilités et Statistiques, 47*(3), 629–649.

Chopin, N. (2004). Central limit theorem for sequential Monte Carlo methods and its application to Bayesian inference. *Annals of Statistics, 32*(6), 2385–2411.

Crisan, D., & Doucet, A. (2002). A survey of convergence results on particle filtering methods for practitioners. *IEEE Transactions on Signal Processing, 50*(3), 736–746.

Del Moral, P. (2004). *Feynman-Kac formulae. Genealogical and interacting particle systems with applications. Probability and its applications*. New York: Springer.

Del Moral, P., & Guionnet, A. (1999). Central limit theorem for nonlinear filtering and interacting particle systems. *Annals of Applied Probability, 9*(2), 275–297.

Del Moral, P., & Guionnet, A. (2001). On the stability of interacting processes with applications to filtering and genetic algorithms. *Annales de l'Institut Henri Poincaré - Probabilités et Statistiques, 37*(2), 155–194.

Douc, R., Garivier, A., Moulines, E., & Olsson, J. (2011). Sequential Monte Carlo smoothing for general state space hidden Markov models. *Annals of Applied Probability, 21*(6), 2109–2145.

Douc, R., & Moulines, E. (2008). Limit theorems for weighted samples with applications to sequential Monte Carlo methods. *Annals of Statistics, 36*(5), 2344–2376.

Douc, R., Moulines, E., & Olsson, J. (2014). Long-term stability of sequential Monte Carlo methods under verifiable conditions. *Annals of Applied Probability, 24*(5), 1767–1802.

Gerber, M., Chopin, N., & Whiteley, N. (2019). Negative association, ordering and convergence of resampling methods. *Annals of Statistics, 47*(4), 2236–2260.

Künsch, H. R. (2001). State space and hidden Markov models. In O. E. Barndorff-Nielsen, D. R. Cox, & C. Klüppelberg (Eds.), *Complex stochastic systems (Eindhoven, 1999). Monographs on statistics and applied probability* (Vol. 87, pp. 109–173). Boca Raton, FL: Chapman and Hall/CRC.

Künsch, H. R. (2005). Recursive Monte Carlo filters: Algorithms and theoretical analysis. *Annals of Statistics, 33*(5), 1983–2021.

Le Gland, F., & Oudjane, N. (2004). Stability and uniform approximation of nonlinear filters using the Hilbert metric and application to particle filters. *Annals of Applied Probability, 14*(1), 144–187.

Oudjane, N., & Rubenthaler, S. (2005). Stability and uniform particle approximation of nonlinear filters in case of non ergodic signals. *Stochastic Analysis and Applications, 23*(3), 421–448.

van Handel, R. (2009). Uniform time average consistency of Monte Carlo particle filters. *Stochastic Processes and Their Applications, 119*(11), 3835–3861.

Whiteley, N. (2013). Stability properties of some particle filters. *Annals of Applied Probability, 23*(6), 2500–2537.

Chapter 12
Particle Smoothing

Summary This chapter describes particle smoothing algorithms, i.e. algorithms to compute the distribution of past states X_{t-k} given data $y_{0:t}$ for a given state-space model.

An important distinction is between on-line algorithms and off-line algorithms. On-line algorithms are able to perform smoothing at every time t, at a cost that stays constant over time. They are said to be "forward-only", as they process the data recursively (in the same way as a particle filter).

Off-line algorithms compute the smoothing distribution of a fixed data-set $y_{0:T}$. They consist of running a particle filter forward in time, from $t = 0$ to $t = T$, then doing some type of backward pass, from time $t = T$ to time 0. That prevents these algorithms to be used in any realistic on-line scenario.

Off-line algorithms may be further decomposed into two classes of algorithms: backward sampling algorithms, and two-filter smoothing algorithms. These two classes correspond to the two decompositions of the smoothing distribution that we derived at the end of Chap. 5.

It will become quickly apparent that smoothing is a more difficult task than filtering. The practical implication is that naive algorithms tend to degenerate quickly (i.e., for k large), while less naive algorithms are either expensive (i.e., have a $\mathcal{O}(N^2)$ complexity) or specialised (to a certain class of functions, or to state-space models fulfilling certain conditions), or both. In particular, most smoothing algorithms are restricted to state-space models such that the process $\{X_t\}$ admits a tractable transition density.

12.1 On-Line Smoothing by Genealogy Tracking

12.1.1 Fixed-Lag Smoothing

We consider first fixed-lag smoothing, i.e. approximating recursively the law of $X_{t-k:t}$ given data $Y_{0:t} = y_{0:t}$, for some fixed integer $k > 0$ (at times $t \geq k$). The

© Springer Nature Switzerland AG 2020

N. Chopin, O. Papaspiliopoulos, *An Introduction to Sequential Monte Carlo*, Springer Series in Statistics, https://doi.org/10.1007/978-3-030-47845-2_12

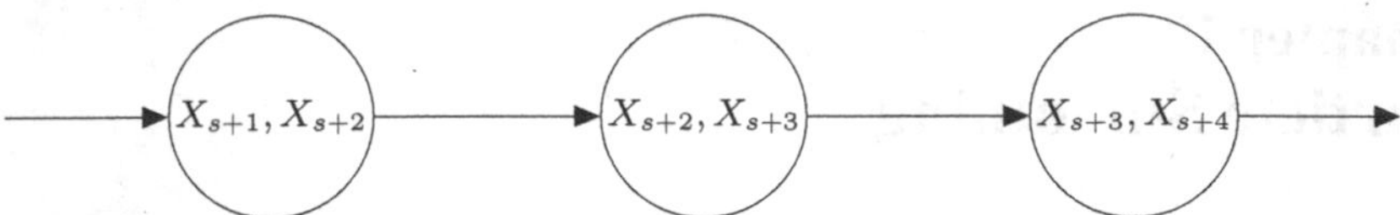

Fig. 12.1 The "lifted up" Markov process for $k = 2$, where every two consecutive variables in the original chain have been merged into overlapping blocks of two. In this specification $Z_t = (X_{t-1}, X_t)$

principle of on-line smoothing is particularly simple in this context: we "lift up" the Markov process $\{X_t\}$ in the definition of the Feynman-Kac model, that is, we replace $\{X_t\}$ by $\{Z_t\}$, with $Z_t = X_{t-k:t}$. In this way, we turn fixed-lag smoothing into a filtering problem.

Formally, the transition kernel of $Z_t = X_{t-k:t}$ is:

$$M_t(z_{t-1}, \mathrm{d}z_t') = \delta_{(x_{t-k:t-1})}(\mathrm{d}x_{t-k:t-1}')P_t(x_{t-1}, \mathrm{d}x_t')$$

where $z_{t-1} = x_{t-1-h:t-1}$, $\mathrm{d}z_t' = \mathrm{d}x_{t-k}' \ldots \mathrm{d}x_t'$. In words, to move from block $Z_{t-1} = X_{t-1-h:t-1}$ to block $Z_t = X_{t-k:t}$, copy components $X_{t-k:t-1}$, and append X_t sampled from kernel $P_t(x_{t-1}, \mathrm{d}x_t)$. This is represented graphically in Fig. 12.1.

If we associate this Markov process with the potential functions $G_t(z_{t-1}, z_t) = f_t(y_t|x_t)$, again where x_t is the last component of z_t, we obtain the Feynman-Kac model that corresponds to fixed-lag smoothing applied to the bootstrap filter. And the corresponding algorithm is simply Algorithm 10.3 applied to this particular Feynman-Kac model. Of course, fixed-lag smoothing versions of the guided and auxiliary filters may be obtained in the same way.

This approach is operationally equivalent to particle filtering on the original state process $\{X_t\}$ followed by *genealogy tracking*, that is for each surviving particle tracking its ancestry (up to time $t - k$). This approach to smoothing apparently comes as a free byproduct of particle filtering.

However, there are at least two important costs associated with this approach. First, by construction, $\{Z_t\}$ forgets its past at a slower rate than $\{X_t\}$. This is especially true if k is large, by large here we mean relative to the speed at which the filter forgets its past. Thus one might expect the particle estimates of functions of X_{t-k} to have large variance in practice for large k. This problem can to some extent be mitigated using additional computations by re-simulating the block of states $X_{t-k:t-1}$ at iteration t (see Doucet and Sénécal 2004, for more details on this approach). A second important cost is that the memory requirement is a multiple of that for filtering, since k past states need to be stored for each particle.

12.1.2 Complete Smoothing

Going further, one may perform complete smoothing by simply taking $k = t$, so to speak; i.e. take $Z_t = X_{0:t}$, with Markov kernel

$$M_t(z_{t-1}, dz'_t) = \delta_{(x_{0:t-1})}(dx'_{0:t-1})P_t(x_{t-1}, dx'_t)$$

again with the notation $z_{t-1} = x_{0:t-1}$, $z'_t = x'_{0:t}$; $\{Z_t\}$ is sometimes called the path process.

The main issue with forward smoothing is that $\{Z_t\}$ is a rather peculiar Markov process, as it does not forget its past at all. In particular, what is typically observed is that the number of distinct values taken by the first components of the simulated trajectories Z_t^n quickly falls down to one. This means that forward smoothing performs very poorly for estimating the smoothing expectation of test functions such as e.g. $\varphi(z_t) = x_0$ that depend on components far away in the past.

Example 12.1 We revisit the family of state-space models introduced in Sect. 2.4.2 which involve count data and linear-Gaussian state dynamics:

$$\begin{aligned} Y_t|X_t = x_t &\sim \mathcal{P}(e^{x_t}) \\ X_t|X_{t-1} = x_{t-1} &\sim \mathcal{N}\left(\mu + \rho(x_{t-1} - \mu), \sigma^2\right) \\ X_0 &\sim \mathcal{N}\left(\mu, \frac{\sigma^2}{1-\rho^2}\right) \end{aligned}$$

with $\mu = 0, \rho = 0.5, \sigma = 0.5$. We run the bootstrap filter until time $T = 100$, with $N = 100$ particles, for data simulated from the model. Figure 12.2 plots the corresponding genealogical tree; i.e. the index of the ancestors at time t of the N particles at time 100. One sees that all the particles originate from a common ancestor since time 40.

Remark 12.1 What Fig. 12.2 represents precisely is the N functions $t \to B_t^n$, for $n = 1, \ldots, N$, where B_t^n stands for the index of the ancestor of X_T^n at time t. These B_t^n may be defined recursively as follows: $B_T^n = n$, then for $t = T-1, \ldots, 0$, $B_t^n = A_{t+1}^{B_{t+1}^n}$.

An interesting property of these N functions is that they do not cross. This is a consequence of the fact that standard resampling algorithms return ordered indices, as we explained in Remark 9.1: $A_t^1 \leq \ldots \leq A_t^N$. Thus, even if we track the complete

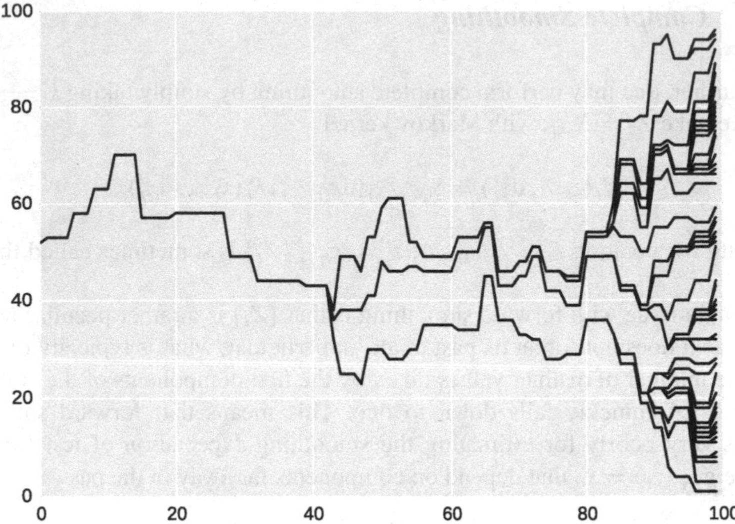

Fig. 12.2 Genealogical tree at time $t = 100$ of a bootstrap filter run with $N = 100$ particles; see Example 12.1

genealogy of a particle filter, we could safely discard any ancestor X_t^n such that n is not in the range $[B_t^1, B_t^N]$. We explain how to make use of this property in the Python corner of this chapter.

12.2 On-Line Smoothing for Additive Functions

We now present an on-line algorithm that does not suffer from the path degeneracy problem. On the other hand, it is limited to additive functions, and has cost $\mathcal{O}(N^2)$. It relies on a particular tool that will be used several times in this chapter: the particle approximation of the backward kernel of the conditional Markov process (process $\{X_t\}$ conditional on $Y_{0:T} = y_{0:T}$).

12.2.1 The Backward Kernel and Its Particle Approximation

Section 5.4.4 provided two expressions for the backward kernel of the conditioned Markov process: the generic expression (5.17), and the practically more useful one based on the assumption that the transition kernel admits a density with respect to a dominating measure, $P_{t+1}(x_t, \mathrm{d}x_{t+1}) = p_{t+1}(x_{t+1}|x_t)\mu(\mathrm{d}x_{t+1})$, in (5.18), which is

$$\overleftarrow{P}_{t|t}(x_{t+1}, \mathrm{d}x_t) = \frac{p_{t+1}(x_{t+1}|x_t)}{\int_{\mathcal{X}} p_{t+1}(x_{t+1}|x_t)\mathbb{P}_t(X_t \in \mathrm{d}x_t|Y_{0:t} = y_{0:t})}\mathbb{P}_t(X_t \in \mathrm{d}x_t|Y_{0:t} = y_{0:t}) . \quad (12.1)$$

In the rest of this chapter we will work with the latter expression.

Assume we have already run a particle filter to approximate the filtering distribution:

$$\sum_{n=1}^{N} W_t^n \delta_{X_t^n}(\mathrm{d}x_t) \approx \mathbb{P}_t(X_t \in \mathrm{d}x_t|Y_{0:t} = y_{0:t}) . \quad (12.2)$$

Notation/Terminology Recall that in Chap. 10, we adapted the notations when necessary so that (12.2) always holds for the considered state-space model; that is, whether the considered algorithm filter is a bootstrap, guided, or auxiliary particle filter. Accordingly, the derivations in this chapter are valid for any particle filter that generates particles X_t^n and normalised weights W_t^n such that (12.2) holds.

If we plug (12.2) in (12.1), we obtain the following approximation for the backward kernel:

$$\overleftarrow{P}^N_{t|t}(x_{t+1}, \mathrm{d}x_t) = \frac{1}{\sum_{m=1}^N W_t^m p_{t+1}(x_{t+1}|X_t^m)} \sum_{n-1}^{N} W_t^n p_{t+1}(x_{t+1}|X_t^n)\delta_{X_t^n}(\mathrm{d}x_t) . \quad (12.3)$$

This is a probability distribution for each x_{t+1} with support on samples generated by the particle filter at time t, $X_t^{1:N}$, but modified weights that depend on x_{t+1}. Any expectation with respect to this backward kernel may be approximated by

$$\frac{\sum_{n=1}^N W_t^n p_{t+1}(x_{t+1}|X_t^n)\varphi(X_t^n)}{\sum_{n=1}^N W_t^n p_{t+1}(x_{t+1}|X_t^n)} \approx \mathbb{E}\left[\varphi(X_t) \mid X_{t+1} = x_{t+1}, Y_{0:t} = y_{0:t}\right] . \quad (12.4)$$

12.2.2 *Principle of On-Line Smoothing for Additive Functions*

We consider in this section the specific problem of computing smoothing expectations of additive functions, i.e. functions of the complete trajectory of the form:

$$\varphi_t(x_{0:t}) = \psi_0(x_0) + \sum_{s=1}^{t} \psi_s(x_{s-1}, x_s) . \tag{12.5}$$

Such functions play an important role in parameter estimation, since algorithms for computing the maximum likelihood estimate of θ require expectations of such functions, see Exercises 12.5, 12.6 and Chap. 14 for more details.

An important theoretical result regarding such functions is that, if we use forward smoothing by genealogy tracking (as described in Sect. 12.1), we obtain estimates whose variance grows quadratically with time; see Poyiadjis et al. (2011) for a proof. This means we need to take $N = \mathcal{O}(T^2)$ to obtain reasonable performance until time T. However, in this framework it is possible to obtain smoothing estimates whose variance grows only linearly with time. The price to pay is that such estimates cost $\mathcal{O}(N^2)$ to compute. This $\mathcal{O}(N^2)$ approach is based on the following recursion.

Proposition 12.1 *For $t \geq 0$ and an additive function of the form* (12.5)*, let*

$$\Phi_t(x_t) := \mathbb{E}\left[\varphi_t(X_{0:t}) \mid X_t = x_t, Y_{0:t} = y_{0:t}\right],$$

then

$$\mathbb{E}\left[\varphi_t(X_{0:t}) \mid Y_{0:t} = y_{0:t}\right] = \mathbb{E}\left[\Phi_t(X_t) \mid Y_{0:t} = y_{0:t}\right]$$

and the Φ_t's may be computed recursively as: $\Phi_0(x_0) = \psi_0(x_0)$, and for $t > 0$

$$\Phi_t(x_t) = \mathbb{E}\left[\Phi_{t-1}(X_{t-1}) + \psi_t(X_{t-1}, x_t) \mid X_t = x_t, Y_{0:t-1} = y_{0:t-1}\right] . \tag{12.6}$$

The proof is left as Exercise 12.1; in fact, this result is a special case of a more general property of expectations of additive functions for Markov processes, see Exercise 12.2.

By the Markov property, and for any test function φ and any $t \leq T$, we have that

$$\mathbb{E}\left[\varphi(X_{t-1}) \mid X_t = x_t, Y_{0:t-1} = y_{0:t-1}\right] = \mathbb{E}\left[\varphi(X_{t-1}) \mid X_t = x_t, Y_{0:T} = y_{0:T}\right]$$

therefore the expectation in (12.6) is with respect to the backward kernel of the conditional Markov process. If we replace this (typically intractable) backward kernel by its particle approximation derived in the previous section, we obtain Algorithm 12.1. Note that this algorithm requires that the considered model admits a tractable transition density.

Algorithm 12.1: $\mathcal{O}(N^2)$ on-line smoothing for additive functions

---

At iteration $t \in 0 : T$ of a particle filtering algorithm:

for $n = 1$ **to** N **do**

  if $t = 0$ **then**

    $\Phi_0^n(X_0^n) \leftarrow \psi_0(X_0^n)$

  else

    $\Phi_t^n(X_t^n) \leftarrow \frac{\sum_{m=1}^N W_{t-1}^m p_t(X_t^n|X_{t-1}^m)\{\Phi_{t-1}^N(X_{t-1}^m)+\psi_t(X_{t-1}^m,X_t^n)\}}{\sum_{m=1}^N W_{t-1}^m p_t(X_t^n|X_{t-1}^m)}$

return $\sum_{n=1}^N W_t^n \Phi_t^N(X_t^n)$ (as an approximation of $\mathbb{E}[\varphi_t(X_{0:t})|Y_{0:t} = y_{0:t}]$)

This algorithm has complexity $\mathcal{O}(N^2)$, since it computes for each of the N particles a sum over N terms. Compared to naive forward smoothing by genealogy tracking, we trade CPU time for reducing memory (since only standard PF output is necessary) and stability (of the variance of smoothing estimates). This trade-off is often beneficial when T is large.

In its current description, the validity of Algorithm 12.1 is not completely straightforward, since it seems to 'pile up' approximations recursively: to approximate Φ_t, we compute a Monte Carlo approximation of the expectation of function Φ_{t-1}, which was itself Monte Carlo approximated, as an expectation involving Φ_{t-2}, and so on. However, we will show in Sect. 12.4.2 that, despite being forward-only, this algorithm is strongly related to backward smoothing algorithms, and this connection will make it possible to give a better understanding of the output of this algorithm.

Example 12.2 We return to Example 12.1, and consider the additive function

$$\varphi_t(x_{0:t}) = -\frac{T+1}{2\sigma^2} + \frac{1-\rho^2}{2\sigma^4}(x_0-\mu)^2 + \frac{1}{2\sigma^4}\sum_{s=1}^t \left\{x_s - \mu - \rho(x_{s-1}-\mu)\right\}^2; \tag{12.7}$$

the expectation of which is the score of the model with respect to parameter σ^2; see Exercise 12.7. Figure 12.3 compares the inter-quartile range of the

(continued)

Example 12.2 (continued)
estimates of $\mathbb{E}[\varphi_t(X_{0:t})|Y_{0:t} = y_{0:t}]$ as a function of time, as obtained from 25 runs of $\mathcal{O}(N)$ forward smoothing by genealogy tracking (Sect. 12.1.2), and the $\mathcal{O}(N^2)$ forward smoothing algorithm for additive functionals (Algorithm 12.1). The number of particles has been chosen so that the CPU time of both algorithms match. These results match the theory quite closely: the variability of smoothing estimates increases over time in both cases, linearly for the $\mathcal{O}(N^2)$ algorithm, quadratically for the $\mathcal{O}(N)$ algorithm. (Note that the log-scale has been used on both axes.) On the other hand, in this particular example, one needs to wait for quite a long time before the $\mathcal{O}(N^2)$ approach becomes competitive.

This example indicates that, although forward smoothing is expected to perform poorly in general, it may actually be a reasonable approach for approximating the smoothing expectation of additive functions, at least until a certain time t, where the $\mathcal{O}(N^2)$ starts to deliver a better CPU vs variance trade-off.

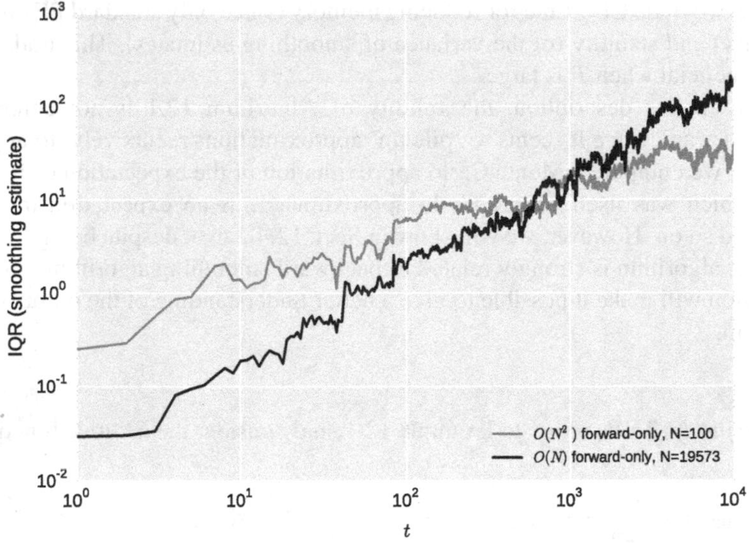

Fig. 12.3 On-line smoothing: inter-quartile range of smoothing estimates (based on 25 runs) of φ_t as a function of time t (log-scale on both axes)

12.3 Backward Smoothing: Complete Trajectories

This section and the next describes the first class of off-line smoothing algorithms, which we refer to as backward smoothing algorithms, and which are based on the backward decomposition of the smoothing distribution.

12.3.1 Principle: The Smoothing Skeleton

We use 'backward algorithms' as an umbrella term for the smoothing algorithms based on the backward decomposition of the smoothing distribution, which was obtained in Sect. 5.4.2:

$$\mathbb{P}_T(\mathrm{d}x_{0:T}|Y_{0:T}=y_{0:T}) = \mathbb{P}_T(\mathrm{d}x_T|Y_{0:T}=y_{0:T}) \prod_{t=0}^{T-1} \overleftarrow{P}_{t|t}(x_{t+1}, \mathrm{d}x_t) . \tag{12.8}$$

The smoothing skeleton is the discrete distribution obtained by replacing each factor by its particle approximation (obtained from the output of a particle filter). In particular each backward kernel is replaced by the particle approximation derived in Sect. 12.2.1:

$$\mathbb{P}_T^N(\mathrm{d}x_{0:T}|Y_{0:T}=y_{0:T}) := \left\{\sum_{n=1}^N W_T^n \delta_{X_T^n}(\mathrm{d}x_T)\right\} \prod_{t=0}^{T-1} \overleftarrow{P}_{t|t}^N(x_{t+1}, \mathrm{d}x_t) . \tag{12.9}$$

To get a better understanding of the skeleton, we develop its expression for $T = 1$:

$$\mathbb{P}_1^N(\mathrm{d}x_{0:1}|Y_{0:1}=y_{0:1}) := \sum_{m=1}^N \sum_{n=1}^N \left\{\frac{W_0^m W_1^n p_1(X_1^n|X_0^m)}{\sum_{l=1}^N W_0^l p_1(X_1^n|X_0^l)}\right\} \delta_{X_1^n}(\mathrm{d}x_1)\delta_{X_0^m}(\mathrm{d}x_0) . \tag{12.10}$$

We see that the support of the skeleton is the set of the N^2 paths that can be formed by pairing any particle from time 0 to any particle from time 1; the weight attached to each path is given inside the curly brackets. More generally, the support of the skeleton consists of the N^{T+1} trajectories that pass through a single particle X_t^n at each time $t = 0, \ldots, T$.

The smoothing skeleton is based on a backward decomposition; yet, surprisingly, it may also be obtained through the following forward recursion.

Proposition 12.2 *For any $t \geq 1$, one has:*

$$\mathbb{P}_t^N(\mathrm{d}x_{0:t}|Y_{0:t} = y_{0:t}) =$$
$$\mathbb{P}_t^N(\mathrm{d}x_{0:t-1}|Y_{0:t-1} = y_{0:t-1}) \otimes \sum_{n=1}^{N} \frac{W_t^n p_t(X_t^n|x_{t-1})}{\sum_l W_{t-1}^l p_t(X_t^n|x_{t-1})} \delta_{X_t^n}(\mathrm{d}x_t) \tag{12.11}$$

where $\otimes$ denotes the outer product between measures. (The right hand side is not a probability measure.)

Proof Expand the last backward kernel in (12.9), as done in (12.10), and change the order of summation. □

Practically, each trajectory in the smoothing skeleton built up to time $t-1$, is extended to N new trajectories by pairing with each X_t^n for $n = 1 : N$. The trajectory that finishes at value X_{t-1}^m at time $t-1$, and extended to finish at X_t^n at time t has its weight multiplied by

$$\frac{W_t^n p_t(X_t^n|X_{t-1}^m)}{\sum_l W_{t-1}^l p_t(X_t^n|X_{t-1}^l)} .$$

We shall use this forward recursion later, in Sect. 12.4.2.

Given the very large size (N^{T+1}) of its support, it is prohibitively expensive to compute exactly expectations with respect to the skeleton. However, we may approximate such expectations using Monte Carlo. The next two sections explain how to do that.

12.3.2 Sampling from the Smoothing Skeleton: Forward Filtering Backward Sampling (FFBS)

The algorithm known as FFBS (forward filtering backward smoothing) in the literature amounts to generating i.i.d. trajectories from the smoothing skeleton, exploiting the backward decomposition (12.9) and the fact that the particle approximations $\mathbb{P}_T^N(\mathrm{d}x_T|Y_{0:T} = y_{0:T})$ and $\overleftarrow{P}_{t|t}^N(x_{t+1}, \mathrm{d}x_t)$ define discrete probability distributions. Since the support of these distributions consists of the particle positions, simulation comes down to simulating genealogies by sampling at each time integers from 1 to N. This is described in Algorithm 12.2.

Algorithm 12.2: FFBS

Input: Output of a particle filter: particles $X_0^{1:N}, \ldots, X_T^{1:N}$ and weights $W_0^{1:N}, \ldots, W_T^{1:N}$.

Output: sequence $B_{0:T}$ of indices in $1:N$; the simulated trajectory is then $(X_0^{B_0}, \ldots, X_T^{B_T})$.

$B_T \sim \mathcal{M}(W_T^{1:N})$

for $t = (T-1)$ **to** 0 **do**

  $\widehat{w}_t^n \leftarrow W_t^n p_{t+1}(X_{t+1}^{B_{t+1}}|X_t^n)$ and $\widehat{W}_t^n = \widehat{w}_t^n / \sum_{m=1}^N \widehat{w}_t^m$ for $n = 1, \ldots, N$

  $B_t \sim \mathcal{M}(\widehat{W}_t^{1:N})$

The cost of generating M trajectories with Algorithm 12.2 is $\mathcal{O}(TMN)$, since one must recompute the N modified weights $\widehat{W}_t^n$ at each time t and for each generated trajectory. From these M trajectories, one may obtain the following estimates

$$\frac{1}{M}\sum_{m=1}^{M} \varphi(X_0^{B_0^m}, \ldots, X_T^{B_T^m}) \approx \mathbb{P}_T^N(\varphi|Y_{0:T} = y_{0:T}) \approx \mathbb{P}_T(\varphi|Y_{0:T} = y_{0:T})$$

which suffer from two levels of error: the Monte Carlo error due to i.i.d. sampling from the skeleton, which vanishes as $M \to +\infty$, and the approximation error of the skeleton itself, which vanishes as $N \to +\infty$. Common practice is to take $M = N$, leading to a $\mathcal{O}(TN^2)$ complexity.

12.3.3 Using Rejection to Speed Up FFBS

The quadratic cost for generating N trajectories in FFBS is the result of repeating an $\mathcal{O}(N)$ operation for each of the N trajectories. Specifically, at each time t we need to simulate from the distribution proportional to

$$\propto \sum_{n=1}^{N} W_t^n p_{t+1}(X_{t+1}^{B_{t+1}}|X_t^n)\delta_{X_t^n}(\mathrm{d}x_t)\,. \tag{12.12}$$

If we make the extra assumption that $p_{t+1}(x_{t+1}|x_t) < C_t(x_t)$ for some known function C_t, then

$$\sum_{n=1}^{N} W_t^n p_{t+1}(X_{t+1}^{B_{t+1}}|X_t^n)\delta_{X_t^n}(\mathrm{d}x_t) \leq \sum_{n=1}^{N} W_t^n C_t(X_t^n)\delta_{X_t^n}(\mathrm{d}x_t).$$

This inequality suggests to sample from (12.12) using rejection sampling, as defined in Algorithm 8.1, based on fixed proposal distribution $\mathcal{M}(\widehat{W}_t^{1:N})$, where $\widehat{W}_t^n \propto W_t^n C_t(X_t^n)$.

At this point, it is worth recalling briefly the discussion in Chap. 9 regarding multinomial resampling. The standard approach (Algorithm 9.3) to generate N samples from a multinomial distribution with support of size M has complexity $\mathcal{O}(M+N)$. In rejection sampling however, we do not know in advance how many draws must be generated. Fortunately, we have seen in the Python corner of Chap. 9 how to address this problem: we may use the alias algorithm, which requires a $\mathcal{O}(N)$ initialisation, and then produces draws on the fly, at cost $\mathcal{O}(1)$ per draw.

Algorithm 12.3, which we call FFBS-reject, implements this idea. It is tempting to state that the complexity of this algorithm is also $\mathcal{O}(N)$ per time step. This statement is valid under the strong condition that $p_t(x_t|x_{t-1})$ is lower bounded; see Exercise 12.3. However, in general the picture is less clear. The cost of generating N accepted indices at each time step is random (as always when using a rejection sampler), and its expectation depends on N in a non-trivial way.

In practice, we strongly recommend to perform preliminary experiments (for small values of T and N) to assess the rejection rates at each step. If these rates are not too low, then FFBS-reject is likely to be much faster than standard FFBS.

Algorithm 12.3: FFBS-reject

Input: Output of particle filter: particles $X_t^{1:N}$ and normalised weights $W_t^{1:N}$ at times $0,\ldots,T$; functions C_t such that $p_{t+1}(x_{t+1}|x_t) \le C_t(x_t)$; properly initialised samplers for multinomial distributions $\mathcal{M}(W_T^{1:N})$, and for $t<T$, $\mathcal{M}(\widehat{W}_t^{1:N})$, where $\widehat{W}_t^n \propto W_t^n C_t(X_t^n)$, which may generate a single value at $\mathcal{O}_P(1)$ cost (see Python corner of Chap. 9).

Output: sequence $B_{0:T}$ of indices in $1:N$; the simulated trajectory is then $(X_0^{B_0},\ldots,X_T^{B_T})$.

$B_T \sim \mathcal{M}(W_T^{1:N})$ ▷ `Using` $\mathcal{O}_P(1)$ `sampler`

for $t=(T-1)$ **to** 0 **do**

  repeat

    $B_t \sim \mathcal{M}(\widehat{W}_t^{1:N})$ ▷ `Using` $\mathcal{O}_P(1)$ `sampler`

    $U \sim \mathcal{U}([0,1])$

  until $U \le p_{t+1}(X_{t+1}^{B_{t+1}}|X_t^{B_t})/C_t(X_t^{B_t})$

12.4 Backward Smoothing: Marginal Distributions

We have seen that, in full generality, exact computation of expectations with respect to the smoothing skeleton costs $\mathcal{O}(N^{T+1})$, which is impractical. However, it is possible to perform such exact computations at cost $\mathcal{O}(N^2)$, if we restrict to marginal expectations of a single state, X_t, or a pair of successive states, (X_t, X_{t+1}). In this way, we obtain smoothing estimates that suffer from only one level of approximation (the approximation error of the skeleton), rather than two levels like FFBS. The corresponding derivations are most similar to those of the exact filter and smoother for finite state-space models (Chap. 6).

12.4.1 Computing Marginal Distributions of the Skeleton

The algorithm to compute marginal distributions from the smoothing skeleton is based on a direct application of its backward decomposition:

$$\mathbb{P}_T^N(\mathrm{d}x_{0:T}|Y_{0:T} = y_{0:T}) := \left\{\sum_{n=1}^{N} W_T^n \delta_{X_T^n}(\mathrm{d}x_T)\right\} \prod_{t=0}^{T-1} \overleftarrow{P}_{t|t}^N(x_{t+1}, \mathrm{d}x_t)\,.$$

From the equation above, one clearly sees that the marginal distribution of component X_T is

$$\sum_{n=1}^{N} W_T^n \delta_{X_T^n}(\mathrm{d}x_T)\,.$$

By replacing $\overleftarrow{P}_{T-1|T-1}^N(x_T, \mathrm{d}x_{T-1})$ by its expression, (12.9), we also obtain the marginal distribution of (X_{T-1}, X_T), a discrete distribution whose support is of size N^2:

$$\sum_{m,n=1}^{N} \frac{W_T^n W_{T-1}^m p_T(X_T^n|X_{T-1}^m)}{\sum_{l=1}^{N} W_{T-1}^l p_T(X_T^n|X_{T-1}^l)} \delta_{X_{T-1}^m}(\mathrm{d}x_{T-1}) \delta_{X_T^n}(\mathrm{d}x_T)\,.$$

Then, by marginalising out component X_T, one obtains the marginal distribution of X_{T-1} as a mixture of N components. This is done by simply summing over index n, for each m.

By repeating the same calculations at time $T-2, T-3, \ldots$, one obtains all the subsequent marginal distributions; see Algorithm 12.4.

Algorithm 12.4: Backward smoothing for pairwise marginals

Input: Output of a particle filter: particles $X_0^{1:N}, \ldots, X_T^{1:N}$ and weights $W_0^{1:N}, \ldots, W_T^{1:N}$.

Output: Smoothing weights $W_{t|T}^n$ and pairwise smoothing weights $W_{t-1:t|T}^{m,n}$.

for $t = T$ **to** 0 **do**
  for $m = 1$ **to** N **do**
    if $t = T$ **then**
      $W_{t|T}^m \leftarrow W_T^m$
    else
      $W_{t|T}^m \leftarrow \sum_{n=1}^N W_{t:t+1|T}^{m,n}$
  if $t > 0$ **then**
    for $n = 1$ **to** N **do**
      $S_t^n \leftarrow \sum_{l=1}^N W_{t-1}^l p_t(X_t^n|X_{t-1}^l)$
      for $m = 1$ **to** N **do**
        $W_{t-1:t|T}^{m,n} \leftarrow W_{t|T}^n W_{t-1}^m p_t(X_t^n|X_{t-1}^m)/S_t^n$

The complexity of Algorithm 12.4 is $\mathcal{O}(N^2)$ per time step; notice in particular that we wrote the algorithm to ensure that the denominator in the expression of $W_{t-1:t|T}^{m,n}$ is computed only once for each n. The output of this algorithm may be used as follows.

Output of Marginal Backward Smoothing

Pairwise marginal smoothing approximations, for $t = 1, \ldots, T$

$$\sum_{m,n=1}^{N} W_{t-1:t|T}^{m,n} \delta_{(X_{t-1}^m, X_t^n)}(\mathrm{d}x_{t-1:t}) \approx \mathbb{P}_T(\mathrm{d}x_{t-1:t}|Y_{0:T} = y_{0:T})$$

(continued)

and marginal approximations, for $t = 0, \ldots, T$,

$$\sum_{n=1}^{N} W_{t|T}^n \delta_{X_t^n}(\mathrm{d}x_t) \approx \mathbb{P}_T(\mathrm{d}x_t | Y_{0:T} = y_{0:T})$$

It is more common to describe Algorithm 12.4 as a way to approximate marginal distributions for single components X_t; but as we can see marginals for pairs (X_{t-1}, X_t) are obtained at no extra cost.

The general usefulness of Algorithm 12.4 is a bit debatable, even if one is interested mainly in smoothing marginals. It does remove an extra level of Monte Carlo error (compared to FFBS algorithms), but its complexity is $\mathcal{O}(N^2)$. Hence, FFBS-reject, when implementable, may provide a better error vs CPU trade-off even for approximating smoothing marginal expectations.

On the other hand, for models when FFBS-reject is not feasible, it does provide a faster and more accurate approximation of smoothing marginals than the standard, $\mathcal{O}(N^2)$ version of FFBS. In addition, it allows us to make a nice connection with forward-only smoothing for additive functions, as we discuss now.

12.4.2 *On the Connection Between the Smoothing Skeleton and Forward-Only Smoothing for Additive Functions*

We may use the marginal smoothing algorithm of the previous section to compute expectations of additive functions $\varphi_T(x_{0:T}) = \psi_0(x_0) + \sum_{t=1}^{T} \psi_t(x_{t-1}, x_t)$, since that expectation is a sum of pairwise marginal smoothing expectations. Reciprocally, one may use the forward-only on-line algorithm of Sect. 12.2 for additive functions to obtain the marginal smoothing expectation of $\psi_t(X_{t-1}, X_t)$, by simply taking $\psi_s = 0$ for all $s \neq t$. In both approaches, we observe that estimates are obtained as weighted sums over the quantities $\psi_t(X_{t-1}^m, X_t^n)$. In fact, the connection between both approaches runs deeper: they compute the same things, i.e. expectations with respect to the smoothing skeleton.

Proposition 12.3 *The output of Algorithm 12.1 at iteration t, for function $\varphi_t(x_{0:t}) = \psi_0(x_0) + \sum_{s=1}^{t} \psi_s(x_{s-1}, x_s)$, equals the expectation of the same function φ_t with respect to the smoothing skeleton at time t, $\mathbb{P}_t^N(\mathrm{d}x_{0:t} | Y_{0:t} = y_{0:t})$, as defined in* (12.9).

Proof For notational convenience, we take $t = T$, and assume that $\varphi_T(x_{0:T}) = \psi_0(x_0)$, i.e. $\psi_s = 0$ for $s > 0$; generalising the result to $\varphi_T(x_{0:T}) = \psi_t(x_t)$ follow the same lines, and generalising to sum of such functions follows from the linearity of expectations.

By the law of total expectation, one has:

$$\begin{aligned}
\mathbb{P}_T^N(\psi_0(X_0)|Y_{0:T} = y_{0:T}) &= \mathbb{P}_T^N\left(\Phi_1^N(X_1)\,\middle|\, Y_{0:T} = y_{0:T}\right) \\
&= \ldots \\
&= \mathbb{P}_T^N\left(\Phi_T^N(X_T)\,\middle|\, Y_{0:T} = y_{0:T}\right)
\end{aligned}$$

where $\Phi_t^N(x_t) := \mathbb{P}_T^N(\psi_0(X_0)|X_t = x_t, Y_{0:T} = y_{0:T})$; i.e. the expectation of $\psi_0(X_0)$ conditional on $X_t = x_t$, relative to the distribution defined by the skeleton. This function Φ_t^N is simply the particle version of function Φ_t (again, for the special case where $\psi_s = 0$ for $s > 0$), which was defined in Proposition 12.1, and admits a similar recursion, by the tower rule:

$$\Phi_t^N(x_t) = \mathbb{P}_T^N(\Phi_{t-1}^N(X_{t-1})|X_t = x_t, Y_{0:T} = y_{0:T}),$$

which we recognise, given the structure of the skeleton, as an expectation with respect to (the particle approximation of) the backward kernel:

$$\Phi_t^N(x_t) = \sum_{n=1}^N \frac{W_{t-1}^n p_t(x_t|X_{t-1}^n)}{\sum_{p=1}^N W_{t-1}^p p_t(x_t|X_{t-1}^p)} \Phi_{t-1}^N(X_{t-1}^n)\,.$$

which is precisely how the $\Phi_t^N(X_t^n)$ are computed recursively in Algorithm 12.1. □

Thus the proper way to understand and analyse Algorithm 12.1 (forward-only smoothing for additive functions) is as a method for computing exactly and recursively (forward in time) expectations of additive functions with respect to the smoothing skeleton. In particular, it only involves the approximation error of the skeleton itself (as an empirical approximation of the true smoothing distribution).

12.4.3 Discussion of Backward Algorithms

We see now that all the smoothing algorithms discussed in this section and the previous one are simple operations (sampling, or marginalisation) effectuated on the smoothing skeleton. Thus, to study the convergence (as $N \to +\infty$) of these algorithms, one must study first the convergence of the skeleton to the true smoothing distribution. See the bibliographical notes at the end of the chapter for references to papers following this approach.

12.5 Two-Filter Marginal Smoothers

We now present the second class of off-line smoothing algorithms, namely the two-filter smoother and its variants. Note that these algorithms may be used only to approximate marginal smoothing distributions (and not the joint smoothing distribution), as we now explain.

12.5.1 General Idea

Two-filter smoothing is commonly described as a particle approximation of the following identity (established in Sect. 5.4.2):

$$\mathbb{P}(X_t \in \mathrm{d}x_t | Y_{0:T} = y_{0:T}) = \frac{p_T(y_{t+1:T}|x_t)}{p_T(y_{t+1:T}|y_{0:t})} \mathbb{P}(X_t \in \mathrm{d}x_t | Y_{0:t} = y_{0:t}) . \tag{12.13}$$

However, we find it more convenient and more general to work with pairwise smoothing marginal distributions; that is, to work with identity

$$\mathbb{P}(X_{t:t+1} \in \mathrm{d}x_{t:t+1} | Y_{0:T} = y_{0:T}) = \frac{p_T(y_{t+1:T}|x_{t+1})}{p_T(y_{t+1:T}|y_{0:t})} \mathbb{P}(X_t \in \mathrm{d}x_t | Y_{0:t} = y_{0:t}) P_{t+1}(x_t, \mathrm{d}x_{t+1}) , \tag{12.14}$$

which may be derived essentially in the same way.

Now let's assume that $P_{t+1}(x_t, \mathrm{d}x_{t+1})$ admits probability density $p_{t+1}(x_{t+1}|x_t)$, and let's introduce a probability distribution $\gamma_{t+1}(\mathrm{d}x_{t+1})$, with density $\gamma_{t+1}(x_{t+1})$. We may then rewrite the expression above as:

$$\mathbb{P}(X_{t:t+1} \in \mathrm{d}x_{t:t+1} | Y_{0:T} = y_{0:T}) \propto \frac{p_{t+1}(x_{t+1}|x_t)}{\gamma_{t+1}(x_{t+1})} \mathbb{P}(X_t \in \mathrm{d}x_t | Y_{0:t} = y_{0:t}) \gamma_{t+1|T}(\mathrm{d}x_{t+1}) \tag{12.15}$$

where

$$\gamma_{t+1|T}(\mathrm{d}x_{t+1}) \propto \gamma_{t+1}(\mathrm{d}x_{t+1}) p_T(y_{t+1:T}|x_{t+1}) .$$

Two-filter smoothing consists of running two particle algorithms: one standard particle filter, which approximates the filtering distribution $\mathbb{P}(X_t \in \mathrm{d}x_t | Y_{0:t} = y_{0:t})$; and an "information filter", which approximates recursively the distribution $\gamma_{t|T}(\mathrm{d}x_t)$. The results of the two algorithms are then combined using the above identity, in order to approximate the pairwise marginal distribution of (X_t, X_{t+1}) given $Y_{0:T} = y_{0:T}$.

The two next sections discuss how to implement the information filter, and how to choose the distributions $\gamma_t(\mathrm{d}x_t)$. The subsequent sections present the actual algorithms.

12.5.2 The Information Filter: The Principle

In Chap. 5, see (5.15), we derived the following recursion:

$$p_T(y_{t+1:T}|x_t) = \int_{\mathcal{X}} f_{t+1}(y_{t+1}|x_{t+1}) p_T(y_{t+2:T}|x_{t+1}) P_{t+1}(x_t, \mathrm{d}x_{t+1}) .$$

This looks suspiciously similar to the forward recursion of Feynman-Kac models. This implies that we may be able to approximate this function using a particle algorithm. There are two caveats however: first, this recursion works backwards; second, the quantity above is not necessarily a probability density with respect to x_t: it may integrate to infinity.

For the first point, and mostly to avoid confusing notations, we consider this time a Feynman-Kac backwards model, with a Markov chain that starts at time T, $X_T \sim \overleftarrow{\mathbb{M}}_T(\mathrm{d}x_T)$, then evolves according to $\overleftarrow{M}_t(x_{t+1}, \mathrm{d}x_t)$ until time 0, and a collection of potential functions $\overleftarrow{G}_T(x_T)$, $\overleftarrow{G}_t(x_{t+1}, x_t)$ for $t = T-1, \ldots, 0$. The corresponding $\mathbb{Q}$-distributions are defined as follows, for $t = T$ to $t = 0$:

$$\mathbb{Q}_t(\mathrm{d}x_{t:T}) \propto \overleftarrow{\mathbb{M}}_T(\mathrm{d}x_T) \overleftarrow{G}_T(x_T) \prod_{s=t}^{T-1} \overleftarrow{M}_s(x_{s+1}, \mathrm{d}x_s) \overleftarrow{G}_s(x_{s+1}, x_s) \tag{12.16}$$

with a normalising constant chosen so that $\mathbb{Q}_t$ integrates to one, and empty products equal one.

For the second point, we introduce a sequence of probability distributions $\gamma_t(\mathrm{d}x_t)$, and we seek to construct a Feynman-Kac model such that $\mathbb{Q}_t(\mathrm{d}x_t)$ matches

$$\begin{aligned} \gamma_{t|T}(\mathrm{d}x_t) &\propto \gamma_t(\mathrm{d}x_t) p_T(y_{t:T}|x_t) \\ &\propto \gamma_t(\mathrm{d}x_t) f_t(y_t|x_t) p_T(y_{t+1:T}|x_t). \end{aligned}$$

The following lemma is easy to establish from first principles.

Lemma 12.4 *Taking*

$$\overleftarrow{G}_T(x_T) = \frac{f_T(y_T|x_T)\gamma_T(\mathrm{d}x_T)}{\overleftarrow{\mathbb{M}}_T(\mathrm{d}x_T)} \tag{12.17}$$

$$\overleftarrow{G}_t(x_{t+1}, x_t) = \frac{f_t(y_t|x_t)\gamma_t(\mathrm{d}x_t)P_{t+1}(x_t, \mathrm{d}x_{t+1})}{\gamma_{t+1}(\mathrm{d}x_{t+1})\overleftarrow{M}_t(x_{t+1}, \mathrm{d}x_t)} \tag{12.18}$$

for $t = 0, \ldots, T-1$, *and assuming that the Radon-Nikodym derivatives above are well defined, the corresponding Feynman-Kac model, defined as* (12.16), *is such that*

$$\begin{aligned}\mathbb{Q}_t(\mathrm{d}x_t) &\propto \gamma_t(\mathrm{d}x_t)p_T(y_{t:T}|x_t)\\ &\propto \gamma_t(\mathrm{d}x_t)f_t(y_t|x_t)p_T(y_{t+1:T}|x_t) \qquad (12.19)\\ &\propto \int_{\mathcal{X}} \mathbb{Q}_{t+1}(\mathrm{d}x_{t+1})\overleftarrow{M}_t(x_{t+1}, \mathrm{d}x_t)\overleftarrow{G}_t(x_{t+1}, x_t). \qquad (12.20)\end{aligned}$$

The practical implication of the above lemma is that we can implement a particle filter that sequentially approximates $\gamma_{t|T}(\mathrm{d}x_t)$, and as a by-product the quantity $p_T(y_{t+1:T}|x_t)$. The corresponding algorithm (known as the information filter) is summarised as Algorithm 12.5.

Input of the Information Filter

- A sequence of distributions $\gamma_t(\mathrm{d}x_t)$, for $t = 0 : T$.
- A distribution $\overleftarrow{\mathbb{M}}_T(\mathrm{d}x_T)$ (a) which may be simulated from; and (b) such that the Radon-Nikodym derivative $\gamma_T(\mathrm{d}x_T)/\overleftarrow{\mathbb{M}}_T(\mathrm{d}x_T)$ may be computed pointwise.
- Markov kernels $\overleftarrow{M}_t(x_{t+1}, \mathrm{d}x_t)$, for $t = 0, \ldots, (T-1)$ (a) which may be simulated from, and (b) such that the Radon-Nikodym derivatives

$$\frac{f_t(y_t|x_t)\gamma_t(\mathrm{d}x_t)P_{t+1}(x_t, \mathrm{d}x_{t+1})}{\gamma_{t+1}(\mathrm{d}x_{t+1})\overleftarrow{M}_t(x_{t+1}, \mathrm{d}x_t)}$$

  are well defined and may be computed point-wise.
- the number of particles M.

Algorithm 12.5: Information filter

`Operations involving index` m `must be performed for` $m = 1, \ldots, M$

$\overleftarrow{X}_T^m \sim \overleftarrow{\mathbb{M}}_T(\mathrm{d}x_T)$
$w_T^m \leftarrow \overleftarrow{G}_T(x_T^m)$
$\overleftarrow{W}_T^m \leftarrow w_T^m / \sum_{n=1}^M w_T^n$
for $t = T-1$ **to** 0 **do**
  $A_t^{1:M} \sim \texttt{resampling}(\overleftarrow{W}_{t+1}^{1:M})$
  $\overleftarrow{X}_t^m \sim \overleftarrow{M}_t(\overleftarrow{X}_{t+1}^{A_t^m}, \mathrm{d}x_t)$
  $w_t^m \leftarrow \overleftarrow{G}_t(\overleftarrow{X}_{t+1}^{A_t^m}, \overleftarrow{X}_t^m)$
  $\overleftarrow{W}_t^m \leftarrow w_t^m / \sum_{n=1}^M w_t^n$

Of course, Algorithm 12.5 is simply Algorithm 10.1 written backwards, and with arrows on certain symbols. It may be generalised in the same way, i.e. using adaptive resampling (see Sect. 10.2).

12.5.3 The Information Filter: The Choice of γ_t and $\overleftarrow{M}_t$

How should we choose the γ_t's and the $\overleftarrow{M}_t$'s? To fix ideas, consider the case where, relative to the law of process $\{X_t\}$, we are able to compute the marginal distribution $\mathbb{M}_t(\mathrm{d}x_t)$ of each X_t, and to simulate from the backward kernel $\overleftarrow{P}_{t+1}(x_{t+1}, \mathrm{d}x_t)$ (the law of X_t given X_{t+1}).

Then one may take $\gamma_t(\mathrm{d}x_t) = \mathbb{M}_t(\mathrm{d}x_t)$, $\overleftarrow{M}_t(x_{t+1}, \mathrm{d}x_t) = \overleftarrow{P}_t(x_{t+1}, \mathrm{d}x_t)$, and, using the fact that

$$\mathbb{M}_t(\mathrm{d}x_t) P_{t+1}(x_t, \mathrm{d}x_{t+1}) = \mathbb{M}_{t+1}(\mathrm{d}x_{t+1}) \overleftarrow{P}_t(x_{t+1}, \mathrm{d}x_t),$$

one obtains

$$\overleftarrow{G}_t(x_{t+1}, x_t) = f_t(y_t|x_t).$$

The corresponding algorithm may be interpreted as a bootstrap filter that assimilates the data in the reverse direction (from y_T to y_0).

More generally, if we set $\gamma_t(\mathrm{d}x_t) \propto \mathbb{M}_t(\mathrm{d}x_t)\overleftarrow{\eta}_t(x_t)$ for some arbitrary function $\overleftarrow{\eta}_t : \mathcal{X} \to \mathbb{R}^+$, we obtain the following alternative expression for $\overleftarrow{G}_t$:

$$\overleftarrow{G}_t(x_{t+1}, x_t) = \frac{f_t(y_t|x_t)\overleftarrow{\eta}_t(x_t)}{\overleftarrow{\eta}_{t+1}(x_{t+1})} \times \frac{\overleftarrow{P}_t(x_{t+1}, \mathrm{d}x_t)}{\overleftarrow{M}_t(x_{t+1}, \mathrm{d}x_t)}, \tag{12.21}$$

assuming of course that the Radon-Nikodym derivative above is well defined. Then we may interpret the information filter as an auxiliary particle filter, with proposal kernel $\overleftarrow{M}_t(x_{t+1}, \mathrm{d}x_t)$, and auxiliary weight function $\overleftarrow{\eta}_t$. The same local optimality results apply (see Theorem 10.2), and are stated without proof.

Theorem 12.5 (Local Optimality in Information Filter) *For a given state-space model, suppose that* $(\overleftarrow{G}_s, \overleftarrow{M}_s)$ *have been chosen to satisfy* (12.17), (12.18) *for* $s \geq t+1$. *Among all pairs* $(\overleftarrow{M}_t, \overleftarrow{G}_t)$ *that satisfy* (12.18) *and functions* $\overleftarrow{\eta}_{t+1}$ *such that* (12.21) *applies, the Markov kernel*

$$\overleftarrow{M}_t^{\mathrm{opt}}(x_{t+1}, \mathrm{d}x_t) = \frac{f_t(y_t|x_t)}{\int_{\mathcal{X}} f_t(y_t|x_t') \overleftarrow{P}_{t+1}(x_{t+1}, \mathrm{d}x_t')} \overleftarrow{P}_{t+1}(x_{t+1}, \mathrm{d}x_t)$$

and the function

$$\overleftarrow{\eta}_{t+1}^{\mathrm{opt}}(x_{t+1}) = \int_{\mathcal{X}} f_t(y_t|x_t') \overleftarrow{P}_t(x_{t+1}, \mathrm{d}x_t')$$

minimise $\mathrm{Var}\left[\overleftarrow{G}_t(X_{t+1}^{A_t^m}, X_t^m)/\overleftarrow{\eta}_t(X_t^m)\right]$.

Of course, the caveats regarding the local optimality criterion we discussed in Sect. 10.3.3 also apply in this case too. We discuss now some concrete examples.

Example 12.3 Consider the basic stochastic volatility model, with Markov transition defined as:

$$X_t - \mu = \rho(X_{t-1} - \mu) + U_t, \quad U_t \sim \mathcal{N}(0, \sigma^2) \tag{12.22}$$

and assume additionally that $X_0 \sim \mathcal{N}(\mu, \sigma^2/(1-\rho^2))$ and $|\rho| < 1$. It is easy to check that $\{X_t\}$ is stationary, $X_t \sim \mathcal{N}(\mu, \sigma^2/(1-\rho^2))$, and that $X_t|X_{t+1}$ has exactly the same transition kernel as $X_t|X_{t-1}$; i.e. $X_t|X_{t+1} \sim \mathcal{N}(\mu + \rho(X_{t+1} - \mu), \sigma^2)$. In that case, designing an efficient information filter is clearly equivalent to designing an auxiliary particle filter for the same model.

Models such that $\{X_t\}$ is stationary makes the interpretation of the information filter as a particle filter in reverse particularly transparent. This interpretation carries over the non-stationary case, but caution must be exercised, as shown in the example below.

Example 12.4 Consider a state-space model such that $\{X_t\}$ is a Gaussian random walk,

$$X_t = X_{t-1} + U_t, \quad U_t \sim \mathcal{N}(0, \sigma^2)$$

such as e.g. our first tracking example in Sect. 2.4.1. Then $X_t \sim \mathcal{N}(0, t\sigma^2)$ marginally. For such a model, it is easy to design examples where the standard bootstrap filter (that works from time 0 to time T) gives good performance, while the 'reverse' bootstrap filter (the particular information filter designed at the beginning of this section) works poorly, because particles at time T will be generated initially from $\mathcal{N}(0, T\sigma^2)$, and very few of them will have high density $f_T(y_T|x_T)$ if T is large. One may have to design e.g. a guided PF specifically for the information filter.

Outside models where $\{X_t\}$ is Gaussian, it is often impossible to compute the marginal distributions $\mathbb{M}_t(\mathrm{d}x_t)$, and the backward kernel $\overleftarrow{P}_{t-1}(x_t, \mathrm{d}x_{t-1})$. In that case, unfortunately, Theorem 12.5 provides only limited intuition on how to design the information filter: one has to choose $\overleftarrow{M}_t(x_{t+1}, \mathrm{d}x_t)$ and γ_t so that $\overleftarrow{M}_t(x_{t+1}, \mathrm{d}x_t)$ and the corresponding auxiliary function $\overleftarrow{\eta}_t$ (i.e. the function $\overleftarrow{\eta}_t$ such that $\gamma_t(\mathrm{d}x_t) \propto \mathbb{M}_t(\mathrm{d}x_t)\overleftarrow{\eta}_t(x_t)$) are close to their optimal values, but the relation between γ_t and $\overleftarrow{\eta}_t$ involves marginal distributions we cannot compute.

12.5.4 Two-Filter Smoother: Standard $\mathcal{O}(N^2)$ Version

We return to (12.15), and assume that we have already run two particle algorithms: a standard particle filter, which generated the approximations

$$\sum_{n=1}^{N} W_t^n \delta_{X_t^n}(\mathrm{d}x_t) \approx \mathbb{P}(X_t \in \mathrm{d}x_t | y_{0:t}) ;$$

and an information filter, which generated the approximations

$$\sum_{m=1}^{M} \overleftarrow{W}_{t+1}^m \delta_{\overleftarrow{X}_{t+1}^m}(\mathrm{d}x_{t+1}) \approx \gamma_{t+1|T}(\mathrm{d}x_{t+1}) .$$

Plugging both approximations into (12.15) leads to

$$\frac{1}{\sum_{m,n=1}^{N} \omega^{m,n}} \sum_{m,n=1}^{N} \omega^{m,n} \delta_{(X_t^n, \overleftarrow{X}_{t+1}^m)}(\mathrm{d}x_{t:t+1}) \tag{12.23}$$

where

$$\omega^{m,n} := W_t^n \overleftarrow{W}_{t+1}^m \frac{p_{t+1}(\overleftarrow{X}_{t+1}^m | X_t^n)}{\gamma_{t+1}(\overleftarrow{X}_{t+1}^m)}$$

and the following class of estimators:

$$\frac{\sum_{m,n=1}^{N} \omega^{m,n} \varphi(X_t^n, \overleftarrow{X}_{t+1}^m)}{\sum_{m,n=1}^{N} \omega^{m,n}} \approx \mathbb{E}[\varphi(X_t, X_{t+1}) | Y_{0:T} = y_{0:T}];$$

see Algorithm 12.6.

Algorithm 12.6: Two-filter smoothing

Input: Function φ, output at time t of a particle filter (N particles X_t^n with weights W_t^n) output at time $t+1$ of an information filter (M particles $\overleftarrow{X}_{t+1}^m$ with weights $\overleftarrow{W}_{t+1}^m$).

for $n = 1$ **to** N **do**
  for $m = 1$ **to** M **do**
    $\omega^{m,n} \leftarrow W_t^n \overleftarrow{W}_{t+1}^m p_{t+1}(\overleftarrow{X}_{t+1}^m | X_t^n) / \gamma_{t+1}(\overleftarrow{X}_{t+1}^m)$

return (as an approximation of $\mathbb{E}[\varphi(X_t, X_{t+1}) | Y_{0:T} = y_{0:T}]$)

$$\frac{1}{\sum_{m,n=1}^{N} \omega^{m,n}} \sum_{m,n=1}^{N} \omega^{m,n} \varphi(X_t^n, \overleftarrow{X}_{t+1}^m).$$

The algorithm has the following properties: its complexity is $\mathcal{O}(NM)$, and thus $\mathcal{O}(N^2)$ if we take $M = N$. It requires that the Markov kernel $P_t(x_{t-1}, \mathrm{d}x_t)$ admits a tractable probability density. And finally, we observe that an algorithm usually described as a way to approximate univariate smoothing marginals provides approximations of pairwise marginals as no extra cost.

12.5.5 A $\mathcal{O}(N)$ Variant of Two-Filter Smoothing

Equation (12.23) involves $\mathcal{O}(N^2)$ terms (if we take $M = N$), but it is possible to derive a $\mathcal{O}(N)$ approximation of this sum using importance sampling.

To do so, we interpret (12.23) as a joint distribution for integer variables I, J, taking values respectively in $1 : M$, and $1 : N$, and we perform importance sampling based on an independent proposal: for $l = 1, \ldots, L$, $I_l \sim \mathcal{M}(\overleftarrow{\beta}^{1:M})$, $J_l \sim \mathcal{M}(\beta^{1:N})$, where both $\overleftarrow{\beta}^{1:M}$ and $\beta^{1:N}$ are vectors of probabilities that sum to one; see Algorithm 12.7.

Algorithm 12.7: $\mathcal{O}(N)$ variant of two-filter smoothing

---

Input: integer L, function φ, probabilities $\beta^{1:N}$ (that sum to one) and $\overleftarrow{\beta}^{1:M}$ (idem), output at time t of a particle filter (N particles X_t^n with weights W_t^n) output at time $t+1$ of an information filter (M particles $\overleftarrow{X}_{t+1}^m$ with weights $\overleftarrow{W}_{t+1}^m$).

Generate $I^{1:L} \sim \mathcal{M}(\overleftarrow{\beta}^{1:M})$ ▷ `using Algorithm 9.3`

Generate $J^{1:L} \sim \mathcal{M}(\beta^{1:N})$ ▷ `using Algorithm 9.3`

for $l = 1$ **to** L **do**

  $\omega^l \leftarrow W_t^{J_l} \overleftarrow{W}_{t+1}^{I_l} p_{t+1}(\overleftarrow{X}_{t+1}^{I_l} | X_t^{J_l}) / \beta^{J_l} \overleftarrow{\beta}^{I_l} \gamma_{t+1}(\overleftarrow{X}_{t+1}^{I_l})$

return (as an approximation of $\mathbb{E}[\varphi(X_t, X_{t+1}) | Y_{0:T} = y_{0:T}]$)

$$\frac{1}{\sum_{l=1}^{L} \omega_l} \sum_{l=1}^{L} \omega_l \varphi(X_t^{J_l}, \overleftarrow{X}_{t+1}^{I_l}) . \tag{12.24}$$

Note that the complexity of operations $I^{1:L} \sim \mathcal{M}(\overleftarrow{\beta}^{1:M})$, $J^{1:L} \sim \mathcal{M}(\beta^{1:N})$, is respectively $\mathcal{O}(L + M)$ and $\mathcal{O}(L + N)$, as we can use multinomial resampling (Algorithm 9.3) to generate these variables. Hence, by taking $L = M = N$, we do obtain a $\mathcal{O}(N)$ complexity. The caveat is that we add another layer of Monte Carlo noise to our approximation.

The remaining question is how to choose the probabilities β^n and $\overleftarrow{\beta}^m$. A simple choice is: $\beta^n = W_t^n$, $\overleftarrow{\beta}^m \propto \overleftarrow{W}_{t+1}^m / \gamma_{t+1}(X_{t+1}^m)$ (normalised to sum to one), so that the weight ω^l in Algorithm 12.7 simplifies to: $\omega^l \leftarrow p_{t+1}(\overleftarrow{X}_{t+1}^{I_l} | X_t^{J_l})$.

Alternatively, one may want to use 'optimal' probabilities. Minimising the variance of (12.24) for a given φ does not seem straightforward, but if we instead minimise the Kullback-Leibler divergence between the proposal and the target of our importance sampler, we obtain a more workable optimality criterion.

Proposition 12.6 *Let (X, Y) a pair of random variables with joint probability density $q(x, y)$ (with respect to some dominating measure). The probability density $p(x, y) = p_x(x)p_y(y)$ that minimises the Kullback-Leibler divergence $\mathbb{E}[\log\{q(X, Y)/p(X, Y)\}]$, under the constraint that X and Y are independent under $p(x, y)$, is the product of the marginal distributions of X and Y under q.*

For a proof, see Exercise 12.4.

Thus, one may want to choose the β^n's and the $\overleftarrow{\beta}^m$'s so that they are close to the marginals of (12.23). Starting with the $\overleftarrow{\beta}^m$'s, we have

$$\begin{aligned}\sum_{n=1}^{N} \omega^{m,n} &= \frac{\overleftarrow{W}^m_{t+1}}{\gamma_{t+1}(\overleftarrow{X}^m_{t+1})} \left\{ \sum_{n=1}^{N} W^n_t p_{t+1}(\overleftarrow{X}^m_{t+1}|X^n_t) \right\} \\ &\approx \frac{\overleftarrow{W}^m_{t+1}}{\gamma_{t+1}(\overleftarrow{X}^m_{t+1})} p_{t+1}(\overleftarrow{X}^m_{t+1}|y_{0:t})\end{aligned}$$

which suggests to take the $\overleftarrow{\beta}^m$ as above, but with $p_{t+1}(X^m_{t+1}|y_{0:t})$ replaced by some approximation of the predictive distribution. Several recipes may be used to that effect.

First, one may fit some parametric distribution (e.g. Gaussian) to the particle approximation of the predictive distribution provided by the forward filter.

Second, one may replace it by an approximation of $p_{t+1}(X^m_{t+1}|y_t)$; such an approximation is typically derived when implementing an auxiliary particle filter, and may be used here as well. Third, and using a different approach, one may approximate the marginal probabilities recursively: use some initial values for the β^n and the $\overleftarrow{\beta}^m$ to approximate through importance sampling both marginals of (12.23), update these probabilities, and repeat a few times.

Regarding the β^n's, we have:

$$\begin{aligned}\sum_{m=1}^{M} \omega^{m,n} &= W^n_t \left\{ \sum_{m=1}^{M} \frac{\overleftarrow{W}^m_{t+1} p_{t+1}(\overleftarrow{X}^m_{t+1}|X^n_t)}{\gamma_{t+1}(\overleftarrow{X}^m_{t+1})} \right\} \\ &\approx W^n_t p_T(y_{t+1:T}|X^n_t)\end{aligned}$$

and the same ideas apply: take β^n as above but with $p_T(y_{t+1:T}|X^n_t)$ replaced by some approximation of it; for instance deduced from a particle approximation of the same quantity, obtained from the information filter.

12.5.6 A Block-of-Three Variant for Models Without a Transition Density

As discussed several times before, some practical models are such that $P_t(x_{t-1}, \mathrm{d}x_t)$ does not admit a probability density (with respect to a fixed dominating measure).

We now present a variant of two-filter smoothing that may be applied to some of these models.

We start by remarking that it may be possible to implement an information filter for such models, as the following example shows.

Example 12.5 We return to the bearings-only tracking model (Sect. 2.4.1), which has motion model

$$X_t = \begin{pmatrix} I_2 & I_2 \\ 0_2 & I_2 \end{pmatrix} X_{t-1} + \begin{pmatrix} 0_2 & 0_2 \\ 0_2 & U_t \end{pmatrix}, \quad U_t \sim \mathcal{N}_2(0, \sigma^2 I_2).$$

The two first components of X_t (the position of the target) are a deterministic function of X_{t-1}, and thus the law of $X_t|X_{t-1}$ involves Dirac masses that enforce the constraints $X_t(1) = X_{t-1}(1) + X_{t-1}(3)$, $X_t(2) = X_{t-1}(2) + X_{t-1}(4)$. Nonetheless, the information filter may be implemented, by choosing the γ_t's and the $\overleftarrow{M}_t(x_{t+1}, \mathrm{d}x_t)$ so that the joint distribution $\gamma_{t+1}(\mathrm{d}x_{t+1})\overleftarrow{M}_t(x_{t+1}, \mathrm{d}x_t)$ imposes the same deterministic constraints.

The block-of-three variant of two-filter smoothing applies to state-space models such that $P_t(x_{t-1}, \mathrm{d}x_t)$ does not admit a probability density, but the law of X_{t+2} given X_t does. The following example describes such a model.

Example 12.6 Consider the auto-regressive process of order two $Z_t = \phi_1 Z_{t-1} + \phi_2 Z_{t-2} + U_t$, $U_t \sim \mathcal{N}(0, \sigma^2)$. As seen in Sect. 2.4.3 on stochastic volatility models, we may cast $\{Z_t\}$ a Markov process as follows:

$$X_t = \begin{pmatrix} \phi_1 & \phi_2 \\ 1 & 0 \end{pmatrix} X_{t-1} + \begin{pmatrix} U_t \\ 0 \end{pmatrix}, \quad U_t \sim \mathcal{N}(0, \sigma^2).$$

Take $\phi_1 = \phi_2 = 1$ for simplicity. Since $X_t(2)$ is a deterministic function of X_{t-1}, $P_t(x_{t-1}, \mathrm{d}x_t)$ does not admit a probability density, but the law of X_t given X_{t-2} does:

$$X_t = \begin{pmatrix} 2 & 1 \\ 1 & 1 \end{pmatrix} X_{t-2} + \begin{pmatrix} U_t + U_{t-1} \\ U_{t-1} \end{pmatrix}, \quad U_t, U_{t-1} \sim \mathcal{N}(0, \sigma^2).$$

(continued)

Example 12.6 (continued)
We also observe that the law of X_{t+1} conditional on X_t and X_{t+2} is a Dirac mass:

$$X_{t+1} = \begin{pmatrix} x_{t+2}(2) \\ x_t(1) \end{pmatrix} \text{ conditional on } X_t = x_t, X_{t+2} = x_{t+2}. \tag{12.25}$$

This type of examples motivates working with consecutive blocks of three states, rather than two as before:

$$\begin{aligned}\mathbb{P}(X_{t:t+2} \in \mathrm{d}x_{t:t+2}|Y_{0:T} = y_{0:T}) \propto f_{t+1}(y_{t+1}|x_{t+1})p_T(y_{t+2:T}|x_{t+2})\times \\ \mathbb{P}(X_t \in \mathrm{d}x_t|Y_{0:t} = y_{0:t})P_{t+1}(x_t, \mathrm{d}x_{t+1})P_{t+2}(x_{t+1}, \mathrm{d}x_{t+2}).\end{aligned}$$

We rewrite $P_{t+1}(x_t, \mathrm{d}x_{t+1})P_{t+2}(x_{t+1}, \mathrm{d}x_{t+2})$, the law of $X_{t+1:t+2}$ given $X_t = x_t$, as

$$P_{t+1:t+2}(x_t, \mathrm{d}x_{t+2})\bar{P}_{t+1}(x_t, x_{t+2}, \mathrm{d}x_{t+1}),$$

the law of X_{t+2} given $X_t = x_t$, times the law of X_{t+1} given X_t and X_{t+2}. We assume the former has probability density $p_{t+1:t+2}(x_{t+2}|x_t)$ that may be evaluated pointwise, but we assume nothing of the sort for the latter. Then we obtain

$$\begin{aligned}\mathbb{P}(X_{t:t+2} \in \mathrm{d}x_{t:t+2}|Y_{0:T} = y_{0:T}) \propto \frac{f_{t+1}(y_{t+1}|x_{t+1})p_{t+1:t+2}(x_{t+2}|x_t)}{\gamma_{t+2}(x_{t+2})}\times \\ \mathbb{P}(X_t \in \mathrm{d}x_t|Y_{0:t} = y_{0:t})\gamma_{t+2|T}(\mathrm{d}x_{t+2})\bar{P}_{t+1}(x_t, x_{t+2}, \mathrm{d}x_{t+1}),\end{aligned}$$

which we approximate, using the same plug-in approach as before, i.e. $\sum_{n=1}^N W_t^n \delta_{X_t^n}(\mathrm{d}x_t) \approx \mathbb{P}(\mathrm{d}x_t|Y_{0:t} = y_{0:t})$ (from the forward particle filter), $\sum_{m=1}^M \overleftarrow{W}_{t+2}^m \delta_{X_{t+2}^m}(\mathrm{d}x_{t+2}) \approx \gamma_{t+2|T}(\mathrm{d}x_{t+2})$ (from the information filter), by

$$\propto \sum_{m,n=1}^{N} \omega^{m,n} f_{t+1}(y_{t+1}|x_{t+1})\delta_{(X_t^n, \overleftarrow{X}_{t+2}^m)}(\mathrm{d}(x_t, x_{t+2}))\bar{P}_{t+1}(X_t^n, \overleftarrow{X}_{t+2}^m, \mathrm{d}x_{t+1}) \tag{12.26}$$

with weights

$$\omega^{m,n} := \frac{W_t^n \overleftarrow{W}_{t+2}^m}{\gamma_{t+2}(\overleftarrow{X}_{t+2}^m)} p_{t+1:t+2}(\overleftarrow{X}_{t+2}^m|X_t^n).$$

To obtain a $\mathcal{O}(N)$ importance sampling approximation of (12.26) we define a proposal for I (taking values in $1:N$), J (taking values in $1:M$) and X_{t+1} (taking values in $\mathcal{X}$) such that I and J are independent: for $l = 1, \ldots, L$, generate $I^l \sim \mathcal{M}(\overleftarrow{\beta}^{1:M})$, $J^l \sim \mathcal{M}(\beta^{1:N})$, and conditional on $I_l = m$, $J_l = n$, generate $\widetilde{X}^l_{t+1} \sim \bar{Q}_{t+1}(X^n_t, \overleftarrow{X}^m_{t+2}, dx_{t+1})$, a proposal kernel such that $\bar{P}_{t+1}(x_t, x_{t+2}, dx_{t+1}) \ll \bar{Q}_{t+1}(x_t, x_{t+2}, dx_{t+1})$, i.e. the Radon-Nikodym derivative

$$\frac{\bar{P}_{t+1}(x_t, x_{t+2}, dx_{t+1})}{\bar{Q}_{t+1}(x_t, x_{t+2}, dx_{t+1})}$$

is well defined for any x_t, $x_{t+2} \in \mathcal{X}$. This approach is summarised as Algorithm 12.8.

Algorithm 12.8: Block-of-three, $\mathcal{O}(N)$ version of two-filter smoother

---

Input: integer L, function φ, probabilities $\beta^{1:N}$ (that sum to one) and $\overleftarrow{\beta}^{1:M}$ (idem), proposal kernel $\bar{Q}_{t+1}(x_t, x_{t+2}, dx_{t+1})$, which defines a distribution for X_{t+1}, for any x_t, $x_{t+2} \in \mathcal{X}$; output at time $t \leq T-2$ of a particle filter, $(X_t^{1:N}, W_t^{1:N})$, output at time $t+2$ of an information filter, $(\overleftarrow{X}^{1:M}_{t+2}, \overleftarrow{W}^{1:M}_{t+2})$.

Generate $I^{1:L} \sim \mathcal{M}(\overleftarrow{\beta}^{1:M})$ ▷ `using Algorithm 9.3`

Generate $J^{1:L} \sim \mathcal{M}(\beta^{1:N})$ ▷ `using Algorithm 9.3`

for $l = 1$ **to** L **do**

$\widetilde{X}^l_{t+1} \sim \bar{Q}_{t+1}(X_t^{J_l}, X_{t+2}^{I_l}, dx_{t+1})$

Compute

$$\chi^l \leftarrow \frac{W_t^{J_l}\overleftarrow{W}^{I_l}_{t+2} p_{t+1:t+2}(\overleftarrow{X}^{I_l}_{t+2}|X_t^{J_l}) f_{t+1}(y_{t+1}|\widetilde{X}^l_{t+1})}{\beta^{J_l}\overleftarrow{\beta}^{I_l}\gamma_{t+2}(\overleftarrow{X}^{I_l}_{t+2})} \times \left\{ \left.\frac{\bar{P}_{t+1}(X^n_t, \overleftarrow{X}^m_{t+2}, dx_{t+1})}{\bar{Q}_{t+1}(X^n_t, \overleftarrow{X}^m_{t+2}, dx_{t+1})}\right|_{x_{t+1}=\widetilde{X}^l_{t+1}} \right\}.$$

return (as an approximation of $\mathbb{E}[\varphi(X_t, X_{t+1}, X_{t+2})|Y_{0:T} = y_{0:T}]$)

$$\frac{1}{\sum_{l=1}^L \chi_l} \sum_{l=1}^{L} \chi_l \varphi(X_t^{J_l}, \widetilde{X}^l_{t+1}, X^{I_l}_{t+2}). \tag{12.27}$$

Regarding the choice of the proposal, the same considerations as in the previous section apply: using as an optimality criterion the Kullback-Leibler divergence between the proposal and the target, the optimal choice for kernel $\bar{Q}_{t+1}(x_t, x_{t+2}, \mathrm{d}x_{t+1})$, is the distribution proportional to $\bar{P}_{t+1}(x_t, x_{t+2}, \mathrm{d}x_{t+1})$ $f_{t+1}(y_{t+1}, x_{t+1})$, i.e. the law of X_{t+1} conditional on X_t, X_{t+2} and Y_{t+1}; and the optimal choice for the vectors of probabilities $\beta^{1:N}$ and $\overleftarrow{\beta}^{1:M}$) are the marginal distributions of (12.26). This leads to the same recipes as in the previous section: take $\beta^n \approx W_t^n p_T(y_{t+1:T}|X_t^n)$, $\overleftarrow{\beta}^m \approx \left(\overleftarrow{W}^m_{t+2}/\gamma_{t+2}(\overleftarrow{X}^m_{t+2})\right) p_{t+2}(X^m_{t+2}|Y_{0:t+1} = y_{0:t+1})$.

One could generalise this block-of-three approach to blocks of four states or more; i.e. trying to approximate the smoothing distribution of $X_{t:t+h}$, for $h \geq 3$. The bigger the block is, the larger the dimension of the sampling space will be, and therefore, the higher the variance of the Monte Carlo noise introduced by importance sampling should be. For this reason, we recommend to use the block-of-three (or more generally block-of-k) approach only for models such that $P_t(x_{t-1}, \mathrm{d}x_t)$ does not admit a probability density (but the law of $X_{t+k-1}|X_t$ does).

Example 12.7 Following Example 12.6, we noticed that the law of X_{t+1} conditional on X_t and X_{t+2} is a Dirac mass, see (12.25). We are then forced to choose as a proposal $\bar{Q}_{t+1}(x_t, x_{t+2}, \mathrm{d}x_{t+1})$ for X_{t+1} the same Dirac mass. In addition, if we take $\beta^n = W_t^n$, $\overleftarrow{\beta}^m \propto \overleftarrow{W}^m_{t+2}/\gamma_{t+2}(\overleftarrow{X}^m_{t+2})$ (normalised to sum to one), then weight χ_l reduces to

$$\chi^l \leftarrow p_{t+1:t+2}(\overleftarrow{X}^{I_l}_{t+2}|X_t^{J_l}) f_{t+1}(y_{t+1}|\widetilde{X}^l_{t+1}).$$

Thus, we may apply the block-of-three algorithm to this model; on the other hand, as we have already explained, the block-of-two variant (seen in the previous section) is not applicable, since for this model the distribution of $X_t|X_{t-1}$ does not have a density.

12.6 Numerical Experiment: Comparison of Off-Line Smoothing Algorithms

We return to Example 12.1 and consider again the additive function (12.7). We compute its smoothing expectation with respect to a simulated dataset of size $T = 100$. For the forward pass, we use a bootstrap filter.

We compare the estimates obtained with the following approaches: FFBS (generating N trajectories, at cost $\mathcal{O}(N^2)$) FFBS-reject (generating N trajectories, at cost $\mathcal{O}_P(N)$), standard two-filter smoothing with cost $\mathcal{O}(N^2)$, and its $\mathcal{O}(N)$

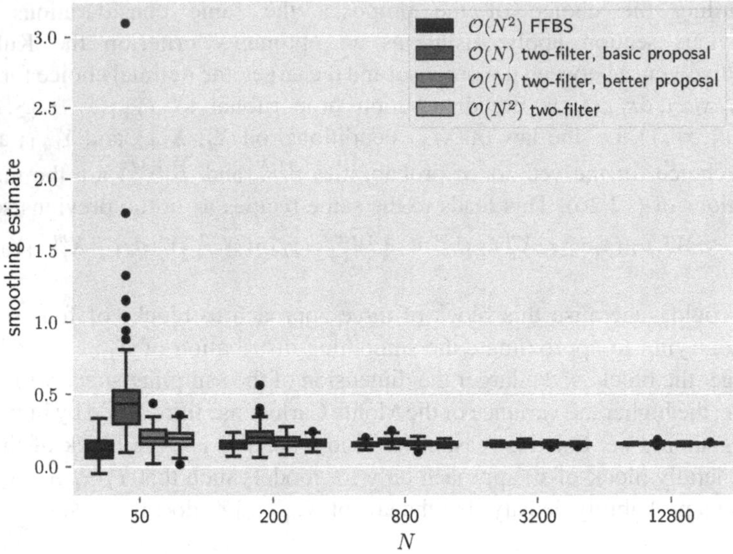

Fig. 12.4 Box-plots of smoothing estimates of φ_T over 100 independent runs as a function of N, as obtained from different off-line smoothing algorithms

variant based on two proposals: the basic proposal (such that the weights reduce to $\omega^l = p_t(\overleftarrow{X}_{t+1}^{I_l}|X_t^{J_l})$, see Sect. 12.5.5), and a proposal based on Gaussian approximations of the predictive distributions; i.e.

$$\overleftarrow{\beta}^m \propto \frac{\overleftarrow{W}_t^m}{\gamma_{t+1}(\overleftarrow{X}_{t+1}^m)} \varphi(\overleftarrow{X}_{t+1}^m; \widehat{\mu}_t, \widehat{\sigma}_t^2)$$

where $\widehat{\mu}_t$ and $\widehat{\sigma}_t^2$ are the (unweighted) mean and variance of the particles X_{t+1}^n (and thus are estimates of the mean and variance of the predictive distribution). The probabilities β^m are set similarly (using particle estimates from the information filter).

Figure 12.4 compares the variability of the smoothing estimates; box-plots for FFBS are not included, as FFBS and FFBS-reject sample from the *same* distribution.

For FFBS-reject, we took $C_t = (2\pi\sigma^2)^{-1/2}$, leading to an average overall acceptance rate about 18%. For the two-filter approaches, we use the fact that the state process is stationary, and set the information filter to a bootstrap filter applied to the data in reverse, as discussed in Example 12.3.

The smoothing estimates have roughly the same variability (except that $\mathcal{O}(N)$ two-filter smoothing with a basic proposal generates more outliers when N is small; see below). However, their CPU cost differ markedly; see Fig. 12.5. $\mathcal{O}(N)$ two-filter smoothing is particularly fast, and takes only two seconds for $N = 12,800$, while

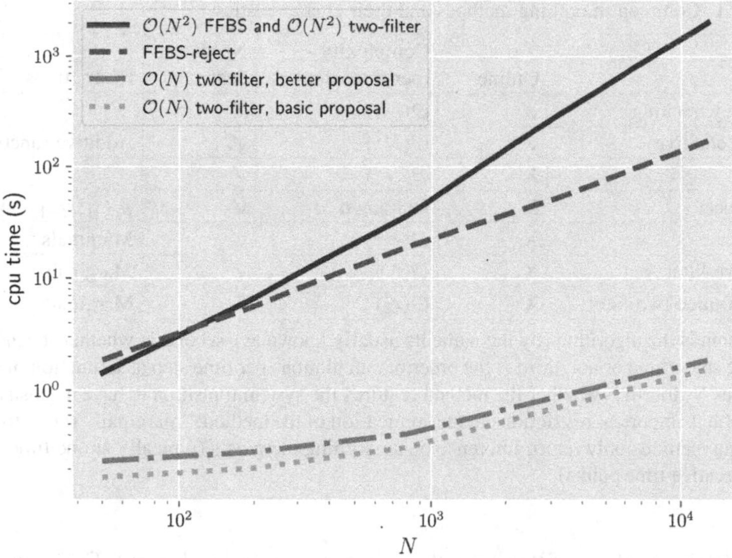

Fig. 12.5 CPU time as a function of N for the various off-line algorithms discussed in text; the CPU times of $\mathcal{O}(N^2)$ FFBS and $\mathcal{O}(N^2)$ two-filter are essentially the same, and cannot be distinguished visually

$\mathcal{O}(N^2)$ algorithms take more than 30 min for the same N. Note also the intermediate performance of FFBS-reject.

Regarding two-filtering smoothing, recall that the $\mathcal{O}(N)$ estimates are *noisy* versions of the $\mathcal{O}(N^2)$ estimate, and therefore should have a larger variance by construction. Thus it is quite remarkable that the extra variability of the $\mathcal{O}(N)$ estimates is so small. The fact that these estimates are unstable for small values of N is not a great concern, given that their low (and linear) CPU cost makes it possible to use it for much larger values of N.

12.7 Conclusion

The numerical results of the previous section should be considered with a pinch of salt. The performance and applicability of each method may vary considerably from one problem to the next.

Still, one reasonable piece of advice is to determine first how difficult it is to implement two-filter smoothing (in particular to design an information filter). If this task is reasonably straightforward, then it seems worth trying to implement its $\mathcal{O}(N)$ version, as $\mathcal{O}(N)$ two-filter smoothing seems hard to beat, at least in terms of CPU time.

Table 12.1 Common smoothing methods and their characteristics

Method	Online	Complexity (per time step)	Needs density	Restrictions
Genealogy tracking	✓	$\mathcal{O}(N)$	X	
Forward additive	✓	$\mathcal{O}(N^2)$	✓	Additive functions
FFBS	X	$\mathcal{O}(N^2)$	✓	
FFBS-reject	X	Unknown	✓	$p_t(x_t\|x_{t-1})$ bounded
Two-filter	X	$\mathcal{O}(N^2)$	✓	Marginals
$\mathcal{O}(N)$ two-filter	X	$\mathcal{O}(N)$	✓	Marginals
Block-of-three two-filter	X	$\mathcal{O}(N)$	X	Marginals

First column is the algorithm (by the name its usually known as), second is whether it can be used for online smoothing or not, third is the order of calculations per time step as a function of number of particles N, fourth is whether the method requires the system transition to have a density of not, and the fifth is important restrictions in the application of the method; "marginals" refers to the fact that certain methods only return inference on smoothing marginals (typically at one time point or two consecutive time points)

Alternatively, if two-filter smoothing is not easy to implement, FFBS would be the default option, and it may be worthwhile trying to implement FFBS-reject, the rejection-based variant (meaning one needs to find a function C_t such that $p_t(x_t|x_{t-1}) \leq C_t(x_t)$). However, one should check that the rejection rate of FFBS-reject is not too high, otherwise the method might become impracticable. Table 12.1 summarises all the smoothing algorithms discussed in this chapter.

Exercises

12.1 *Use the law of iterated conditional expectation and the Markov property to prove Proposition 12.1.*

12.2 *The result in Proposition 12.1 is in fact an instance of the following more general result. Let X_t be a Markov process with forward transition kernel $P_t(x_{t-1}, \mathrm{d}x_t)$ and backward kernel $\overleftarrow{P}_{t-1}(x_t, \mathrm{d}x_{t-1})$, and $\varphi_t(x_{0:t})$ be an additive function*

$$\varphi_t(x_{0:t}) = \psi_0(x_0) + \sum_{s=1}^{t} \psi_s(x_{s-1}, x_s)\,.$$

Then,

$$\mathbb{E}[\varphi_t(X_{0:t})] = \int_{\mathcal{X}} \Phi_t(x_t)\mathbb{P}_t(\mathrm{d}x_t)$$

where

$$\Phi_t(x_t) := \mathbb{E}[\varphi_t(X_{0:t})|X_t = x_t]$$

and

$$\Phi_t(x_t) = \int_{\mathcal{X}} \{\Phi_{t-1}(x_{t-1}) + \psi_t(x_{t-1}, x_t)\} \overleftarrow{P}_{t-1}(x_t, \mathrm{d}x_{t-1}) .$$

Establish this general result. Explain why Proposition 12.1 is a direct corollary of this result.

12.3 *Write down the acceptance rate of the rejection step in Algorithm 12.3 (at time t, for a fixed B_{t+1}); show that it is lower-bounded as soon as $p_{t+1}(x_{t+1}|x_t)$ is lower-bounded (by a positive constant) and C_t is constant. Deduce that the overall CPU cost of this algorithm is $\mathcal{O}_P(NT)$ in this case. (Explain the rationale behind symbol P in the $\mathcal{O}_P$.)*

12.4 *(Recall Exercise 8.12.) Prove Proposition 12.6 on optimal Kullback-Leibler approximation. To that effect, show that*

$$\mathbb{E}\left[\log \frac{q(X,Y)}{p_X(X)p_Y(Y)}\right] = \mathbb{E}\left[\log \frac{q_X(X)}{p_X(X)}\right] + \mathbb{E}\left[\log \frac{q_{Y|X}(Y|X)}{p_Y(Y)}\right]$$

where q_x and $q_{y|x}$ are respectively the marginal distribution of X, and the distribution of Y conditional on $X = x$ (relative to q). Use this identity to find the optimal p_x. Conclude.

12.5 *Identities for the score in latent variable and state-space models.*

1. *Consider a latent variable model with parameters θ, data Y and latent variables X linked in terms of an observation density $f^\theta(y|x)$ and a prior density $p^\theta(x)$. The likelihood is thus given by*

$$p^\theta(y) = \int f^\theta(y|x)p^\theta(x)\mathrm{d}x .$$

We assume that these densities are differentiable as functions of θ and regular enough so that we can "differentiate under the integral sign", that is that the following holds:

$$\nabla p^\theta(y) = \int \nabla \left(f^\theta(y|x)p^\theta(x)\right) \mathrm{d}x .$$

Establish the identity

$$\nabla \log p^\theta(y) = \mathbb{E}\left[\nabla \left(\log f^\theta(y|X) + \log p^\theta(X)\right)|Y = y\right]$$

where the expectation is with respect to the conditional distribution of X given $Y = y$, i.e., according to the density proportional to $f^\theta(y|x)p^\theta(x)$.

2. *Consider now a state-space model with densities $p_0^\theta(x_0)$, $p_t^\theta(x_t|x_{t-1})$, $f_0^\theta(y_0|x_0)$, $f_t^\theta(y_t|x_t)$, all of which are differentiable with respect to θ and regular enough to justify differentiation under the integral sign. Apply the result above to show that*

$$\nabla \log p^\theta(y_{0:t}) = \mathbb{E}[\varphi_t(X_{0:t})|Y_{0:t} = y_{0:t}]$$

for

$$\varphi_t(x_{0:t}) = \psi_0(x_0) + \sum_{s=1}^{t} \psi_s(x_{s-1}, x_s)$$

and

$$\begin{aligned}\psi_0(x_0) &= \nabla\left(\log f_0^\theta(y_0|x_0) + \log p_0^\theta(x_0)\right)\\ \psi_s(x_{s-1}, x_s) &= \nabla\left(\log f_s^\theta(y_s|x_s) + \log p_s^\theta(x_s|x_{s-1})\right).\end{aligned}$$

12.6 *A generic algorithm for maximum likelihood inference in state-space models is the EM algorithm; which we will discuss more in detail in Chap. 14. As in Exercise 12.5 consider a latent variable model with observation density $f^\theta(y|x)$ and a prior density $p^\theta(x)$. The E-step of an EM algorithm requires the computation of the function*

$$Q(\theta, \theta') = \mathbb{E}_{\mathbb{P}^{\theta'}}[\log f^\theta(y|X) + \log p^\theta(X)|Y = y].$$

The expectation is with respect to the conditional distribution of the latent variable X given $Y = y$, assuming that the true parameter value is θ'. We highlight the dependence of the averaging distribution on θ' by writing $\mathbb{E}_{\mathbb{P}^{\theta'}}$. The M-step then maximises this as a function of θ for a given value of θ'. The iteration of these two steps together with setting θ' to the output of the M-step defines the basic EM-algorithm that obtains a local maximiser of $p^\theta(y)$. Consider now a state-space model with components $p_0^\theta(x_0)$, $p_t^\theta(x_t|x_{t-1})$, $f_0^\theta(y_0|x_0)$, $f_t^\theta(y_t|x_t)$, and let $Q(\theta, \theta')$ be the function that corresponds to $p_t^\theta(y_{0:t})$. Show that

$$Q(\theta, \theta') = \mathbb{E}[\varphi_t(X_{0:t})|Y_{0:t} = y_{0:t}]$$

for

$$\varphi_t(x_{0:t}) = \psi_0(x_0) + \sum_{s=1}^{t} \psi_s(x_{s-1}, x_s)$$

and

$$\psi_0(x_0) = \log f_0^\theta(y_0|x_0) + \log p_0^\theta(x_0)\,,$$
$$\psi_s(x_{s-1}, x_s) = \log f_s^\theta(y_s|x_s) + \log p_s^\theta(x_s|x_{s-1})\,.$$

12.7 *Show that for the model described in Example 12.1,*

$$\nabla_{\sigma^2} \log p^{\sigma^2}(y_{0:t}) = \mathbb{E}\left[\varphi_t(X_{0:t}) \mid Y_{0:t} = y_{0:t}\right]$$

where

$$\varphi_t(x_{0:t}) = -\frac{T+1}{2\sigma^2} + \frac{1-\rho^2}{2\sigma^4}(x_0 - \mu)^2 + \frac{1}{2\sigma^4}\sum_{s=1}^{t}\{x_s - \mu - \rho(x_{s-1} - \mu)\}^2\,.$$

Python Corner

In `particles`, on-line smoothing algorithms are implemented in the `collectors` module; collectors are objects that perform computation on the current particle system, and store results, at each iteration t of a SMC algorithm.

Off-line smoothing algorithms operate on the complete *history* of a particle filter: they take as inputs the outputs of the particle filter generated at times $t = 0, \ldots, T$. This history must be recorded in some way. To that effect, `particles` contains a `smoothing` module which implements class `ParticleHistory`. When a `SMC` object is instantiated with option `store_history` set to `True`, e.g.

```
pf = particles.SMC(fk=fk_boot_sv, N=100, store_history=True)
```

the history of the corresponding particle algorithm is recorded in a `pf.hist` object, which is an instance of `ParticleHistory`. Specifically, this object records at each time t the values of $X_t^{1:N}$, $W_t^{1:N}$, and $A_t^{1:N}$. The off-line smoothing algorithms are defined as *methods* of this class. For instance, here is how $\mathcal{O}(N^2)$ FFBS is implemented.

```
import numpy as np
from particles import resampling as rs  # see Chapter 9

class ParticleHistory(object):
    """Particle history."""

    def __init__(self, fk):
        self.A, self.X, self.W, self.lw = [], [], [], []
        self.fk = fk

    def save(self, X=None, lw=None, W=None, A=None):
        """Save one page of the history."""
```

```
            self.X.append(X)
            self.lw.append(lw)
            self.W.append(W)
            self.A.append(A)

        def backward_sampling_ON2(self, M):
            """O(N^2) FFBS, returns M trajectories."""

            idx = np.empty((self.T, M), dtype=int)
            idx[-1, :] = rs.multinomial(M, self.W[-1])
            for m in range(M):
                for t in reversed(range(self.T-1)):
                    lwm = (self.lw[t]
                           + self.fk.logpt(t+1, self.X[t],
                                           self.X[t+1][idx[t+1, m]]) )
                    Wm = rs.exp_and_normalise(lwm)
                    idx[t, m] = rs.multinomial(1, Wm)
          return [self.X[t][idx[t, :]] for t in range(self.T)]
```

(This extract above is actually a simplified version of the current implementation. However the differences between the two are immaterial to this discussion.)

Attributes `X`, `W`, `lw`, and `A` are lists: e.g. `self.W[t]` points to the `NumPy` array that contains $W_t^{1:N}$. Note that a `ParticleHistory` object must have access to the `model` object that represents the considered state-space model. In this way, methods such as `backward_sampling_ON2` have access to the transition density of the model.

Implementing FFBS-reject requires more efforts, unfortunately. Since this method is a rejection sampler based on a multinomial proposal, there is the issue of simulating an arbitrary number of draws from such a distribution. However we have already discussed how to address this in the Python corner of Chap. 9. The other issue is performance. Consider the following extra method for class `ParticleHistory`:

```
    def backward_sampling_lincost_pedagogical(self, M):
        """O(N) FFBS, slow version."""

        idx = np.empty((self.T, M), dtype=int)
        idx[-1, :] = rs.multinomial(M, self.W[-1])
        for t in reversed(range(self.T - 1)):
            gen = rs.MulinomialQueue(M, self.W[t])
            for m in range(M):
                while True:  # SLOW
                    nprop = gen.dequeue(1)
                    lpr_acc = (self.fk.logpt(t+1, self.X[t][nprop],
                                             self.X[t+1][idx[t+1, m]])
                               - self.fk.logC(t+1))
                    if np.log(rand()) < lpr_acc:
                        break
                idx[t, m] = nprop
        return [self.X[t][idx[t, :]] for t in range(self.T)]
```

The piece of code above is from an early version of `particles`, but it turned out to be painfully slow. The culprit is of course the nested loop; and more precisely the fact that the body of the inner loop consists of only a few basic operations. (In contrast, the inner loop of method `backward_sampling_ON2` performs operations on `NumPy` arrays of size $\mathcal{O}(N)$, hence the interpreter overhead remains small in that case.)

The interested reader may have a look at the source code for our actual (faster) implementation. Basically, it operates the M rejection samplers in parallel, using `NumPy` operations. First, a vector of M proposals is generated, and one computes the vector of indices for which these proposals are accepted. Then a vector of M' proposals are generated a second time, where M' is the number of draws that got rejected in the first time. And so on. The price to pay is that the code is twice longer and much less readable than the version above. More generally, rejection samplers are difficult to implement (efficiently) in interpreted languages. (Another option would be to use `Numba` to make the piece of code above run faster, as discussed in Chap. 9, but we have not yet experimented with this.)

Two-filter smoothing algorithms do not pose any specific difficulty (apart from using as an argument an object representing the output of the information filter); again he interested reader is encouraged to have a look at the source code.

It is easy to recover the genealogical tree of the N final particles from a `ParticleHistory` object: using the $A_t^{1:N}$ variables one can compute recursively the ancestors of each particle X_T^n; see below.

```
def compute_trajectories(self):
    """Compute the complete trajectories.

    Compute and add attribute B to *self*, where
    B[t,n] is the index of ancestor at time t of X_T^n,
    and T is the current length of history.
    """

    self.B = np.empty((self.T, self.N), int)
    self.B[-1, :] = self.A[-1]
    for t in reversed(range(self.T-1)):
        self.B[t, :] = self.A[t+1][self.B[t+1]]
```

Reconstructing trajectories in this way is of course more efficient than carrying forward these trajectories, by recopying over and over these trajectories in bigger and bigger arrays.

The memory cost of saving history until time T is $\mathcal{O}(TN)$. One may be interested only in reconstructing the genealogical tree (as above); uses of that genealogical tree will appear later (in Chap. 16). In such a case, one may want to reduce the memory cost by "pruning the tree", i.e. by discarding particles X_t^n that have no ancestors at time T.

There are various ways to do that. One way is to implement particles as nodes that contain a pointer to their ancestors. When a node is no longer pointed to, it gets removed from memory (by some form of garbage collection). This type of programming is a bit advanced, and makes more sense in a language like C++; see Jacob et al. (2015) for more details.

Alternatively, recall (Remarks 9.1 and 12.1) that the ancestor variables $A_{t+1}^{1:N}$ are ordered: $A_{t+1}^1 \leq \ldots \leq A_{t+1}^N$. Hence, the variables B_t^n defined above (the label of ancestor of X_T^n at time t) are also ordered. This means in particular that all particles X_t^n such that $n < B_t^1$ or $n > B_t^N$ may be discarded, as they do not have off-springs at time T. One may periodically prune the tree in this way, by simply computing recursively the indices of the ancestors of particles X_t^1 and X_t^N at a given time t. Note however we are trading CPU cost for memory cost. In that respect, an interesting property of the genealogical tree is that the time to coalescence (the s such that all particles at time t have a single ancestor at time $t-s$) is usually $\mathcal{O}(\log N)$; see Jacob et al. (2015) for a proof. Thus pruning should reduce memory cost down to $\mathcal{O}(T + N \log N)$. Given the memory vs CPU trade-off, it seems reasonable to prune only from time to time.

Bibliographical Notes

On-line smoothing by genealogy tracking, as described in Sect. 12.1.1 was first introduced in Kitagawa (1996). Fixed-lag smoothing (and the possibility to regenerate at every time step all the components of the block) is discussed in Doucet and Sénécal (2004). Recursion (12.6) has a long history, see Section 2 of Del Moral et al. (2010) for references and a historical perspective.

The general version of the FFBS algorithm (which simulates complete trajectories) was proposed in Godsill et al. (2004); a simpler version restricted to marginal smoothing distributions was presented earlier in Hürzeler and Künsch (1998), Isard and Blake (1998) and Doucet et al. (2000). The $\mathcal{O}(N)$ variant of FFBS based on rejection was introduced in Douc et al. (2011). (In that paper, proposed particles are drawn uniformly from the set of *resampled* particles, but as we have seen it is also possible to draw directly proposed particles from the set of weighted particles.) The idea of using rejection within FFBS appeared previously in Hürzeler and Künsch (1998).

The general version of two-filter smoothing, which requires choosing a pseudo-prior $\gamma_t(\mathrm{d}x_t)$ for each state in the information filter, is due to Briers et al. (2010); the initial version of Kitagawa (1996) corresponds to the special case where $\gamma_t(\mathrm{d}x_t)$ has density $\gamma_t(x_t) = 1$; a choice which is valid only when the corresponding pseudo-posterior is proper (i.e. $\int_{\mathcal{X}} p_T(y_{t+1:T}|x_t)\,\mathrm{d}x_t < +\infty$), as discussed in Briers et al. (2010).

The block-of-three version of $\mathcal{O}(N)$ two-filter smoothing is due to Fearnhead et al. (2010). The idea of using importance sampling to derive a $\mathcal{O}(N)$ approximation of a sum of N^2 terms was used previously in Briers et al. (2005).

Convergence of smoothing algorithms have been studied in Douc et al. (2011) and Gerber and Chopin (2017).

Another approach that one may use to approximate smoothing expectations is to run an MCMC algorithm (see Chap. 15) that targets the smoothing distribution. One particular MCMC scheme that may be used to this aim is Particle Gibbs, which we cover in Chap. 16; other MCMC approaches are briefly showcased in the numerical experiment section of Chap. 15. By and large, MCMC methods tend to be less convenient to use than the particle smoothing methods covered in this chapter. One advantage however is that, by using an elegant coupling construction due to Glynn and Rhee (2014), one may obtain *unbiased* estimates of smoothing expectations from MCMC algorithms, as shown by Jacob et al. (2020) and Middleton et al. (2019); see these papers for more details.

Bibliography

Briers, M., Doucet, A., & Maskell, S. (2010). Smoothing algorithms for state-space models. *Annals of the Institute of Statistical Mathematics, 62*(1), 61–89.

Briers, M., Doucet, A., & Singh, S. S. (2005). Sequential auxiliary particle belief propagation. In *Proceedings of the 8th International Conference on Information Fusion* (Vol. 1).

Del Moral, P., Doucet, A., & Singh, S. (2010). Forward smoothing using sequential Monte Carlo. *arXiv e-prints 1012.5390.*

Douc, R., Garivier, A., Moulines, E., & Olsson, J. (2011). Sequential Monte Carlo smoothing for general state space hidden Markov models. *Annals of Applied Probability, 21*(6), 2109–2145.

Doucet, A., Godsill, S., & Andrieu, C. (2000). On sequential Monte Carlo sampling methods for Bayesian filtering. *Statistics and Computing, 10*(3), 197–208.

Doucet, A., & Sénécal, S. (2004). Fixed-lag sequential Monte Carlo. In *2004 12th European Signal Processing Conference* (pp. 861–864).

Fearnhead, P., Wyncoll, D., & Tawn, J. (2010). A sequential smoothing algorithm with linear computational cost. *Biometrika, 97*(2), 447–464.

Gerber, M., & Chopin, N. (2017). Convergence of sequential quasi-Monte Carlo smoothing algorithms. *Bernoulli, 23*(4B), 2951–2987.

Glynn, P. W., & Rhee, C.-H. (2014). Exact estimation for Markov chain equilibrium expectations. *Journal of Applied Probability, 51*(A (Celebrating 50 Years of The Applied Probability Trust)), 377–389.

Godsill, S. J., Doucet, A., & West, M. (2004). Monte Carlo smoothing for nonlinear times series. *Journal of the American Statistical Association, 99*(465), 156–168.

Hürzeler, M., & Künsch, H. R. (1998). Monte Carlo approximations for general state-space models. *Journal of Computational and Graphical Statistics, 7*(2), 175–193.

Isard, M., & Blake, A. (1998). A smoothing filter for condensation. In H. Burkhardt & B. Neumann (Eds.), *Computer Vision — ECCV'98: 5th European Conference on Computer Vision Freiburg, Germany, June, 2–6, 1998 Proceedings, Volume I* (pp. 767–781). Berlin/Heidelberg: Springer.

Jacob, P. E., Lindsten, F., & Schön, T. B. (2020). Smoothing with couplings of conditional particle filters. *Journal of the American Statistical Association, 115*, 721–729.

Jacob, P. E., Murray, L. M., & Rubenthaler, S. (2015). Path storage in the particle filter. *Statistics and Computing, 25*(2), 487–496.

Kitagawa, G. (1996). Monte Carlo filter and smoother for non-Gaussian state space models. *Journal of Computational and Graphical Statistics, 5*(1), 1–25.

Middleton, L., Deligiannidis, G., Doucet, A., & Jacob, P. E. (2019). Unbiased smoothing using Particle Independent Metropolis-Hastings. *arXiv e-prints* 1902.01781.

Poyiadjis, G., Doucet, A., & Singh, S. S. (2011). Particle approximations of the score and observed information matrix in state space models with application to parameter estimation. *Biometrika, 98*(1), 65–80.

Chapter 13
Sequential Quasi-Monte Carlo

Summary So far, the algorithms we have discussed rely on Monte Carlo, that is, on averages of random variables. QMC (quasi-Monte Carlo) is an alternative to Monte Carlo where random points are replaced with low-discrepancy sequences. The advantage is that QMC estimates usually converge faster than their Monte Carlo counterparts.

This chapter explains how to derive QMC particle algorithms, also called SQMC (sequential quasi-Monte Carlo) algorithms.

13.1 A Very Basic Introduction to Quasi-Monte Carlo

13.1.1 Settings

We return to the initial motivation of Monte Carlo, namely the approximation of any quantity that may be written as an expectation; however, we restrict our attention to expectations with respect to the uniform distribution over $[0, 1]^d$:

$$\mathbb{E}[\varphi(U)] = \int_{[0,1]^d} \varphi(u)\,\mathrm{d}u\,. \tag{13.1}$$

This restriction is not so limiting: we have seen in Chap. 8 that computer-generated random variables are almost always obtained as some deterministic transformation of uniform variables (see Exercise 13.1 for more comments on this point). By the same logic, many expectations of interest may be rewritten as (13.1).

The standard Monte Carlo approximation of the quantity above is

$$\frac{1}{N}\sum_{n=1}^{N} \varphi(U^n)$$

where the U^n are N i.i.d. variables generated from $\mathcal{U}([0, 1]^d)$. Quasi-Monte Carlo replaces these random variables by N deterministic points, chosen so as to have

© Springer Nature Switzerland AG 2020

N. Chopin, O. Papaspiliopoulos, *An Introduction to Sequential Monte Carlo*, Springer Series in Statistics, https://doi.org/10.1007/978-3-030-47845-2_13

'low discrepancy'. Informally, this means that these points must cover the space more regularly than random points would. More formally, the standard definition of the discrepancy of points $u^{1:N}$ in $[0, 1]^d$ is:

$$D^\star(u^{1:N}) = \sup_{I \in \mathcal{I}_d} \left| \frac{1}{N} \sum_{n=1}^{N} \mathbb{1}_I(u^n) - \lambda_d(I) \right| \tag{13.2}$$

where $\mathcal{I}_d$ is the set of d-dimensional intervals of the form $\prod_{i=1}^d [0, b_i)$, and $\lambda_d(I)$ is the volume (Lebesgue measure) of I: $\lambda_d(I) = \prod_{i=1}^d b_i$ for $I = \prod_{i=1}^d [0, b_i)$.

This particular discrepancy is usually called the 'star' discrepancy, hence the star in the notation. It plays a central role in QMC theory, thanks to the Koksma-Hlawka inequality:

$$\left| \frac{1}{N} \sum_{n=1}^{N} \varphi(u^n) - \int_{[0,1]^d} \varphi(u)\, \mathrm{d}u \right| \le V(\varphi) D^\star(u^{1:N}) \tag{13.3}$$

where $V(\varphi)$ is the variation of function φ in the sense of Hardy and Krause (see below). This inequality conveniently separates the properties of φ on one hand, and the choice of the points $u^{1:N}$ on the other hand. Thus, the only task that remains in order to use QMC to approximate integrals is to construct points $u^{1:N}$ such that $D^\star(u^{1:N})$ converges as quickly as possible to 0.

We do not give the exact definition of $V(\varphi)$, the variation in the sense of Hardy and Krause, because it is quite technical. We mention however that it equals the total variation when $d = 1$, see Exercise 13.2, and that $V(\varphi) < +\infty$ if all partial derivatives up to order one in each variable of φ are integrable. On the other hand $V(\varphi) = +\infty$ when φ is an indicator function (unless the corresponding set has probability zero or one). Loosely speaking, (13.3) requires that φ is smooth in some sense. (Recall that Monte Carlo does not have such a requirement.) We refer the readers to Chapter 3 of Leobacher and Pillichshammer (2014) for more background on the Koksma-Hlawka inequality, in particular how $V(\varphi)$ may be interpreted as the norm of a certain RKHS (reproducing kernel Hilbert space).

13.1.2 Low-Discrepancy Point Sets and Sequences

To understand how one may construct a set of points with low discrepancy, we start with the one-dimensional case, $d = 1$. Then an obvious choice is the regular grid:

$$u^n = \frac{2n-1}{2N}, \quad n \in 1:N, \tag{13.4}$$

i.e. u^n is the centre of interval $[(n-1)/N, n/N]$. It is easy to show that $D^\star(u^{1:N}) = 1/2N$, and that this particular configuration actually minimises the star discrepancy in dimension one (Exercise 13.3).

Grids are not extensible: to go from N to $N+1$ points, we must re-evaluate φ at $N+1$ new locations. This is cumbersome if we wish to progressively refine our approximation. In other words, $u^{1:N}$ is not a sequence, and it would be more rigorous to denote it by $u^{N,1:N}$. In the QMC terminology, $u^{N,1:N}$ is called a low-discrepancy point set.

It is possible to construct low-discrepancy sequences. A simple example is the van der Corput sequence in base $b \geq 2$, which is obtained by reversing the base-b representation of integers: i.e., for $n = \sum_{k=0}^{K} \phi_k(n) b^k$, with $0 \leq \phi_k(n) < b$ (K being the number of digits in base b), set $u^n = \sum_{k=0}^{K} \phi_k(n) b^{-k-1}$. For $b = 2$, we obtain

$$\frac{1}{2}, \frac{1}{4}, \frac{3}{4}, \frac{1}{8}, \frac{5}{8}, \ldots$$

as one can check from Table 13.1.

Such a sequence is dense in $[0, 1]$ (i.e. it will eventually visit any given interval included in $[0, 1]$), and has discrepancy $D^\star(u^{1:N}) = (\log N)/N$ (see e.g. Chapter 2 of Leobacher and Pillichshammer (2014) for a proof). In other words, it seems that the price to pay for using a sequence is a $\log N$ factor.

For $d \geq 2$, things get more complicated (unsurprisingly). One way to construct a low-discrepancy sequence in $[0, 1]^d$ is to use d van der Corput sequences; that is component i, $i = 1 : d$ of the sequence is a van der Corput sequence in base b_i. The resulting sequence is called a Halton sequence. It is such that $D^\star(u^{1:N}) = \mathcal{O}\left((\log N)^d / N\right)$, provided that the bases b_i of the d Van der Corput sequences are pairwise co-prime. If this condition is not met, the Halton sequence does not visit the whole hyper-cube $[0, 1]^d$. Consider in particular the case where the same base is used for all the components; then the points fall on the hyper-plane $x_1 = \ldots = x_{d_x}$. A standard recipe is to use the i-th prime number for component i.

It is also possible to construct points sets in dimension d which have discrepancy of order $\mathcal{O}\left((\log N)^{d-1}/N\right)$. Again we observe that we can speed up convergence by a $(\log N)$ factor if we are willing to use a point set instead of a sequence. One

Table 13.1 Construction of the Van der Corput sequence in base 2

n	Binary representation	Reversed binary	u^n
1	1	0.1	1/2
2	10	0.01	1/4
3	11	0.11	3/4
4	100	0.001	1/8
5	101	0.101	5/8
6	110	0.011	3/8
...	...	...	...

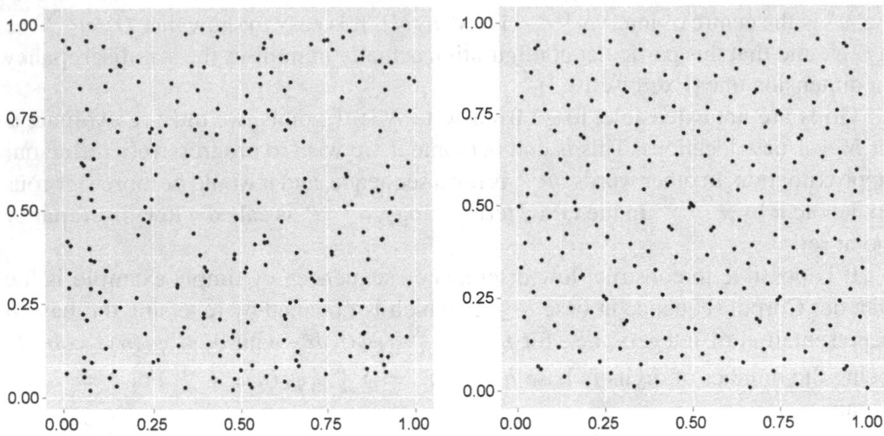

Fig. 13.1 Quasi-Monte Carlo vs Monte Carlo: 256 random variates from $\mathcal{U}([0,1]^2)$ (left) vs 256 first points of Sobol sequence in dimension 2 (right)

example is the Hammersley point set, which is obtained by combining the point set $(0, 1/N, 2/N, \ldots)$ for the first component with a Halton sequence of dimension $d-1$ for the $d-1$ remaining components.

Discussing the different ways to construct low-discrepancy sequences and point sets is quite beyond the scope of this chapter. From now on, we will assume a certain method is used to construct $u^{1:N}$, whose discrepancy converges at rate $\mathcal{O}(N^{-1+\varepsilon})$ for any $\varepsilon > 0$. In fact, we will no longer pay attention to the distinction between point sets and sequences, and stick to the notation $u^{1:N}$. Figure 13.1 compares random points with a QMC sequence. Observe the greater regularity of the latter.

One last comment regarding QMC rates of convergence is that they reveal a curse of dimensionality: asymptotically $(\log N)^d/N \leq N^{-1/2}$, the Monte Carlo error rate, but this inequality holds only for $N \gg 10^{39}$ if we take $d = 10$. To beat MC with QMC, one might need to take N larger and larger when the dimension increases.

In practice, QMC often remains competitive for medium to large dimensions. After all, the theory only provides an upper bound, which might be very conservative in practical cases. But ones does observe empirically that the comparative advantage of QMC tends to deteriorate with the dimension.

13.1.3 Randomised Quasi-Monte Carlo

RQMC (randomised QMC) amounts to re-injecting some randomness in QMC. This looks paradoxical at first: the whole point of QMC was to get rid of random points, so as to obtain faster convergence. Yet randomisation comes with several benefits.

RQMC transforms a QMC point set $u^{1:N}$ into a collection of N random variables, $U^{1:N}$, such that $U^n \sim \mathcal{U}([0,1]^d)$ marginally, for each n. This makes our approximation an unbiased estimate of the quantity of interest, like in Monte Carlo:

$$\mathbb{E}\left[\frac{1}{N}\sum_{n=1}^{N}\varphi(U^n)\right] = \int_{[0,1]^d}\varphi(u)\,du\,.$$

This gives the first benefit of RQMC: We can assess the evaluation error, by simply evaluating our approximation several times, and compute the empirical variance of these evaluations.

A second advantage of RQMC is that it comes with extra guarantees regarding the approximation error, expressed this time in terms of the variance: for instance, Owen (1995, 1997, 1998) established (for a certain type of RQMC construction, known as scrambling), that

$$\mathrm{Var}\left[\frac{1}{N}\sum_{n=1}^{N}\varphi(U^n)\right] = o(N^{-1})$$

for any square-integrable φ; and that if φ is smooth in a certain sense, one may even have (for certain values of N)

$$\mathrm{Var}\left[\frac{1}{N}\sum_{n=1}^{N}\varphi(U^n)\right] = \mathcal{O}(N^{-3+\varepsilon})\,.$$

This means that the error rate is $N^{-3/2+\varepsilon}$, to be compared with $N^{-1+\varepsilon}$ for straight QMC. The fact we gain an extra order of magnitude in the convergence is of course rather striking.

The simplest way to generate RQMC points is to apply a 'random shift' to QMC points $u^{1:N}$:

$$U^n \equiv u^n + V \pmod 1\,,$$

where $V \sim \mathcal{U}([0,1])$, and the modulo operation works component-wise. Clearly, $U^n \sim \mathcal{U}([0,1])$. The scrambling strategy mentioned above applies random permutations to the digits of the base b representation of the $N \times d$ entries of $u^{1:N}$. Again, we refer to the bibliography for more details.

A third advantage of RQMC, which is specific to our settings, is that using a RQMC sequence will allow us to derive an algorithm that still provides an unbiased estimate of the normalising constant, and therefore may be used in the PMCMC framework developed in Chap. 16.

13.1.4 Take-Home Message

From the sake of simplicity, we consider from now on only randomised QMC sequences; i.e. $U^{1:N}$ will denote a collection of N random variables, marginally distributed as $\mathcal{U}([0,1])$, and jointly distributed so that they have low discrepancy (with probability one). (In practice, we shall use scrambled Sobol sequences in our simulations.)

The path to a QMC version of SMC seems now direct: first, rewrite SMC algorithms as a deterministic function of uniform variables, and second, replace these uniform variables with (R)QMC sequences. However, a finer point is that this deterministic function should maintain "low discrepancy" in some sense. We shall revisit this finer point after we have defined the SQMC algorithm.

13.2 SQMC

13.2.1 Preliminaries

For simplicity, we assume from now on that the state space $\mathcal{X}$ is $\mathbb{R}^{d_x}$ or some open subset of $\mathbb{R}^{d_x}$. The considered Feynman-Kac model defines quantities $\mathbb{Q}_t(\varphi)$ that, at time t, are integrals of dimension $d_x(t+1)$. We could rewrite these integrals as expectations with respect to a uniform distribution over $[0,1]^{d_x(t+1)}$, but this seems cumbersome, and does not seem to take into account the sequential nature of Feynman-Kac models.

Instead, let's recall our interpretation of SMC as a sequence of importance sampling steps, where the proposal distribution is partly based on a Monte Carlo approximation from the previous iteration. More precisely, we said in Chap. 10 that step t of Algorithm 10.1 simulates N times from distribution:

$$\sum_{n=1}^{N} W_{t-1}^{n} \delta_{X_{t-1}^{n}}(\mathrm{d}x_{t-1}) M_t(X_{t-1}^{n}, \mathrm{d}x_t) \tag{13.5}$$

and then reweights the simulated values $(X_{t-1}^{A_t^n}, X_t^n)$ according to weight function G_t. The joint distribution (13.5) combines the resampling step (the choice of the ancestor out of N possible values) with the step that simulates X_t^n according to kernel M_t.

This suggests viewing SMC not as one single Monte Carlo exercise, of dimension $d_x(t+1)$ at time t, but as a succession of Monte Carlo exercises of dimension (d_x+1); the $+1$ is for the ancestor variables, as we explain in next section. Moreover, it is the simulation from (13.5) that should be written as a deterministic function of uniform variables.

The next two sections explain how to do that, first for $d_x = 1$, then for $d_x \geq 2$. As a preliminary, we suppose that we know a function Γ_t such that, $\Gamma_t(x_{t-1}, U)$ for

$U \sim \mathcal{U}([0,1]^{d_x})$ has the same distribution as $M_t(x_{t-1}, \mathrm{d}x_t)$ for $t \geq 1$ and $x_{t-1} \in \mathcal{X}$. (At time 0, we assume we know Γ_0 such that $\Gamma_0(U)$ has the same distribution as $\mathbb{M}_0(\mathrm{d}x_0)$.)

Example 13.1 When $\mathcal{X} \subset \mathbb{R}$, we may set $\Gamma_t(x_{t-1}, \cdot)$ to the inverse CDF of distribution $M_t(x_{t-1}, \mathrm{d}x_t)$. When $\mathcal{X} \subset \mathbb{R}^{d_x}$, with $d > 1$, we can generalise this approach by using the Rosenblatt transformation: first component of $\Gamma_t(x_{t-1}, u)$ is the inverse CDF of $X_t(1)|X_{t-1} = x_{t-1}$, evaluated at $u(1)$; second component is the inverse CDF of $X_t(2)|X_{t-1} = x_{t-1}, X_t(1) = x_t(1)$, evaluated at $u(2)$, and where $x_t(1)$ is set to first component of $\Gamma_t(x_{t-1}, u)$, and so on.

Example 13.2 Assume that kernel M_t is such that $X_t|X_{t-1} = x_{t-1} \sim \mathcal{N}(\mu(x_{t-1}), \Sigma(x_{t-1}))$. Let $L(x_{t-1})$ be the lower triangular matrix in the Cholesky decomposition $\Sigma(x_{t-1}) = L(x_{t-1})L(x_{t-1})^T$. Then one way to define Γ_t is

$$\Gamma_t(x_{t-1}, U) = \mu(x_{t-1}) + L(x_{t-1}) \begin{pmatrix} \Phi^{-1}(U(1)) \\ \vdots \\ \Phi^{-1}(U(d)) \end{pmatrix}$$

where Φ^{-1} is the unit Gaussian inverse cumulative distribution function.

13.2.2 Univariate State-Space ($d_x = 1$)

When $d_x = 1$, two uniform variables seem enough to sample from (13.5); one to choose the ancestor, the other to simulate X_t^n, given the ancestor (using function Γ_t). With symbols, $U_t^n \sim \mathcal{U}([0,1]^2)$, $A_t^n = \chi(U_t^n(1))$ for some function χ, and $X_t^n = \Gamma_t(X_{t-1}^{A_t^n}, U_t(2))$.

In SQMC, we use for χ the function that returns for a given input u the index n such that $X_{t-1}^n = F^{-1}(u)$, where F^{-1} is the inverse of the empirical CDF:

$$F(x) = \sum_{n=1}^{N} W_{t-1}^n \mathbb{1}\{X_{t-1}^n \leq x\}; \tag{13.6}$$

see Fig. 13.2. This plot is similar to the inverse CDF plot in Fig. 9.3, except that the points that appear on the x-axis are the N (ordered) ancestors, not the integers 1 to N.

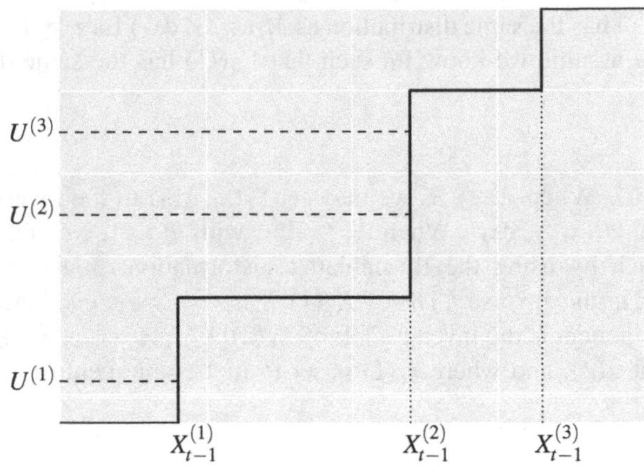

Fig. 13.2 Pictorial description of the inverse CDF algorithm for distribution $\sum_{n=1}^{N} W_{t-1}^{n} \delta_{X_{t-1}^{n}}(\mathrm{d}x_t)$, when $\mathcal{X} \subset \mathbb{R}$. The $X_{t-1}^{(n)}$'s denote the *order statistics* of the ancestors: $X_{t-1}^{(1)} \leq X_{t-1}^{(2)} \leq \ldots \leq X_{t-1}^{(n)}$

Algorithm 13.1 describes how function χ may be computed. It amounts to applying the inverse CDF algorithm to the ordered ancestors. Because of the sort operation, this algorithm has complexity $\mathcal{O}(N \log N)$.

Algorithm 13.1: Function `icdfgen`($X^{1:N}$, $W^{1:N}$, $U^{(1:N)}$) for computing the inverse of CDF $F : x \to \sum_{n=1}^{N} W^n \mathbb{1}\{X^n \leq x\}$ at ordered points $U^{(1)}, \ldots, U^{(N)}$

Input: $X^{1:N}$ ($X^n \in \mathbb{R}$), $W^{1:N}$ (normalised weights), and ordered points $0 < U^{(1)} < \ldots < U^{(N)} < 1$.

Output: $A^{1:N}$ (indices such that $X^{A^n} = F^{-1}(U^{(n)})$)

Function `icdfgen`($X^{1:N}$, $W^{1:N}$, $U^{(1:N)}$):

- $\sigma \leftarrow$ `argsort`($X^{1:N}$) ▷ find σ s.t. $X^{\sigma(1)} \leq \ldots \leq X^{\sigma(N)}$
- $\bar{A}^{1:N} \leftarrow$ `icdf`($W^{\sigma(1:N)}$, $U^{(1:N)}$) ▷ using Algorithm 9.1
- $A^{1:N} \leftarrow \sigma^{-1}(\bar{A}^{1:N})$

The fact we need to order the ancestors is startling at first. In Monte Carlo sampling, ordering the inputs would make no difference. On the other hand, when QMC is involved, the 'regularity' of the functions that transform QMC points becomes important. By using ordered particles, we ensure that the function $u \to X_{t-1}^{\chi(u)}$ (which returns the resampled ancestors) is more regular than the one which would obtain without ordering.

A more formal way to say this is that this particular transformation ensures that the resampled ancestors have the same low discrepancy as the inputs; Exercise 13.4 gives some extra intuition on this point.

Algorithm 13.2 summarises the SQMC algorithm in dimension one.

Algorithm 13.2: SQMC (dimension $d_x = 1$)

Generate RQMC sequence $U_0^{1:N}$ of length N and dimension 1.
$X_0^n \leftarrow \Gamma_0(U_0^n)$ for $n \in 1:N$
$w_0^n \leftarrow G_0(X_0^n)$ and $W_0^n = w_0^n / \sum_{m=1}^N w_0^m$ for $n \in 1:N$
for $t = 1$ **to** T **do**
  Generate a RQMC sequence of length N and dimension 2, and sort it according to its first component; call $U_t^{1:N}$ the result. (Thus $U_t^1(1) < \ldots < U_t^N(1)$.)
  $A_t^{1:N} \leftarrow$ `icdfgen`$(X_{t-1}^{1:N}, W_{t-1}^{1:N}, U_t^{1:N}(1))$ ▷ `using Algorithm 13.1`
  $X_t^n \leftarrow \Gamma_t(X_{t-1}^{A_t^n}, U_t^n(2))$ for $n = 1, \ldots, N$
  $w_t^n \leftarrow G_t(X_{t-1}^{A_t^n}, X_t^n)$ and $W_t^n \leftarrow w_t^n / \sum_{m=1}^N w_t^m$ for $n \in 1:N$

13.2.3 Multivariate State-Space ($d_x \geq 2$)

For $d_x \geq 2$, the CDF $x \to \sum_{n=1}^N W_{t-1}^n \mathbb{1}\{X_t^n \leq x\}$ is obviously not invertible. What we could do instead is to transform the N ancestors (which are points in $\mathbb{R}^{d_x}$) into N scalars. Then we could apply our inverse CDF approach to these N scalars. But we must find a transformation that maintains (in some sense) the low discrepancy of the N original ancestors.

For that purpose, one may use the Hilbert curve. The Hilbert curve of dimension $d \geq 2$ is a space-filling curve, that is, a continuous function $H : [0, 1] \to [0, 1]^d$ which fills entirely the hyper-cube $[0, 1]^d$: $H([0, 1]) = [0, 1]^d$. It may be obtained as the limit of a certain sequence of functions, represented in Fig. 13.3. Function H

Fig. 13.3 The Hilbert curve in two dimensions is defined as the limit of the above sequence

is not a bijection: there exists $x \in [0, 1]^d$ such that the equation $H(u) = x$ has two solutions or more. However, the set of such x's is of measure 0. As a consequence, it is possible to construct a pseudo-inverse $h : [0, 1]^d \to [0, 1]$ such that $H \circ h(x) = x$ for all $x \in [0, 1]^d$.

Function h is not the only way to transform $[0, 1]^{d_x}$ into $[0, 1]$. However, as far as we know, it is the only one which has been proven to maintain discrepancy. That is, for a sequence (X^n) in $[0, 1]^{d_x}$ such that $D^\star(X^{1:N}) \to 0$, the transformed sequence $(h(X^n))$ is also such that $D^\star(h(X^{1:N})) \to 0$. (For a more precise and more general statement, see Theorem 3 of Gerber and Chopin (2015).)

In practice, the N ancestors X_{t-1}^N lie in $\mathcal{X} \subset \mathbb{R}^{d_x}$, so, to apply h, one must first apply a certain preliminary transformation $\xi : \mathcal{X} \to [0, 1]^{d_x}$. If $\mathcal{X} = \mathbb{R}^{d_x}$, a convenient choice is a (properly rescaled) component-wise logistic transform: i.e. component i of vector $\xi(x)$ is $1/(1 + \exp((m_i - x(i))/s_i)$ for some user-chosen location and scale parameters $m_i, s_i > 0$.

Algorithm 13.3 summarises the general SQMC algorithm. The description assumes $d_x \geq 2$, but notice how we recover the $d_x = 1$ algorithm by replacing function $h \circ \xi$ by the identity function.

Input of SQMC

- A Feynman-Kac model such that:
  - the state-space $\mathcal{X}$ is a subset of $\mathbb{R}^{d_x}$, $d_x \geq 1$.
  - its weight function G_t may be evaluated pointwise (for all t);
- A function $\Gamma_0 : [0,1]^{d_x} \to \mathcal{X}$, such that $\Gamma_0(U)$ for $U \sim \mathcal{U}([0,1]^{d_x})$ has the same distribution as $\mathbb{M}_0$.
- Functions $\Gamma_t : \mathcal{X} \times [0,1]^{d_x} \to \mathcal{X}$, such that $\Gamma_t(x_{t-1}, U)$ for $U \sim \mathcal{U}([0,1]^{d_x})$ has the same distribution as $M_t(x_{t-1}, \mathrm{d}x_t)$ (for all $t \geq 1$ and $x_{t-1} \in \mathcal{X}$).
- The number of particles N.
- If $d_x \geq 2$, a bijection ξ between $\mathcal{X}$ and $[0,1]^{d_x}$.

Algorithm 13.3: SQMC (any dimension d_x)

Generate RQMC sequence $U_0^{1:N}$ of length N and dimension d_x.
$X_0^n \leftarrow \Gamma_0(U_0^n)$ for $n = 1, \ldots, N$
$w_0^n \leftarrow G_0(X_0^n)$ and $W_0^n \leftarrow w_0^n / \sum_{m=1}^N w_0^m$ for $n = 1, \ldots, N$
for $t = 1$ **to** T **do**
  Generate a RQMC sequence of length N and dimension $d_x + 1$, and sort it according to its first component; call $U_t^{1:N}$ the result. (Thus $U_t^1(1) < \ldots < U_t^N(1)$.)
  $H_{t-1}^n \leftarrow h \circ \xi(X_{t-1}^n)$ for $n = 1, \ldots, N$
  $A_t^{1:N} \leftarrow$ `icdfgen`$(H_t^{1:N}, W_{t-1}^{1:N}, U_t^{1:N}(1))$ ▷ `using Algorithm 13.1`
  $X_t^n \leftarrow \Gamma_t(X_{t-1}^{A_t^n}, U_t^n(2 : d_x + 1))$ for $n = 1, \ldots, N$
  $w_t^n \leftarrow G_t(X_{t-1}^{A_t^n}, X_t^n)$ and $W_t^n \leftarrow w_t^n / \sum_{m=1}^N w_t^m$ for $n = 1, \ldots, N$

The output of SQMC may be used exactly in the same way as in SMC.

Output of SQMC
Exactly as the output of the generic PF:

$$\frac{1}{N}\sum_{n=1}^{N}\delta_{X_t^n} \quad \text{approximates } \mathbb{Q}_{t-1}(\mathrm{d}x_t)$$

$$\mathbb{Q}_t^N(\mathrm{d}x_t) = \sum_{n=1}^{N} W_t^n \delta_{X_t^n} \quad \text{approximates } \mathbb{Q}_t(\mathrm{d}x_t)$$

$$L_t^N = \prod_{s=0}^{t} \ell_s^N \quad \text{approximates } L_t = \prod_{s=0}^{t} \ell_s$$

13.2.4 Further Considerations

Stating precisely the convergence of SQMC is beyond the scope of this book. We will simply say that, under appropriate conditions, the MSE of a particle estimate $\sum_{n=1}^{N} W_t^n \varphi(X_t^n)$ has been shown to be $o(N^{-1})$ (while it is $\mathcal{O}(N^{-1})$ for SMC). The exact rate of convergence remains an open problem. More importantly, SQMC greatly outperforms SMC in a range of practical problems, at least when the dimension d_x of the state-space is not too large. See Sect. 13.4 for a numerical illustration and a discussion.

The estimate L_t^N remains an unbiased estimate of L_t in SQMC. Thus we may use SQMC as a direct substitute for SMC inside PMCMC algorithms such as PMMH; see Chap. 16. This usually improves the mixing, as the likelihood estimate provided by SQMC tends to have lower variance than the one generated by standard SMC.

If we consider a Feynman-Kac model such that $G_t = 1$ for all t, then $\mathbb{Q}_t(\mathrm{d}x_t)$ reduces to $\mathbb{M}_t(\mathrm{d}x_t)$, the marginal distribution of X_t, for a given Markov process $\{X_t\}$. This means we can use SQMC to sequentially approximate expectations with respect to these marginal distributions. In that case, SQMC collapses to a variant of the array-RQMC algorithm, which predates SQMC.

One point that remains to be discussed is how SQMC may be adapted to perform smoothing, as discussed in Chap. 12 for SMC. This is the topic of the following section.

13.3 Particle Smoothing and SQMC

We discuss briefly how some of the smoothing algorithms presented in Chap. 12 may be turned into QMC algorithms; that is algorithms that produce smoothing estimates that converge faster than their Monte Carlo counter-parts. For more details and theoretical justification, we refer to Gerber and Chopin (2017).

Two-filter smoothing (Sect. 12.5) requires running two particle filters: one forward and one backward. Smoothing estimates are then obtained (at $\mathcal{O}(N^2)$ cost) as a deterministic function of the outputs of both filters. Provided both filters are implemented using SQMC, the corresponding smoothing estimates converge, as expected, faster than the Monte Carlo rate.

FFBS (Sect. 12.3) requires first, running a particle filter, then, simulating full trajectories using a backward pass. We may use a hybrid strategy, where SQMC is used in the first step, and the standard backward pass (Algorithm 12.2) is used in the second step. The so-obtained smoothing estimates converge as $N \to +\infty$, but the Monte Carlo error due to the backward pass tends to dominate, and the improvement brought by SQMC is negligible.

A better strategy is use quasi-Monte Carlo 'all the way', that is, in both steps. To that effect, we now present a QMC version of the backward pass of FFBS. Recall that FFBS reconstructs trajectories backwards in time: at times $t = T-1, T-2, \ldots$, and given a successor $X_{t+1}^{B_{t+1}^n}$, we select randomly one particle X_t^n, according to a certain vector of (modified) weights. This is most similar to the resampling step, and we may use the same ideas to derive a QMC version of FFBS: i.e. we use the inverse CDF algorithm to select randomly a particle at time t, and we use as an input for the inverse CDF a RQMC variate. These remarks lead to Algorithm 13.4.

Algorithm 13.4: FFBS (QMC version)

Input: Integer $M \geq 1$, output of a SQMC particle filter: particles $X_0^{1:N}, \ldots, X_T^{1:N}$, weights $W_0^{1:N}, \ldots, W_T^{1:N}$, and permutations σ_t such that $h \circ \xi(X_t^1) \ldots \leq h \circ \xi(X_t^N)$, for $t = 0 : T$. (Permutation σ_t is computed at iteration t of SQMC inside the body of function `icdfgen`, see Algorithm13.1.)

Output: M sequences of indices $B_{0:T}^m$ in $1 : N$; the t-component of the m-th smoothing trajectory is $X_t^{B_t^m}$.

Generate a RQMC sequence $U_{0:T}^{1:M}$ of dimension $(T+1)$ and length M.

$C_T^{1:N} \leftarrow$ `cumsum`$(W_T^{\sigma_T(1:N)})$

$B_T^{1:M} \leftarrow \sigma_T^{-1}\left(\texttt{searchsorted}(C_T^{1:N}, U_T^{1:M})\right)$ ▷ `Python corner Chap 9`

for $t = T - 1 \to 0$ **do**

  for $m = 1 \to M$ **do**

    for $n = 1 \to N$ **do**

      $\widehat{w}_t^n \leftarrow W_t^n p_{t+1}(X_{t+1}^{B_{t+1}^m} | X_t^n)$

    $C_t^{1:N} \leftarrow$ `cumsum`$\left(\widehat{w}_t^{\sigma_t(1:N)} / \sum_{n=1}^N \widehat{w}_t^n\right)$

    $B_t^m \leftarrow \sigma_t^{-1}\left(\texttt{searchsorted}(C_t^{1:N}, U_t^m)\right)$ ▷ `Python corner Chap 9`

We have focused on $\mathcal{O}(N^2)$ smoothing algorithms; obtaining better a convergence rate for these algorithms is particularly interesting, since increasing N is more expensive. It is not clear how to derive QMC versions of the $\mathcal{O}(N)$ smoothing algorithms presented in Chap. 12.

13.4 Numerical Experiments

We now present several numerical experiments that illustrate the increased accuracy brought by SQMC. We define, for a given estimate, the SQMC gain as the ratio of the variance (based on 100 runs) of that estimate when obtained with plain SMC, over the variance of the same estimate when obtained with SQMC. A gain of 10 means that the SMC algorithm would need about ten times more particles to match the MSE of the corresponding SQMC algorithm.

13.4.1 Filtering and Likelihood Evaluation in a Toy Example

We consider the bootstrap formalism of the following popular toy example (due to Gordon et al. 1993): $X_0 \sim \mathcal{N}(0, 1)$, and

$$X_t = \delta_1 X_{t-1} + \delta_2 \frac{X_{t-1}}{1 + X_{t-1}^2} + \delta_3 \cos(\delta_4 t) + U_t$$

$$Y_t = \gamma X_t^2 + V_t$$

where $U_t \sim \mathcal{N}(0, \sigma^2)$, $V_t \sim \mathcal{N}(0, 1)$. The data $y_{0:T}$ were simulated from the model (taking $T + 1 = 100$, $\delta_1 = 0.5$, $\delta_2 = 25$, $\delta_3 = 8$, $\delta_4 = 1.2$, $\gamma = 0.05$, and $\sigma^2 = 10$). This model is a popular benchmark, as it is highly non-linear, and the observation density may be bimodal.

The left panel of Fig. 13.4 plots the SQMC gain for the filtering expectation, $\mathbb{E}[X_t | Y_{0:t} = y_{0:t}]$, for two different values of N. We obtain gains larger than 10^5 at certain times. The right panel plots the range of log-likelihood estimates obtained over 100 runs, as a function of N.

These results are (in our experience) representative of the performance of SQMC in univariate state-space models; see also e.g. Figs. 14.1 and 14.2, and surrounding comments. We now turn to higher-dimensional models.

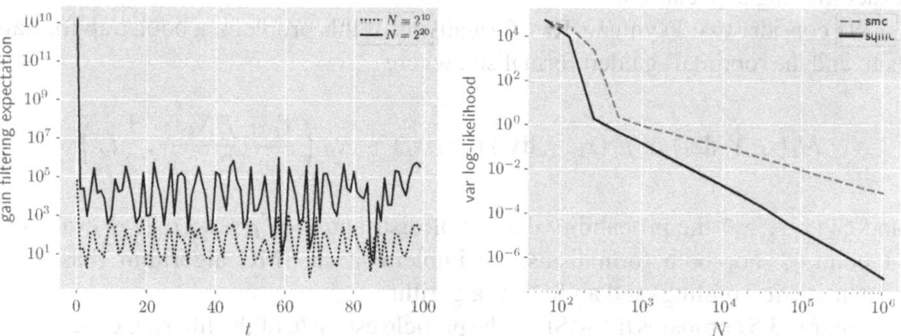

Fig. 13.4 Toy example. Left: SQMC gain for the filtering expectation, as a function of t, for different values of N. Right: empirical variance (over 100 runs) of the log-likelihood estimate as a function of N; light (resp. dark) grey is SMC (resp. SQMC)

13.4.2 Impact of Dimension of the State Space

Numerical experiments in the literature suggests that SQMC suffers from a curse of dimensionality; i.e. its performance gain (relative to SMC) seems to decrease sharply with the dimension of the state space. There are several factors that could explain this:

1. The inherent curse of dimensionality of QMC: the convergence rate of QMC estimates deteriorates with the dimension.
2. Regularity of the Hilbert curve: The SQMC algorithm relies on the inverse of the Hilbert curve, which is $1/d_x$-Hölder. In words, this function is less and less regular as the dimension increases. (Recall that the performance of QMC estimates also depends on the regularity of the function that transforms the uniform variates.)
3. Curse of dimensionality of importance sampling: SQMC (like SMC) relies on importance sampling, which also suffers from a curse of dimensionality (Chap. 8): the larger the dimension of the state space, the greater the discrepancy between the proposal distribution and the target distribution.

To distinguish between these various effects, we adapt slightly the first numerical experiment of Chopin and Gerber (2018). We consider a family of linear Gaussian models, where: $X_0 \sim \mathcal{N}_d(0, I_{d_x})$,

$$\begin{aligned} X_t &= F X_{t-1} + V_t, \qquad & V_t &\sim \mathcal{N}_d(0, I_d), \\ Y_t &= X_t + W_t, \qquad & W_t &\sim \mathcal{N}_d(0, I_d), \end{aligned}$$

with $F_{ij} = \alpha^{|i-j|}$, $1 \leq i, j \leq d$, and $\alpha = 0.3$. We simulate $T = 50$ data-points from the model, for $d_x = 5$, 10, 15 and 20. We use the Kalman filter to compute the exact filtering expectations.

We consider two Feynman-Kac formalisms of this problem: a bootstrap formalism, and the 'optimal' guided formalism, where

$$M_t(x_{t-1}, \mathrm{d}x_t) \propto P_t(x_{t-1}, \mathrm{d}x_t) f_t(y_t|x_t) \sim \mathcal{N}_d\left(\frac{Y_t + F X_{t-1}}{2}, \frac{1}{2} I_d\right)$$

and $G_t(x_{t-1}, x_t)$ the probability density of distribution $\mathcal{N}_d(F x_{t-1}, 2I_d)$ evaluated at point y_t. For both formalisms, we implement an SMC algorithm (based on systematic resampling) and an SQMC algorithm.

Figure 13.5 compares the MSE of the particle estimate of the filtering expectation of the first component obtained from the four considered algorithms (run with $N = 10^4$): SMC-bootstrap, SQMC-bootstrap, SMC-guided, and SQMC-guided. SMC-guided is used as a reference: we plot MSE ratios (MSE of the reference algorithm divided by MSE of the considered algorithm) for the three other algorithms. The violin plots represent the variability of these ratios over the $(T + 1)$ considered filtering expectations.

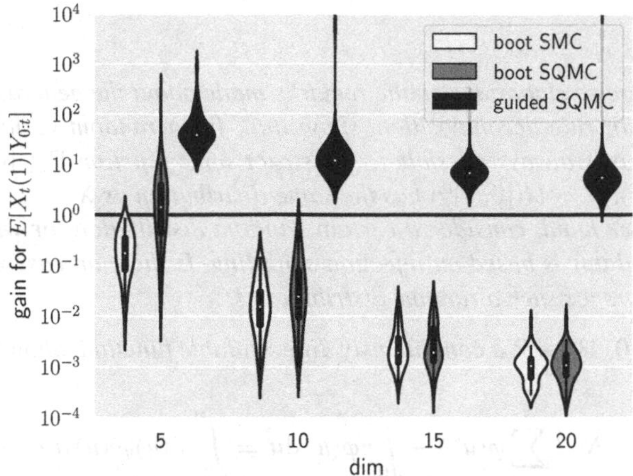

Fig. 13.5 Violin plots of MSE ratios (MSE of reference algorithm divided by MSE of considered algorithm) when estimating the filtering expectations $\mathbb{E}[X_t(1)|Y_{0:t}]$ for $t = 0, \ldots, T = 50$. Each violin plot represents the variability of the $(T+1)$ ratios for these $(T+1)$ estimates. The reference algorithm is guided SMC

We make two observations from Fig. 13.5. First, the guided algorithms outperform significantly the bootstrap algorithms; in fact the gap of performance between the two types of algorithms widens as d_x increases. This is directly related to point 3 above: the larger d_x is, the greater is the discrepancy between the proposal and the target of each importance sampling step of the bootstrap algorithms.

Second, the MSE reduction brought by SQMC is much larger for the guided formalism, and remains non-negligible even for $d_x = 20$ (while it has vanished for $d_x \geq 10$ for bootstrap algorithms).

This phenomenon has an intuitive explanation: the point of QMC is to distribute particles across space in a more regular manner than Monte Carlo. However, if most particles get a negligible weight, this regular distribution should be much less useful (in reducing the variance of estimates). This seems to be what happens for the bootstrap algorithms in large dimension.

The bottom line is that the performance of SQMC may depend more strongly on the choice of the proposal kernels than on the dimension of the state-space. Practically, if we manage to construct good kernel proposals that ensure that the ESS does not drop too much at each iteration, then there is a good chance that SQMC may provide a significant performance gain, even if the dimension is not small.

Exercises

13.1 *This exercise elaborates on the remarks made about the generality of* (13.1). *Using the chain rule decomposition, show that, for a random variable X taking values in* $\mathbb{R}^d$*, it is always possible to construct some function* $\Gamma : [0, 1]^d \to \mathbb{R}^d$ *such that* $\Gamma(U)$*,* $U \sim \mathcal{U}([0, 1]^d)$ *has the same distribution as X.*

On the other hand, consider a certain random distribution for which the only available simulator is based on rejection sampling. Is there an easy way to derive QMC estimators for such a random distribution?

13.2 *Let* $\varphi : [0, 1] \to \mathbb{R}$ *a continuously differentiable function. Show that*

$$N^{-1} \sum_{n=1}^{N} \varphi(u^n) - \int_0^1 \varphi(u)\,\mathrm{d}u = \int_0^1 \delta(u)\varphi'(u)\,du$$

where $\delta(u) = u - N^{-1} \sum_{n=1}^{N} \mathbb{1}(u^n \leq u)$*. Deduce from the definition of the star discrepancy* (13.2) *the Koksma-Hlawka inequality in dimension* $d = 1$*. In that case,* $V(\varphi) = \int_0^1 |\varphi'(u)|\,\mathrm{d}u$*, the total variation of function* φ*.*

13.3 *Show that, for the function* δ *defined in the previous exercise, for an arbitrary point set* $u^{1:N}$*, and for a sufficiently small* $\varepsilon > 0$*, it is possible to find u, v such that* $|\delta(u) - \delta(v)| \geq N^{-1} - \varepsilon$*. (Take for instance u and v immediately before and after some point* u^n*.) Deduce that* $D^\star(u^{1:N}) \geq 1/2N$ *for all point sets* $u^{1:N}$*. Show that, when* $u^{1:N}$ *is the grid defined in* (13.4)*, function* δ *is piece-wise linear, with pieces that vary from* $-1/2N$ *to* $1/2N$*. Deduce that this grid indeed minimises the star discrepancy in one dimension, as claimed at the beginning of Sect. 13.1.2.*

13.4 *One may wonder why the uniform distribution is given such a strong focus in QMC theory. Consider the following generalisation of the star discrepancy, when* $d = 1$*:*

$$\sup_{b\in\mathbb{R}} \left| \frac{1}{N} \sum_{n=1}^{N} \mathbb{1}\{x^n \leq b\} - F(b) \right|$$

where the x^n*'s are N points in* $\mathbb{R}$*, and where* $F : \mathbb{R} \to [0, 1]$ *is the CDF of a certain real-valued probability distribution. (The quantity is in fact the Kolmogorov-Smirnov statistic that one uses in goodness-of-fit testing.)*

Consider first the case where F is continuous and strictly increasing, and define $u^n = F(x^n)$*. Show that the quantity above equals the star discrepancy of* $u^{1:N}$*. Discuss.*

Consider then the case where F is piecewise constant. This time, assume that the x^n *have been obtained as* $x^n = F^{-1}(u^n)$*, where* F^{-1} *is the generalised inverse of F. Show that the quantity above is smaller than the star discrepancy of the* u^n*'s. Discuss the implication for the resampling of ancestors, as described in Sect. 13.2.2.*

13.5 *Consider the Markov kernel M_t such that $X_t|X_{t-1} = x_{t-1}$ follows a Student distribution of dimension $d_x > 1$ (with parameters that depend on x_{t-1}). The Rosenblatt transformation of a Student distribution is hard to compute. Instead we may use the property that, if $Y \sim \mathcal{N}_d(0, \Sigma)$, $Z \sim \chi^2(\kappa)$, then $Y/\sqrt{Z/\kappa}$ follows a Student distribution. Explain how you may use this property to rewrite the considered model so that the underlying Markov chain lies in $\mathbb{R}^{d_x+1}$, and use the Rosenblatt transform that corresponds to this extended Markov chain to define functions Γ_t to be used in the SQMC algorithm.*

Python Corner

We have seen that, for a given Feynman-Kac model, SQMC requires the specification of functions Γ_t such that $\Gamma_t(x_{t-1}, U)$ has the same distribution as $M_t(x_{t-1}, dx_t)$ when $U \sim \mathcal{U}([0, 1]^{d_x})$ (and likewise for Γ_0 and $\mathbb{M}_0$). In `particles`, this is achieved by adding methods `Gamma0` and `Gamma` to a given Feynman-Kac class, as follows:

```
import particles
from particles import distributions as dists

class Bootstrap_SV(particles.FeynmanKac):
    ... # As in Chapter 5

    def Gamma0(self, u):
        sigma0 = self.sigma / np.sqrt(1. - self.rho**2)
        return dists.Normal(scale=sigma0).ppf(u)

    def Gamma(self, t, xp, u):
        return dists.Normal(loc=self.mu + self.rho * (xp - self.mu),
                            scale=self.sigma).ppf(u)
```

Method `ppf` returns the quantile function of a given distribution.

Once this is done, one may run SQMC (Algorithm 13.3) by setting option `qmc=True`:

```
# as in Chapter 5
_, y = dists.Normal().rvs(size=100)  # artificial data
fk_boot_sv = Bootstrap_SV(mu=-1., sigma=0.15, rho=0.9, data=y)

mypf = particles.SMC(fk_boot_SV, N=100, qmc=True)
mypf.run()
```

We have explained in the Python corners of Chaps. 5 and 10 how to automatically define, from a given state-space model, the Feynman-Kac models that correspond to a bootstrap or guided filter: i.e. by calling classes `Bootstrap` and `GuidedPF`. These classes also define automatically methods `Gamma0` and `Gamma`. If the state-space is univariate, these methods return the inverse CDF of the corresponding distributions. If the state-space is of dimension 2 or more, and if the transition kernel is such that, for a given x_{t-1}, $P_t(x_{t-1}, dx_t)$ is a multivariate Gaussian $\mathcal{N}_d(\mu, \Sigma)$, then Γ computes $u \rightarrow \mu + C(\Phi^{-1}(u(1)), \ldots, \Phi^{-1}(u(d_x))$, where C is the Cholesky lower triangle of Σ, $CC^T = \Sigma$; as discussed in Example 13.2. Other multivariate distributions will be implemented in the future.

Here are some pointers for anyone willing to generate QMC sequences in Python. At the time of writing, there is unfortunately no extensive QMC library available in Python. Fortunately, there is such a library in R (`randtoolbox`), and any R library may be accessed in Python through the `rpy2` library:

```
import rpy2
import rpy2.robjects as robjects
from rpy2.robjects.packages import importr
randtoolbox = importr('randtoolbox')

u = randtoolbox.sobol(1000, dim=2, scrambling=1)
#  returns a scrambled QMC sequence of size 1000 and dimension 2
```

For more information, see the on-line documentation of `rpy2` (http://rpy2.bitbucket.org/ and `randtoolbox` (https://cran.r-project.org/web/packages/randtoolbox).

Another possibility is to simply call directly the Fortran routines of `randtoolbox`. Interfacing Fortran and Python is straightforward thanks to the `f2py` utility provided with `NumPy`. This is what is currently done in `particles`.

Bibliographical Notes

The SQMC algorithm was introduced in Gerber and Chopin (2015); see also Gerber and Chopin (2017) for the extension to smoothing. The array-RQMC algorithm mentioned in Sect. 13.2.4 was proposed in Lécot and Tuffin (2004), L'Ecuyer et al. (2006).

The following references are good introductions to QMC: the paper of Owen (2003), Chap. 5 of Glasserman (2004) and Chaps. 5 and 6 of Lemieux (2009). To readers willing to delve a bit more into the theory of QMC, we highly recommend the book of Leobacher and Pillichshammer (2014).

Bibliography

Chopin, N., & Gerber, M. (2018). Sequential quasi–Monte Carlo: Introduction for non-experts, dimension reduction, application to partly observed diffusion processes. In *Monte Carlo and quasi–Monte Carlo methods. Springer proceedings in mathematics & statistics* (Vol. 241, pp. 99–121). Cham: Springer.

Gerber, M., & Chopin, N. (2015). Sequential quasi Monte Carlo. *Journal of the Royal Statistical Society: Series B (Statistical Methodology), 77*(3), 509–579.

Gerber, M., & Chopin, N. (2017). Convergence of sequential quasi-Monte Carlo smoothing algorithms. *Bernoulli, 23*(4B), 2951–2987.

Glasserman, P. (2004). *Monte Carlo methods in financial engineering. Applications of mathematics (New York)* (Vol. 53). New York: Springer. Stochastic Modelling and Applied Probability.

Gordon, N. J., Salmond, D. J., & Smith, A. F. M. (1993). Novel approach to nonlinear/non-Gaussian Bayesian state estimation. *IEE Proceedings F - Communications, Radar and Signal Processing, 140*(2), 107–113.

Lécot, C., & Tuffin, B. (2004). Quasi-Monte Carlo methods for estimating transient measures of discrete time Markov chains. In *Monte Carlo and quasi-Monte Carlo methods 2002* (pp. 329–343). Berlin: Springer.

L'Ecuyer, P., Lécot, C., & Tuffin, B. (2006). Randomized quasi-Monte Carlo simulation of Markov chains with an ordered state space. In *Monte Carlo and quasi-Monte Carlo methods 2004* (pp. 331–342). Berlin: Springer.

Lemieux, C. (2009). *Monte Carlo and quasi-Monte Carlo sampling. Springer series in statistics.* Berlin: Springer.

Leobacher, G., & Pillichshammer, F. (2014). *Introduction to quasi-Monte Carlo integration and applications. Compact textbooks in mathematics.* Cham: Birkhäuser/Springer.

Owen, A. B. (1995). Randomly permuted (t, m, s)-nets and (t, s)-sequences. In *Monte Carlo and quasi-Monte Carlo methods in scientific computing. Lecture notes in statististics* (Vol. 106, pp. 299–317). New York: Springer.

Owen, A. B. (1997). Scrambled net variance for integrals of smooth functions. *Annals of Statistics, 25*(4), 1541–1562.

Owen, A. B. (1998). Scrambling Sobol' and Niederreiter-Xing points. *Journal of Complexity, 14*(4), 466–489.

Owen, A. B. (2003). Quasi-Monte Carlo sampling. In H. W. Jensen (Ed.), *Monte Carlo ray tracing: Siggraph 2003 course 44* (pp. 69–88). San Diego, CA: SIGGRAPH.

Chapter 14
Maximum Likelihood Estimation of State-Space Models

Summary We have seen in Chap. 2 that most state-space models depend on some parameter θ; typically $\theta \in \mathbb{R}^{d_\theta}$. In certain applications, θ is known, while in others it must be learnt from data, either in one go (from some data-set $y_{0:T}$), or sequentially (as datapoints $y_0, y_1, \ldots$ arrive).

This chapter covers maximum likelihood estimation of state-space models. Maximum likelihood estimation is popular in many areas of Statistics, but it is a bit cumbersome for state-space models. For one thing, the likelihood is intractable for such models. Another, perhaps more fundamental problem is the likelihood is rarely a well behaved function (at least as maximisation is concerned). In particular, if one wishes to fully account for parameter uncertainty, one may be better off by adopting a Bayesian approach (covered in Chap. 16).

Methods to compute the maximum likelihood estimator usually combine some particular instance of a general optimisation scheme with some particular implementation of a particle algorithm. Covering all possible combinations, down to the smallest detail, would be tedious. Rather, we provide a general overview, and stress the main ideas. One such idea is that there is a strong connection between maximum likelihood estimation and smoothing. In fact, most approaches will require running one of the smoothing algorithms covered in Chap. 12.

Notation/Terminology In this section we reinstate θ in the kernels, emission densities etc., hence write $p_t^\theta(x_t|x_{t-1})$, $f_t^\theta(y_t|x_t)$, etc., a notation we first introduced in Chap. 2.

© Springer Nature Switzerland AG 2020

N. Chopin, O. Papaspiliopoulos, *An Introduction to Sequential Monte Carlo*, Springer Series in Statistics, https://doi.org/10.1007/978-3-030-47845-2_14

14.1 Generalities on Maximum Likelihood Estimation

14.1.1 Asymptotics

Consider data $y_{0:T}$ and a generic statistical model with likelihood function $\theta \to p_T^\theta(y_{0:T})$. The model is assumed to be identifiable, that is, if $\theta \neq \theta'$, then the functions $y_{0:T} \to p_T^\theta(y_{0:T})$ and $y_{0:T} \to p_T^{\theta'}(y_{0:T})$ differ (in the usual measure-theoretic sense: the set of $y_{0:T}$ where the two functions differ is of positive measure).

A MLE (maximum likelihood estimator) is conventionally defined as:

$$\widehat{\theta}_T \in \arg\max_{\theta \in \Theta}\{\log p_T^\theta(y_{0:T})\} \tag{14.1}$$

i.e. as a global maximiser of the likelihood (or equivalently of its logarithm).

The following result gives some insight on why maximising the likelihood makes sense. (Recall Exercise 8.12.)

Proposition 14.1 *Let $\{f^\theta, \theta \in \Theta\}$ be an identifiable parametric family of probability densities (with respect to a common measure), such that $D_{\mathrm{KL}}(f^\theta || f^{\theta'}) < \infty$ for all $\theta, \theta' \in \Theta$. Let $Y_0, Y_1, \ldots$ be a sequence of i.i.d. random variables, distributed according to density f^{θ_0}, for some $\theta_0 \in \Theta$. Then, for any $\theta \neq \theta_0$,*

$$\mathbb{P}\left(p_T^{\theta_0}(Y_{0:T}) > p_T^\theta(Y_{0:T})\right) \to 1 \quad as\ T \to +\infty$$

where $p_T^\theta(y_{0:T}) = \prod_{t=0}^T f^\theta(y_t)$ is the likelihood of the corresponding model.

Proof The inequality above is equivalent to:

$$\frac{1}{T+1}\sum_{t=0}^{T} \log \frac{f^{\theta_0}(Y_t)}{f^\theta(Y_t)} > 0\,.$$

The left-hand side converges to the Kullback-Leibler divergence

$$D_{\mathrm{KL}}(f^{\theta_0} || f^\theta) := \int_{\mathcal{Y}} f^{\theta_0}(y) \log\left\{\frac{f^{\theta_0}(y)}{f^\theta(y)}\right\} dy$$

almost surely by the law of large numbers. $D_{\mathrm{KL}}(f^{\theta_0}||f^\theta) > 0$ as soon as $\theta \neq \theta_0$, since the model is identifiable. Since almost sure convergence implies convergence in probability, the probability of the inequality above converges to one. □

By using the large deviation theorem, we may prove further than the probability above converges exponentially fast (Exercise 14.1). The result above is stated for an i.i.d. model, but, by looking at the proof, we see that it may be generalised to any model such that a law of large number may be stated (e.g. models for stationary processes).

The proposition above says that the likelihood function always ends up being larger at θ_0 than at any other $\theta \neq \theta_0$. Thus we expect the likelihood to concentrate on a smaller and smaller region around θ_0; in particular its mode, the MLE, should be consistent, $\widehat{\theta}_T \to \theta_0$ almost surely. (To properly formalise the "should" in the previous sentence, see Exercise 14.2.)

Provided that the log-likelihood is differentiable, the MLE is a solution of the equation:

$$\nabla\{\log p_T^\theta(y_{0:T})\} = 0\,. \tag{14.2}$$

Notation/Terminology We use the standard symbol ∇ to denote the gradient of a function. Most gradients in this book are with respect to variable θ. In the few cases there are not, we make this explicit by putting the variable as a subscript, e.g. ∇_ξ; if that variable is a scalar, this notation means of course the corresponding partial derivative.

If we make additional assumptions on the regularity of the log-likelihood, we may apply the following Taylor expansion around the MLE:

$$\log p_T^\theta(y_{0:T}) \approx \log p^{\widehat{\theta}_T}(y_{0:T}) + \frac{1}{2}(\theta - \widehat{\theta}_T)^T H(\widehat{\theta}_T)(\theta - \widehat{\theta}_T) \tag{14.3}$$

where $H_T(\theta)$ is the Hessian (matrix of second derivatives) of the log-likelihood; $H_T(\widehat{\theta}_T)$ should be negative, since $\widehat{\theta}_T$ maximises the log-likelihood. Furthermore, the MLE is typically asymptotically normal:

$$\sqrt{T+1}\left(\widehat{\theta}_T - \theta_0\right) \Rightarrow \mathcal{N}\left(0, \Sigma_0\right)$$

and the asymptotic variance Σ_0 may be consistently estimated by $-H_T(\widehat{\theta}_T)/(T+1)$.

In terms of numerical optimisation, (14.3) implies that, around the mode, the log-likelihood is 'well-behaved', that is, concave. Thus, standard numerical procedures should converge quickly to the mode, provided they are initialised to some point in the vicinity of the mode.

However, one must keep in mind that all the discussion above relies on regularity assumptions and asymptotic approximations. In practice, a non-trivial model may very well generate a "pathological" log-likelihood function for a fixed sample size T; pathologies include non-concavity, multiple local modes, regions where the function is very flat, and so on. The bibliography contains a few pointers to more formal and detailed reviews of MLE asymptotic theory.

14.1.2 Computation

Numerical optimisation is a vast subject. For this discussion we shall focus on three classes of algorithms: (a) derivative-free optimisers; (b) gradient-based optimisers; and (c) the EM algorithm (discussed separately in the next section). In each case, the aim is to maximise some objective function h.

Derivative-free optimisers are algorithms that require being able to compute the objective function h pointwise, but not its gradient. The most commonly used derivative-free optimiser is the simplex (Nelder-Mead) algorithm; in fact it is implemented as the default optimiser in many pieces of software.

The simplex algorithm is not a very popular way to compute the MLE, because it is typically slower than gradient-based algorithms. However it remains a useful fall-back option when the gradient is difficult to compute. To work properly however, the simplex algorithm must be supplied with a 'good' starting point; otherwise it may converge to a local minimum.

Gradient-based optimisers are effectively root-finding algorithms; that is they seek solutions $\theta_\star$ of the equation

$$\nabla h(\theta) = 0 \tag{14.4}$$

which is equivalent to (14.2). They require evaluating the gradient of the objective function pointwise. A popular gradient-based algorithm is gradient ascent, which amounts to the iteration:

$$\theta_n = \theta_{n-1} + \gamma_{n-1} \nabla h(\theta_{n-1}) \,, \qquad \gamma_{n-1} > 0 \,,$$

i.e., at each iteration n we move along the direction given by the gradient. Regarding the γ_n coefficients, two popular choices are (1) setting $\gamma_n = \gamma$ for some user-chosen constant; and (2) use some form of line search to find the γ_n that maximises the increase of the objective function along the gradient direction.

Gradient-based algorithms converge quickly when initialised close to a local mode. On the other hand (and this is the main drawback of such methods) they may become unstable or even diverge when initialised too far from a solution of (14.4). In particular, the algorithm may repetitively 'overshoot', that is move too far in the gradient direction, beyond the closest solution.

Finally, we mention in passing the Newton-Raphson algorithm which relies on first and second derivatives:

$$\theta_n = \theta_{n-1} - H(\theta_{n-1})^{-1} \nabla h(\theta_{n-1})$$

where $H(\theta_{n-1})$ is the Hessian matrix of h at θ_{n-1}. This update equation may be justified by linearising ∇h around θ_{n-1}:

$$\nabla h(\theta_\star) = 0 \approx \nabla h(\theta_{n-1}) + H(\theta_{n-1})(\theta_\star - \theta_{n-1}) \,.$$

The relative performance of Newton-Raphson versus gradient ascent is highly case-dependent. On one hand, the cost of a single iteration is higher for Newton-Raphson, as it requires computing the Hessian, and solving a linear system of d_θ equations. (To compute accurately $A^{-1}b$, for some $d \times d$ matrix A, and vector b, it is better to solve the linear system $Ax = b$, rather than computing the inverse A^{-1} directly. The complexity of solving such a system is $\mathcal{O}(d^3)$.)

On the other hand, Newton-Raphson tends to converge quicker locally, at least when the objective function is concave, and such that the Hessian is not nearly singular for any θ.

For the rest of the discussion, we will focus on gradient ascent as the method of choice for gradient-based optimisation.

14.1.3 The EM Algorithm

The EM (Expectation Maximisation) algorithm may be applied to any model with a 'missing data' structure: i.e., the likelihood, $p^\theta(y)$, may be written as the marginal density of observed variable Y, relative to some joint distribution (X, Y):

$$p^\theta(y) = \int_{\mathcal{X}} p^\theta(x, y)\,\mathrm{d}x\,, \quad \text{where } p^\theta(x, y) = f^\theta(y|x)p^\theta(x)\,\mathrm{d}x,$$

and $f^\theta(y|x)$ is the density of Y given $X = x$, $p^\theta(x)$ is the density of X. (These densities are defined with respect to appropriate, θ-independent dominating measures, μ and ν; see Exercise 14.3.) We will also assume the existence of conditional densities $p^\theta(x\,|\,y)$ to focus on the most relevant aspects of the following developments.

Assuming this particular structure, and given a starting point θ_0, the EM algorithm computes recursively:

$$\theta_n = \arg\max_{\theta\in\Theta} Q(\theta, \theta_{n-1}), \tag{14.5}$$

where

$$Q(\theta, \theta') = \mathbb{E}_{\mathbb{P}^{\theta'}}\left[\log p^\theta(X, y)\,\middle|\, Y = y\right]$$

is an expectation with respect to probability density $p^{\theta'}(x\,|\,y) = p^{\theta'}(x, y)/p^{\theta'}(y)$, the density of X conditional on $Y = y$, when the parameter is θ'.

The EM algorithm is an example of the so-called minorisation algorithms for maximising a function. This minorisation property is established below. To appreciate the main ingredient in the proof, recall Exercise 8.12.

Proposition 14.2 *For an appropriate function $c(\theta)$ (whose computation is not necessary for implementing the EM algorithm),*

$$Q(\theta, \theta') + c(\theta') \leq \log p^{\theta}(y), \quad \forall \theta \in \Theta,$$

with equality if and only if $\theta = \theta'$.

Proof The Kullback-Leibler divergence between $p^{\theta}(x|y)$ and $p^{\theta'}(x|y)$ can be developed using basic probability calculus as:

$$\mathbb{E}_{\mathbb{P}^{\theta'}}\left[\log p^{\theta'}(X|y) \,\middle|\, Y = y\right] + \log p^{\theta}(y) - Q(\theta, \theta').$$

We define $c(\theta')$ to be minus the first term above. Using the fact that the Kullback-Leibler divergence is non-negative we obtain the inequality

$$Q(\theta, \theta') + c(\theta') \leq \log p^{\theta}(y),$$

with strict equality if and only if $\theta = \theta'$, due to the main property of the divergence. □

The idea behind minorisation algorithms is that once we have obtained a function with the minorising property established above, we can immediately build an algorithm that increases the marginal likelihood at every step. To see this notice that if $Q(\theta, \theta') > Q(\theta', \theta')$ then

$$\log p^{\theta}(y) \geq Q(\theta, \theta') + c(\theta') > Q(\theta', \theta') + c(\theta') = \log p^{\theta'}(y).$$

Therefore, increasing the "Q-function" is sufficient for increasing the marginal likelihood.

The EM algorithm is particularly appealing when maximisation (14.5) may be carried out analytically. The following proposition covers a large number of practical cases.

Proposition 14.3 *Assume that the parametric family of joint densities $p^{\theta}(x, y)$ is a natural exponential family, i.e. $p^{\theta}(x, y) = \exp\{\theta^T s(x, y) - \kappa(\theta) - \xi(x, y)\}$, for some measurable functions $s, \xi : \mathcal{X} \times \mathcal{Y} \to \mathbb{R}$, and $\kappa : \Theta \to \mathbb{R}$. Provided κ is differentiable, maximising* (14.5) *amounts to finding θ such that*

$$\nabla\kappa(\theta) = \mathbb{E}_{\mathbb{P}^{\theta_{n-1}}}\left[s(X, y) \mid Y = y\right]. \tag{14.6}$$

The proof is left as a simple exercise (Exercise 14.4). This proposition explains why the EM update has often a closed-form, familiar expression. Indeed, if both X and Y were observed, the MLE would be solution of:

$$\nabla\kappa(\theta) = s(x, y)$$

which is quite similar to (14.6). The following example illustrates this point in the context of state-space models.

Example 14.1 Consider the following univariate linear Gaussian model:

$$X_t = X_{t-1} + U_t, \quad U_t \sim \mathcal{N}(0, \sigma_X^2)$$
$$Y_t = X_t + V_t, \qquad V_t \sim \mathcal{N}(0, \sigma_Y^2)$$

and $X_0 \sim \mathcal{N}(0, \sigma_X^2)$. Clearly, this model falls in the above framework, with $y = y_{0:T}$, $x = x_{0:T}$, and (letting $x_{-1} = 0$ for convenience):

$$p^\theta(x, y) = \left(\frac{1}{2\pi\sigma_X\sigma_Y}\right)^{T+1} \exp\left\{-\frac{1}{2\sigma_X^2}\sum_{t=0}^{T}(x_t - x_{t-1})^2 - \frac{1}{2\sigma_Y^2}\sum_{t=0}^{T}(y_t - x_t)^2\right\}.$$

We recognise a natural exponential family, where

$$\theta = \begin{pmatrix} \frac{1}{\sigma_X^2} \\ \frac{1}{\sigma_Y^2} \end{pmatrix}, \quad s(x, y) = \begin{pmatrix} -\frac{1}{2}\sum_{t=0}^{T}(x_t - x_{t-1})^2 \\ -\frac{1}{2}\sum_{t=0}^{T}(y_t - x_t)^2 \end{pmatrix},$$

and

$$\kappa(\theta) = \frac{T+1}{2}\left\{\log\left(\frac{1}{\sigma_X^2}\right) + \log\left(\frac{1}{\sigma_Y^2}\right)\right\},$$

hence, applying (14.6), we see that the EM update takes the following form: $\theta_n = (1/\sigma_{X,n}^2, 1/\sigma_{Y,n}^2)$, with

$$\sigma_{X,n}^2 = \mathbb{E}_{\mathbb{P}^{\theta_{n-1}}}\left[\frac{1}{T+1}\sum_{t=0}^{T}(X_t - X_{t-1})^2 \,\middle|\, Y_{0:T} = y_{0:T}\right]$$

$$\sigma_{Y,n}^2 = \mathbb{E}_{\mathbb{P}^{\theta_{n-1}}}\left[\frac{1}{T+1}\sum_{t=0}^{T}(y_t - X_t)^2 \,\middle|\, Y_{0:T} = y_{0:T}\right].$$

Of course, if both the X_t's and the Y_t's would be observed, we would obtain the same formulae for the MLE, but without the expectations. In practice, to compute the above expectations, we need to run the Kalman smoother (Chap. 7) at each iteration.

The above example may easily be generalised to a larger class of linear Gaussian models, see Exercises 14.5 and 14.6. In fact, the EM algorithm is often the preferred method to compute the MLE for such models. This remark applies to finite state-space models as well, see Exercises 14.7 and 14.8.

If maximisation (14.5) does not admit a closed-form solution, one may maximise numerically function $\theta \rightarrow Q(\theta, \theta^{n-1})$ at each iteration, using gradient ascent (or Newton-Raphson) steps. Note that exact maximisation is not required: it is sufficient to find some θ_n such that $Q(\theta_n, \theta_{n-1}) \geq Q(\theta_{n-1}, \theta_{n-1})$ to ensure that the algorithm increases monotonically the likelihood.

Like the algorithms we have previously discussed, the EM algorithm may converge to a local mode of the likelihood, depending on the position of the starting point θ_0 in the parameter space. However, compared to e.g. gradient ascent, it tends to be much more stable numerically. Typically, it requires a small number of iterations to get close to a mode, and it does not exhibit the erratic behaviour that gradient ascent occasionally shows when poorly initialised. As a result, the EM algorithm is very often favoured by practitioners, when feasible.

14.2 Issues with Maximum Likelihood Estimation of State-Space Models

Applying maximum likelihood estimation to a state-space model raises three specific issues: first, conditions under which the asymptotic behaviour of the MLE is known are quite restrictive; second, the likelihood function of a state-space model is often ill-behaved; and third, the likelihood function cannot be computed exactly. We discuss these three issues in turn.

14.2.1 Conditions Ensuring a Standard Asymptotic Behaviour for the MLE

Developing the asymptotic theory of maximum likelihood estimation in state-space models is beyond the scope of the book. It seems worth to point out however that currently known results on the asymptotic behaviour of the MLE for such models rely on very restrictive assumptions, such as: (a) Θ is compact (see comments in Exercise 14.2); and (b) the conditions (G) and (M) discussed in Sect. 11.4 hold uniformly in θ (e.g. the observation density $f_t^\theta(y_t|x_t)$ bounded, with bounds that do not depend on θ). Constructing a model that fulfils these conditions is a contrived exercise.

Of course, the fact that these assumptions are so restrictive may be mostly due to the challenging nature of the asymptotic theory of state-space models. However, it remains that, for most practical models, there is no guarantee that the MLE has the expected asymptotic behaviour.

14.2.2 State-Space Models Generate Nasty Likelihoods

Regarding the second issue, consider the following example.

Example 14.2 We apply the basic univariate stochastic volatility model (introduced in Sect. 2.4.3): $X_0 \sim \mathcal{N}\left(\mu, \sigma^2/(1-\rho^2)\right)$,

$$Y_t|X_t = x_t \quad\quad \sim \mathcal{N}(0, e^{x_t})$$
$$X_t|X_{t-1} = x_{t-1} \sim \mathcal{N}\left(\mu + \rho(x_{t-1} - \mu), \sigma^2\right)$$

to the following real data: the 200 first log-returns observed in 1997 for the GBP vs USD exchange rate. We let $\sigma = 0.18$, and plot the contours of the log-likelihood as a function of μ and $\rho \in (-1, 1)$.

The log-likelihood is T-shaped: it is nearly flat with respect to μ when $\rho \approx 1$, while it is quite flat with respect to ρ for $\mu \approx -1.3$. In addition, it is bimodal with a global mode such that $\rho \approx -0.8$. If we add more data, this mode eventually vanishes, while the other mode where $\rho \approx 0.9$ takes over. The likelihood remains T-shaped however.

"Nasty" log-likelihood functions are not a rare occurrence in state-space modelling. By construction, data points are not very informative in this context. In addition, state-space models are often richly parameterised, sometimes to the point that some parameters are hardly identifiable (i.e., the likelihood function is flat with respect to them).

Example 14.3 Consider the theta-logistic model which is used in population ecology, as discussed in Sect. 2.4.5: X_t is the logarithm of the population size, $X_0 \sim \mathcal{N}(0, 1)$,

$$X_t = X_{t-1} + \tau_0 - \tau_1 \exp(-\tau_2 X_{t-1}) + U_t, \quad U_t \sim \mathcal{N}(0, \sigma_X^2)$$
$$Y_t = X_t + V_t, \quad V_t \sim \mathcal{N}(0, \sigma_Y^2),$$

and $\tau_0, \tau_1, \tau_2 > 0$. We simulate data from this model (using $T = 100$, $\tau_0 = 0.15$, $\tau_1 = 0.12$, $\tau_2 = 0.10$, $\sigma_X = 0.47$ and $\sigma_Y = 0.39$, as in Peters et al. (2010)). Note that $T = 100$ is a typical sample size in population ecology. We plot the corresponding log-likelihood as a function of τ_1 and τ_2 (while keeping the other parameters fixed to the true values). We observe that the

(continued)

Example 14.3 (continued)
log-likelihood is not concave and does not vanish at the boundaries of the parameter space. In addition, it seems difficult to estimate simultaneously τ_1 and τ_2.

We draw two lessons from Figs. 14.1 and 14.2. First, state-space models may generate log-likelihood functions that are multi-modal, non-concave, flat in certain directions, and so on; in short, functions that are challenging to optimise numerically (even if we could compute these functions exactly, but that point will be covered in the next section). This is particularly true for methods, such as gradient ascent, which tend to be unstable when applied to non-concave functions.

Second, and perhaps more importantly, maximum likelihood estimation may not always be sensible when dealing with state-space models. In situations such as above, reporting a point estimate (plus possibly a confidence region), does not seem to properly account for parameter uncertainty. A more satisfactory solution would be to consider a Bayesian approach, and report the full posterior distribution of θ; see Chap. 16.

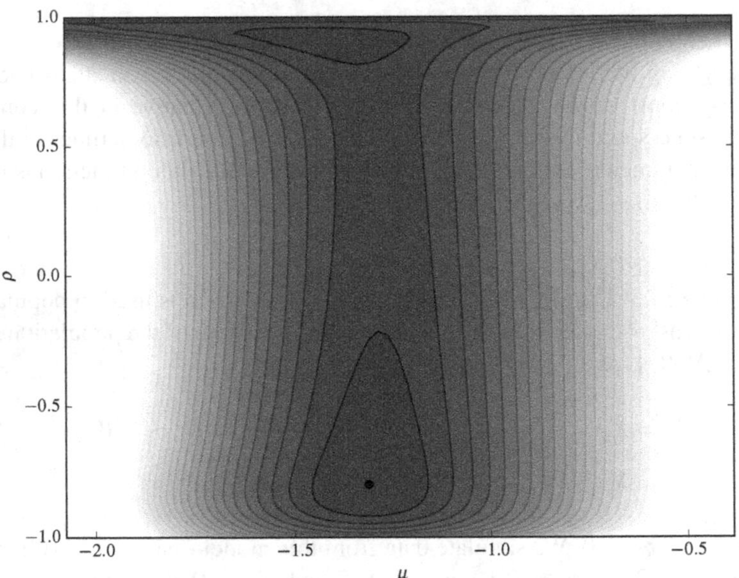

Fig. 14.1 Contour plot of the log-likelihood function of the stochastic model described in Example 14.2. The black dot marks the maximum value; contours correspond to log-likelihood values max log $p_T^\theta(y_{0:T}) - k$, for $k = 1, \ldots, 20$. The likelihood was evaluated by running, for each θ, a bootstrap SQMC algorithm with $N = 10^4$ particles

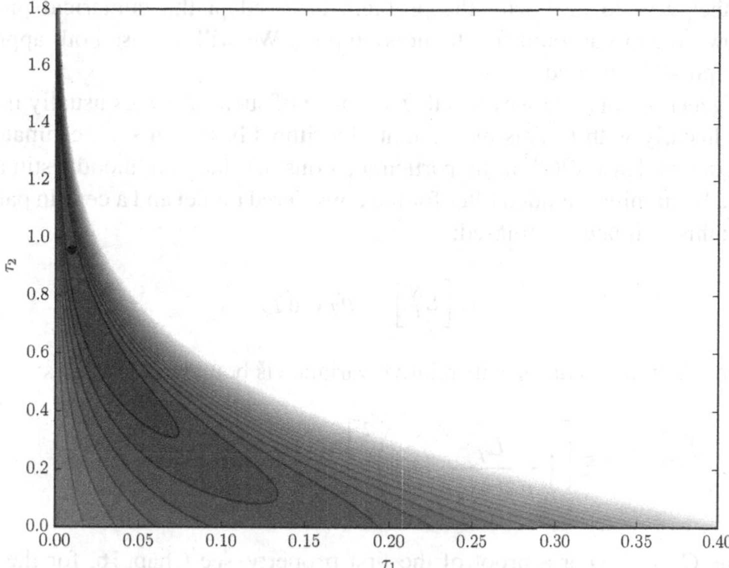

Fig. 14.2 Contour plot of the log-likelihood function of the theta-logistic model described in Example 14.3, based on simulated data ($T = 100$). The black dot marks the maximum value; contours correspond to log-likelihood values $\max \log p_T^\theta(y_{0:T}) - k$, for $k = 1, \ldots, 20$. The likelihood was evaluated by running, for each θ, a guided SQMC algorithm with $N = 10^4$ particles

Of course, having so much parameter uncertainty may be a sign that the model requires to be simplified somehow, e.g. by imposing constraints on the parameters. But, in order to take such decisions, one must detect the high parameter uncertainty in the first place, and maximum likelihood estimation does not seem to be the best approach to do so.

14.2.3 Noisy Estimates

The numerical optimisation procedures described in the previous section require the exact computation of quantities such as the log-likelihood, its gradient, or certain smoothing expectations. For most state-space models, however, the best thing we can do is to obtain a noisy estimate of these quantities, by running a particle filter.

One way to deal with this issue is to use a brute-force approach: set the number of particles N to a large value, so that the noise of the particle estimate of interest becomes negligible, and use it as a direct replacement of the estimated quantity. This approach remains reasonable in simple cases. Figure 14.1 was actually obtained by performing independent runs of the SQMC algorithm ($N = 10^4$) over a grid of parameter values. Note how smooth are the contours.

Another way to deal with the problem is to adapt the numerical procedure somehow, so as to account for the noisy inputs. We will discuss both approaches for each possible method.

One general remark however is that variance of such estimates usually increases at least linearly with t. This means that algorithms based on such estimates have complexity at least $\mathcal{O}(T^2)$. In particular, consider the likelihood estimate L_T^N obtained by running a guided filter for the considered model and a certain parameter value θ; this estimate is unbiased:

$$\mathbb{E}\left[L_T^N\right] = p_T^\theta(y_{0:T})$$

and, under certain conditions, its relative variance is bounded as follows:

$$\mathbb{E}\left[\left(\frac{L_T^N}{p_T^\theta(y_{0:T})} - 1\right)^2\right] \leq \left(1 + \frac{C}{N}\right)^T - 1 \qquad (14.7)$$

for some $C > 0$. (For a proof of the first property, see Chap. 16, for the second property, see Cérou et al. (2011).)

For N fixed, the upper bound in (14.7) blows up exponentially in T. On the other hand, if we take $N = \mathcal{O}(T)$, it converges to a finite limit. These two asymptotic regimes seem to be representative of the behaviour of the likelihood estimate in practice.

Example 14.4 We return to the basic linear Gaussian model used in the numerical example in Sect. 10.5: $X_0 \sim \mathcal{N}(0, \sigma_X^2/(1-\rho^2))$, $X_t|X_{t-1} = x_{t-1} \sim \mathcal{N}(\rho x_{t-1}, \sigma_X^2)$ and $Y_t|X_t = x_t \sim \mathcal{N}(x_t, \sigma_Y^2)$, with $\rho = 0.9$, $\sigma_X = 1$, and $\sigma_Y = 0.2$. The same dataset (simulated from the model) is used throughout.

As a first experiment, we fix N and run 10^3 bootstrap filters until time $T = 100$. Naively, one may want to illustrate (14.7) by plotting the empirical variance (over the 200 runs) of the relative error $L_t^N/p_t^\theta(y_{0:t})$ as a function of t. Unfortunately, such a plot proves to be unreadable and pointless. First, the actual variance is expected to grow exponentially fast. Second, the random variable $L_t^N/p_t^\theta(y_{0:t})$ appears to be very heavy-tailed. This implies that its empirical variance is not a reliable estimate of the actual variance.

Instead, we plot in Fig. 14.3 the inter-quantile ranges for the log-likelihood error, $\log(L_t^N/p_t^\theta(y_{0:t}))$ at iteration t, as a function of t, for $N = 100$ (left) and $N = 1000$ (right). (The exact likelihood is computed using the Kalman filter.) With these plots, we see at least that the spread of the log-error does

(continued)

Example 14.4 (continued)
grow with t (when N is fixed). Note also how large is this spread for $N = 100$ and $t = T = 100$.

As a second experiment, we run 10^3 bootstrap filters for different values of (T, N), with $N = 10T$, and $T = 10^k$, $k = 1, \ldots 4$. Figure 14.4 plots the box-plots of the log-error of the so obtained likelihood estimates. This time, the spread of the error remains stable as T (and N) increases.

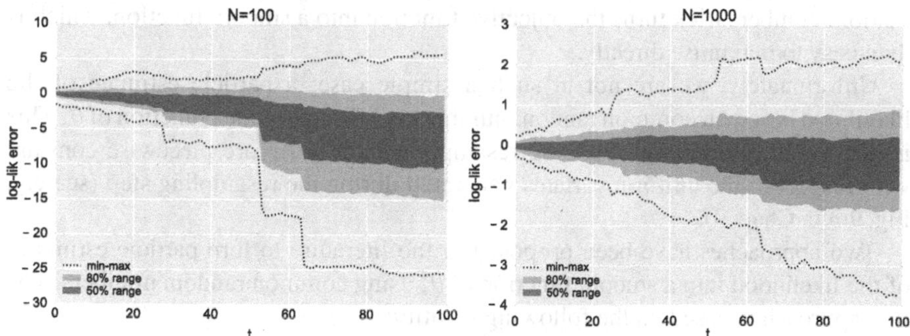

Fig. 14.3 Inter-quantile ranges (over 10^3 independent runs) of $(\log L_t^N - \log p_t^\theta(y_{0:t}))$ for a bootstrap filter (left: $N = 100$, right: $N = 1000$) and the model described in Example 14.4

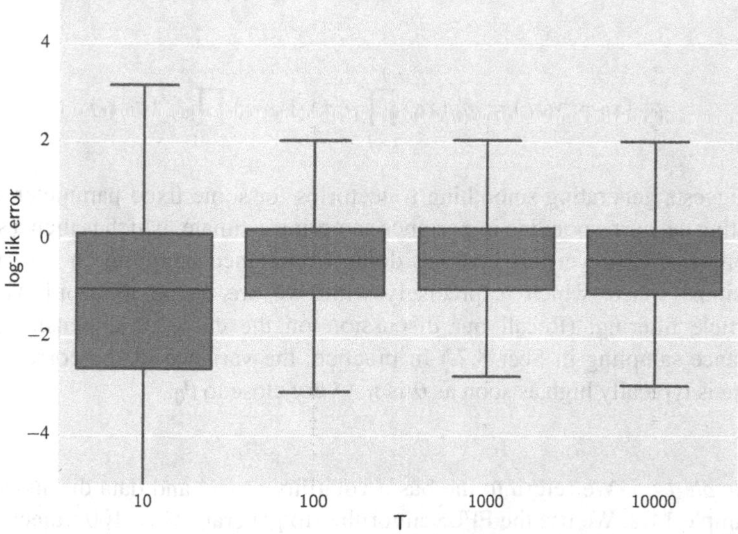

Fig. 14.4 Box-plots of the log-error, $\log(L_T^N / p_T^\theta(y_{0:T}))$ as a function of T, when $N = 10T$, for the model discussed in Example 14.4

14.3 Derivative-Free Approaches to Compute the MLE in a State-Space Model

Maximising the likelihood of a state-space model is a noisy optimisation problem: i.e., optimisation of a function (here, the likelihood) that is evaluated with some noise.

A popular way to perform 'noisy optimisation' is to use 'common random numbers': at each function evaluation, the random number generator is reset to the same seed. This is equivalent to writing the objective function as a function of random variables whose distribution is θ-independent, and to "freeze" these variables to their realisation at the first evaluation. In simple cases, the common random numbers trick turns the objective function into a smooth function, which is then easy to maximise directly.

Unfortunately, we are not in such a simple case: a particle estimate of the likelihood based on common random numbers is a discontinuous function of θ. This is due to the resampling steps: the resampling indices A_t^n are piecewise constant with respect to the uniform variates generated during the resampling step (see e.g. Fig. 9.3 in Chap. 9).

Two approaches have been proposed in the literature to turn particle estimates of the likelihood into a smooth function of θ, using common random numbers. The first approach is based on the following identity:

$$\frac{p_T^\theta(y_{0:T})}{p_T^{\theta_0}(y_{0:T})} = \mathbb{E}_{\mathbb{P}_T^{\theta_0}}\left[\left.\frac{p_T^\theta(X_{0:T}, y_{0:T})}{p_T^{\theta_0}(X_{0:T}, y_{0:T})}\right| Y_{0:T} = y_{0:T}\right] \tag{14.8}$$

where

$$p_T^\theta(x_{0:T}, y_{0:T}) = p_0^\theta(x_0)\prod_{t=1}^{T} p_t^\theta(x_t|x_{t-1})\prod_{t=0}^{t} f_t^\theta(y_t|x_t)\,.$$

This suggests generating smoothing trajectories for some fixed parameter θ_0, and computing the corresponding importance sampling estimate, which is then a smooth function of θ. However this involves doing importance sampling on a very large dimensional space, which is precisely what we are trying to avoid when we do particle filtering. (Recall our discussion on the curse of dimensionality of importance sampling in Sect. 8.7.) In practice, the variance of the corresponding estimate is typically high as soon as θ is not very close to θ_0.

Example 14.5 We return to the basic volatility model and data discussed in Example 14.2. We use the FFBS algorithm to generate $N = 100$ trajectories from the smoothing distribution, corresponding to $\theta_0 = (-1, 0.9, 0.3)$. We

(continued)

Example 14.5 (continued)
plot the ESS of the importance weights in (14.8) as a function of σ, when σ varies but μ and ρ are kept fixed (to the same values as for θ_0); see Fig. 14.5. We see that the ESS drops very quickly as soon as we move away from θ_0, and especially so when T is large.

The second approach is restricted to the univariate case ($\mathcal{X} \subset \mathbb{R}$), and amounts to replacing the inverse CDF used in the resampling step by a continuous approximation. More precisely, resampled particle $X_{t-1}^{A_t^n}$ is replaced by $\tilde{F}^{-1}(U_t^n)$, $U_t^n \sim \mathcal{U}([0,1])$, where $\tilde{F}$ is the continuous, piecewise linear function that goes through the mid-points of the steps of CDF $x \rightarrow \sum_{n=1}^N W_t^n \mathbb{1}\{X_{t-1}^n \leq n\}$. Figure 14.6 illustrates the idea. In that way, the likelihood estimate typically becomes continuous with respect to θ when common random numbers are used. (This is not true for all models, see Exercise 14.9.)

This approach is difficult to generalise to the multivariate case. In addition, the SQMC algorithm works particularly well in the univariate case, and provides likelihood estimates with a very low variance, as seen in Example 14.2, without introducing another level of approximation (due to the linear interpolation).

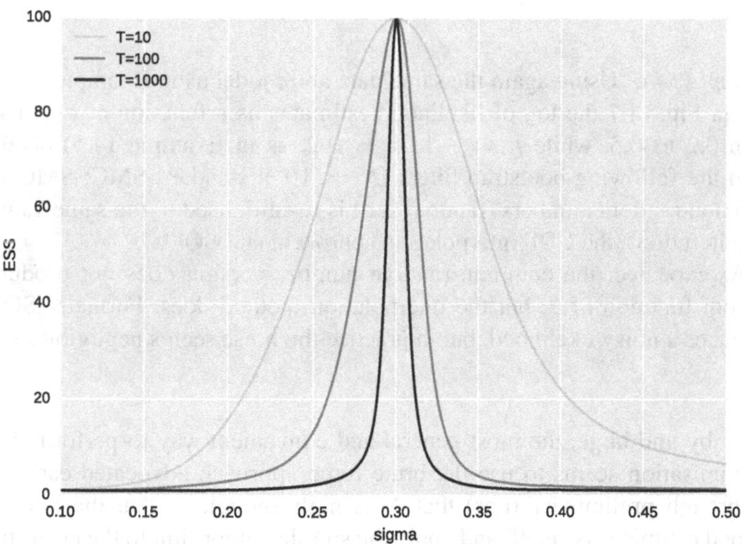

Fig. 14.5 ESS of the importance weights in (14.8) as a function of σ, for the data and the stochastic volatility model discussed in Example 14.2. Parameter is $\theta = (\mu, \rho, \sigma)$, μ and ρ are kept fixed to values -1 and 0.9, respectively, and θ_0 corresponds to $\sigma_0 = 0.3$. The three lines correspond to different values of T: for a given T, the data is the $(T+1)$ log-returns observed since January 1997

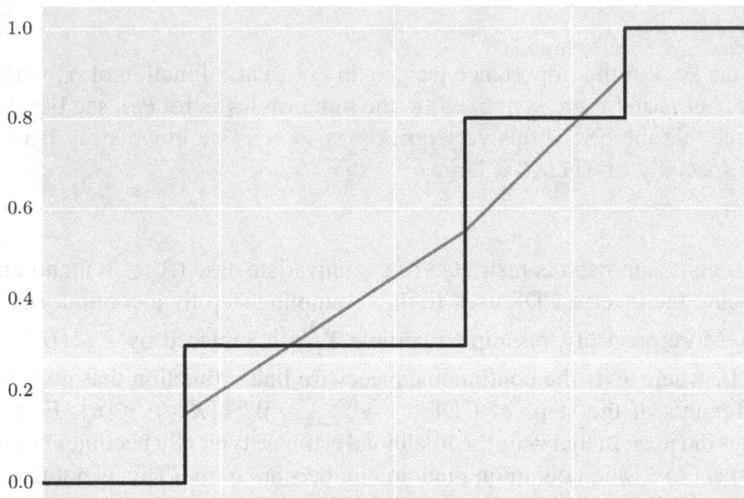

Fig. 14.6 The empirical CDF $x \to \sum_{n=1}^{N} W_t^n \mathbb{1}\{X_t^n \leq x\}$ of $N = 3$ weighted particles (black line), and the corresponding piecewise linear approximation $\widetilde{F}$ that goes through the mid-points of the vertical segments. At the extremities, we define the inverse $\widetilde{F}^{-1}$ to be equal to the smallest or largest particles; i.e. $\widetilde{F}^{-1}(u) = X_t^{(1)}$ for $u < W_t^{(1)}/2$, $\widetilde{F}^{-1}(u) = X_t^{(N)}$ for $u > W_t^{(N)}/2$, where the $X_t^{(n)}$ (and the $W_t^{(n)}$) stand for the ordered particles (and their weights)

Example 14.6 Using again the same data and model as in Example 14.2 we plot in Fig. 14.7 the log of likelihood estimates as a function of σ (varying from 0.2 to 0.5, while $\mu = -1$, $\rho = 0.9$, as in Example 14.5) obtained from the following bootstrap filters ($N = 10^4$): standard SMC; SMC with common random numbers (random seed is re-initialised to the same value at each iteration); the CDF interpolation approach; and SQMC.

As expected, the common random number version does not produce a smooth function of θ, but the interpolation method does. Formally SQMC produces a noisy likelihood, but in practice the noise seems negligible.

Thus, by and large, the most general and convenient way to perform gradient-free optimisation seems to use the brute force approach advocated earlier: make sure (through preliminary runs) that N is high enough so that the noise of the likelihood estimates is small, and apply the simplex algorithm to the corresponding noisy function (which to each θ associates a particle estimate of $\log p_T^\theta(y_{0:T})$). Empirically, the simplex algorithm seems reasonably robust to noisy inputs. The main issue with this approach is that the simplex algorithm is slow to converge

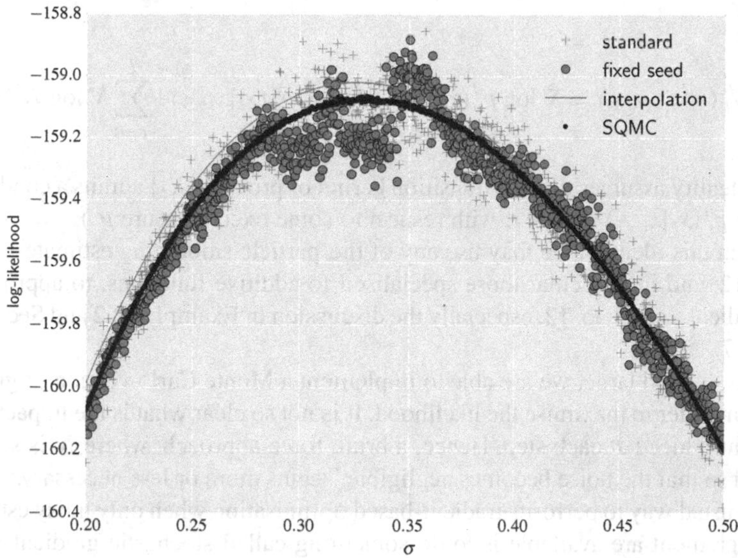

Fig. 14.7 Comparison of likelihood estimates obtained from various bootstrap filters: standard SMC; common random numbers (random seed is fixed to the same value for all the evaluations); interpolation method; and SQMC. Same data and model as Examples 14.2 and 14.5

when the dimension of θ is large. (This issue would remain if the log-likelihood would be computed exactly).

While this approach may be expensive and is not particularly elegant, it has the advantages of being simple, and of not requiring that the considered model admits a tractable transition density, contrary to the approaches we discuss below.

14.4 Gradient Ascent Applied to State-Space Models

We have seen that applying gradient ascent to maximise the likelihood amounts to the iteration:

$$\theta_n = \theta_{n-1} + \gamma_n \nabla \log p_T^\theta(y_{0:T}) \,. \tag{14.9}$$

Fisher's identity (established in Exercise 12.5) expresses the gradient above as a smoothing expectation:

$$\nabla \log p_T^\theta(y_{0:T}) = \mathbb{E}\left[\nabla \log p_T^\theta(X_{0:T}, y_{0:T}) \,\middle|\, Y_{0:T} = y_{0:T}\right]$$

where

$$\nabla \log p_T^\theta(x_{0:T}, y_{0:T}) = \nabla \log p_0^\theta(x_0) + \sum_{t=1}^{T} \nabla \log p_t^\theta(x_t|x_{t-1}) + \sum_{t=0}^{T} \nabla \log f_t^\theta(y_t|x_t) .$$

(This identity assumes that the transition kernel of process $\{X_t\}$ admits a conditional density $p_t^\theta(x_t|x_{t-1})$ at time t, with respect to some fixed measure μ.)

Given this identity, we may use any of the particle smoothing estimates seen in Chap. 12, and in particular those specialised to additive functions, to approximate the gradient, see Chap. 12, especially the discussion in Example 12.2 and Sects. 12.6 and 12.7.

Thus, by and large, we are able to implement a Monte Carlo version of gradient ascent in order to maximise the likelihood. It is not so clear what is the impact of the noise introduced at each step. Hence, a brute force approach, where N is set large enough so that the noise becomes negligible, seems more or less necessary.

The usual way to perform gradient-based optimisation when only noisy estimates of the gradient are available is to do something called stochastic gradient ascent: iterate (14.9), but with the gradient replaced by a noisy estimate, and with γ_n's such that (a) $\sum \gamma_n = \infty$; (b) $\sum \gamma_n^2 < \infty$. These are standard conditions for stochastic approximation schemes. Unfortunately, stochastic gradient descent requires unbiased estimates of the gradient, and the particle smoothing estimates developed in Chap. 12 are not unbiased. However, stochastic gradient descent will prove useful in the on-line context, see Sect. 14.9.

14.5 EM Algorithm for State-Space Models

Assume that the considered state-space model is such that the joint density $p_T^\theta(x_{0:T}, y_{0:T})$ exists and belongs to a natural exponential family:

$$p_T^\theta(x_{0:T}, y_{0:T}) = \exp\left\{\theta^T s_T(x_{0:T}, y_{0:T}) - \kappa_T(\theta) - \xi_T(x_{0:T}, y_{0:T})\right\} .$$

Note that, given the structure of state-space models, functions s_T and ξ_T are necessarily additive; e.g.

$$s_T(x_{0:T}, y_{0:T}) = \psi_0(x_0, y_0) + \sum_{t=1}^{T} \psi_t(x_{t-1}, x_t, y_t) .$$

We may apply directly Proposition 14.3 to derive the corresponding EM update:

$$\theta_n = (\nabla\kappa)^{-1}\left(\mathbb{E}_{\mathbb{P}^{\theta_{n-1}}}\left[s_T(X_{0:T}, y_{0:T}) \mid Y_{0:T} = y_{0:T}\right]\right)$$

where $(\nabla\kappa)^{-1}$ is the inverse of $\nabla\kappa$ (assuming this function is invertible). Again, we find that the update formula involves the smoothing expectation of an additive function, which may be approximated using one of the several particle smoothing algorithms presented in Chap. 12.

Formally, the so-obtained algorithm belongs to the class of MCEM (Monte Carlo EM) algorithms. For such an algorithm, the monotonicity obtained earlier does not hold. It is sometimes advocated to increase the number of particles used at each iteration of a MCEM algorithm (linearly, or super-linearly). Formally, it is the only way to ensure actual convergence. Practically, it seems reasonable to allocate less CPU effort to the first iterations, as such iterations perform big steps in the parameter space, and seems able to do such steps even when provided with rough estimates. The caveat however is that particle smoothing estimates are sometimes quite unstable for small values of N; see e.g. the box-plots for $N = 50$ in Fig. 12.4. Thus, one should avoid starting the EM algorithm with too small a number of particles.

For simplicity, we focused on the case where the joint density has an exponential form. In some cases, the parameter θ may be decomposed into two blocks, $\theta = (\theta_1, \theta_2)$, such that the joint density is exponential in θ_1 (for a fixed θ_2), but not in θ. One then may perform several coordinate ascent steps: maximise with respect to θ_1 (while keeping θ_2 fixed) using the corresponding tractable formula, then maximise with respect to θ_2 (while keeping θ_1 fixed) using numerical optimisation.

14.6 Numerical Experiment

We now compare the different methods discussed so far on a neuroscience experiment taken from Temereanca et al. (2008): the observation y_t at time t is the number of activated neurons over $M = 50$ experiments, and is modelled as $Y_t|X_t \sim \mathcal{B}in(50, \text{logit}^{-1}(X_t))$, where $\text{logit}^{-1}(x) = 1/(1+e^{-x})$, and

$$X_t = \rho X_{t-1} + \sigma U_t, \quad U_t \sim \mathcal{N}(0, 1)$$

for $t \geq 1$. For $t = 0$, $X_0 \sim \mathcal{N}(0, \sigma^2)$. We use the same dataset as in Temereanca et al. (2008), for which $T + 1 = 3000$. The parameter is thus $\theta = (\rho, \sigma^2)$, and $\Theta = [0, 1] \times \mathbb{R}^+$. We only consider the bootstrap formalism throughout.

The log-likelihood function is a bit challenging to maximise, see the contour plot in Fig. 14.8: first, the parameter space is constrained, and the actual MLE is very close to the boundaries of Θ (lower right corner). Second, the contours have an elongated shape which may be difficult to explore; see below.

We note in passing that these contours were obtained by approximating the log-likelihood (using SQMC and $N = 100$) on a grid. Incidentally, this grid computation turns out to be the simplest way to find the MLE in this particular example; and also the fastest on a multi-core machine, as it is trivially parallelisable. Of course, this

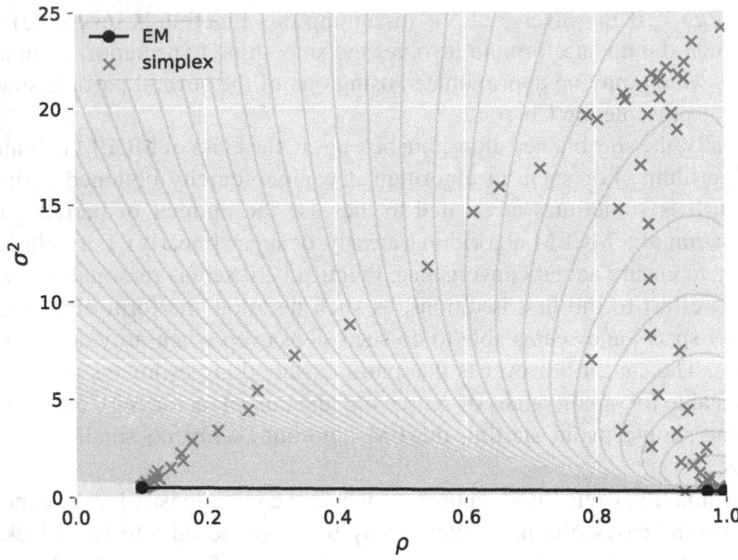

Fig. 14.8 Log-likelihood function of neuroscience experiment example (gray contours), and iterates obtained from EM and Nelder-Mead, when started from the lower left corner. The MLE is in lower right corner. See text for more details

basic approach works well here because of the small dimension of Θ. Moreover, the approximation error of a grid approach cannot be smaller than the resolution of the grid. Still, it is worth noting that such a brute force approach might be expedient in simple cases.

Figure 14.8 also plots the iterates of the following algorithms: EM, and Nelder-Mead, when started in the lower left corner ($\rho_0 = 0.1, \sigma_0^2 = 0.5$). The EM algorithm requires only two iterations to be in the vicinity of the MLE (subsequent points cannot be distinguished on the plot). In addition, EM is simple to implement here, as the maximisation step leads to explicit formulae, the solution of which respects the constraints of the parameter space by construction (see Exercise 14.10). The only drawback of EM here is that each iteration is a bit expensive: we use FFBS-reject (with $N = 100$) to approximate the smoothing expectations that appear in the maximisation step, but the rejection rate is low at all iterations, and particularly so at the first ones, when the parameter estimate is far from the MLE.

On the other hand, each iteration of the Nelder-Mead (simplex) approach is quite fast (we used SQMC with $N = 500$ to approximate the likelihood), but the number of iterations required to reach convergence is much larger, as the algorithm makes a long 'detour' because of the particular shape of the log-likelihood function. We also observed that for other starting points, or lower values of N, the algorithm may not converge. One issue is that when the algorithm 'wanders' too far away from the MLE, the variance of the likelihood estimate becomes large, making the algorithm unstable. Another issue is the constraints imposed on the parameter space:

to implement these constraints, the log-likelihood evaluation function returns $-\infty$ when the argument $\theta = (\rho, \sigma^2)$ is outside $\Theta = [0, 1] \times \mathbb{R}^+$. But in some cases, this makes the algorithm get stuck when it gets close to the boundary of the parameter space. By and large, the simplex algorithm requires far more trial and error than EM in this particular example.

The same could be said for gradient ascent (results not shown). In fact, we did not manage to find a (fixed) value for γ, in (14.9), that makes the algorithm converge in less than 400 iterations. (Bear in mind that each iteration in gradient ascent requires computing smoothing expectations, which is potentially expensive). This seems in line with the common wisdom that gradient-based algorithms may not be competitive with other algorithms when the dimension of Θ is small.

14.7 Final Recommendations

When feasible, the EM approach usually represents the best trade-off in terms of performance versus user's convenience. Thus it would be our default recommendation. When EM not feasible (for instance because the maximisation of function $Q(\theta, \cdot)$ is difficult in some sense), the relative performance and ease of use of gradient-free optimisation and gradient ascent seem to be model-dependent. At the very least, the gradient-free approach is typically easier to implement and to tune, at least when the dimension of θ is not too large.

Whatever the chosen method, we recommend to run it several times with different starting points. This is a good way to assess simultaneously (a) local convergence; and (b) the random fluctuations of the output (which results from the Monte Carlo noise introduced at each step). A good candidate for the starting point is an easy-to-compute consistent estimate, when available; see Exercise 14.11.

14.8 Obtaining Confidence Intervals

Once the MLE $\widehat{\theta}_T$ is obtained, it is often useful to compute the Hessian of the log-likelihood at $\widehat{\theta}_T$, so as to derive confidence intervals. This Hessian may be approximated by using Louis' formula, which is the second-order analogue of Fisher's identity:

$$\nabla^2 \log p_T^\theta(y_{0:T}) = -\left\{\nabla \log p_T^\theta(y_{0:T})\right\}\left\{\nabla \log p_T^\theta(y_{0:T})\right\}^T \\ + \mathbb{E}_{\mathbb{P}^\theta}\left[\phi_{\text{Louis}}(X_{0:T}) \mid Y_{0:T} = y_{0:T}\right]$$

where

$$\phi_{\text{Louis}}(x_{0:T}) = \nabla^2 \log p^\theta(x_{0:T}, y_{0:T}) + \left\{\nabla \log p^\theta(x_{0:T}, y_{0:T})\right\} \left\{\nabla \log p^\theta(x_{0:T}, y_{0:T})\right\}^T$$

and the first and second derivatives of $\log p^\theta(x_{0:T}, y_{0:T})$ have the usual additive expressions, e.g.

$$\nabla \log p^\theta(x_{0:T}, y_{0:T}) = \nabla p_0^\theta(x_0) + \sum_{t=1}^{T} \nabla p_t^\theta(x_t|x_{t-1}) + \sum_{t=0}^{T} \nabla f_t^\theta(y_t|x_t)\,.$$

Note however that test function ϕ_{Louis} itself is not additive. Still it makes it possible to approximate the Hessian by particle smoothing, provided that the densities $p_t^\theta(x_t|x_{t-1})$ and $f_t^\theta(y_t|x_t)$ exist and admit tractable first and second derivatives.

14.9 On-Line Maximum Likelihood Estimation of State-Space Models

On-line maximum likelihood estimation aims at computing, at each time t, an estimate $\widehat{\theta}_t$ based on the data $y_{0:t}$, such that the sequence $\{\widehat{\theta}_t\}$ has the same asymptotic behaviour as the MLE. The corresponding on-line algorithms are constrained to do a single pass on the data; i.e. at iteration t they only have access to y_t (plus perhaps a small number of recent values). As a result, their complexity is linear with respect to time.

In some problems, on-line estimation may be of genuine interest, whereas in other problems, we may use an on-line algorithm due to real time constraint, or because T is so large that off-line algorithms are too expensive. (Recall that the complexity of off-line algorithms is generally $\mathcal{O}(T^2)$.)

On-line maximum likelihood algorithms are variations of off-line algorithms based on stochastic approximation. For simplicity, we focus on gradient ascent, but some of the ideas may be applied to obtain an on-line EM algorithm as well.

Consider the following stochastic gradient scheme:

$$\widehat{\theta}_{t+1} = \widehat{\theta}_t + \gamma_t \nabla \log p^{\widehat{\theta}_t}(y_t|y_{0:t-1})$$

where the γ_t's are chosen so that $\sum \gamma_t = \infty$, $\sum \gamma_t^2 < \infty$ (e.g. $\gamma_t = (1+t)^{-\alpha}$, $\alpha \in (1/2, 1]$)). This update may be viewed as a stochastic approximation version of the initial gradient update (14.9), since

$$\log p_t^\theta(y_{0:t}) = \sum_{s=0}^{t} \log p_s^\theta(y_s|y_{0:s-1})\,.$$

(We are assuming that process $\{Y_t\}$ is ergodic; the method we are about to describe does not work properly for models where $\{Y_t\}$ is not.)

To actually compute the gradient of the log-likelihood term $p^{\widehat{\theta}_t}(y_t|y_{0:t-1})$, we would need to run a particle filter from time 0 to time t. Instead, we write

$$\nabla \log p^{\widehat{\theta}_t}(y_t|y_{0:t-1}) = \nabla \log p_t^{\widehat{\theta}_t}(y_{0:t}) - \nabla \log p_t^{\widehat{\theta}_t}(y_{0:t-1})$$

and approximate both terms by an ad hoc version of Fisher's identity, which would correspond to a model where parameter θ is set to $\widehat{\theta}_t$ at time t; e.g. the first term is approximated by

$$\mathbb{E}_{\mathbb{P}^{\widehat{\theta}_{0:t}}}\left[\nabla \log p_0^{\widehat{\theta}_0}(x_0) + \sum_{s=1}^{t} \nabla \log p_s^{\widehat{\theta}_s}(x_s|x_{s-1}) + \sum_{s=0}^{t} \nabla \log f_s^{\widehat{\theta}_s}(y_s|x_s) \,\middle|\, Y_{0:t} = y_{0:t}\right]$$

where the expectation is with respect to the conditional distribution of $X_{0:t}$, given $Y_{0:t} = y_{0:t}$, for a model where parameter θ would be equal to $\widehat{\theta}_s$ at time s; i.e. at time s transition density is $p_s^{\widehat{\theta}_s}(x_s|x_{s-1})$, observation density is $f_s^{\widehat{\theta}_s}(y_s|x_s)$.

It is now possible to recursively approximate the so-obtained approximation of the gradient of $\log p^{\widehat{\theta}_t}(y_t|y_{0:t-1})$, by running an on-line smoother that changes the parameter θ at every time step, as described above.

The so-obtained on-line algorithm is a bit ad hoc, but it seems to work well empirically, and it has been proved to be convergent in the finite case (where the smoothing expectations above may be computed exactly).

The main caveat with on-line algorithms is the initialisation. A reasonable recipe in many cases is to perform off-line on a first batch of data (say the first 100 datapoints) in order to initialise the on-line algorithm.

Exercises

14.1 *The basic result of large deviations theory is*

$$\lim_{T\to+\infty} \log \mathbb{P}\left(\frac{1}{T}\sum_{t=1}^{T} X_t \geq 0\right) = -c$$

for i.i.d. variables $X_1, \ldots, X_T$, where $c \in (0, +\infty)$ as soon as (a) $\mathbb{P}(X_t > 0) > 0$; (b) $\mathbb{E}(X_t) < 0$ and (c) $\mathbb{E}[\exp(aX_1)] < \infty$ for some $a > 0$.

Use this result to show that the probability that appears in Proposition 14.1 converges exponentially fast to one. In particular, show that the conditions above holds. (Hint: consider Hölder's inequality.)

14.2 *Maximum likelihood estimation is a particular example of the more general framework of M-estimation, where the estimator is* $\hat{\theta}_n = \arg\max_\theta M_n$, M_n *is a sequence of functions such that* $M_n(\theta) \rightarrow M(\theta)$ *almost surely, for all* θ, *and the limit function* M *is such that* θ_0 *is the unique maximiser. (Functions* M *and* M_n *depend on datapoints* $Y_1, \ldots, Y_n$, *and are therefore random.) Show that, if (a)* $\sup_{\theta\in\Theta} |M_n(\theta) - M(\theta)| \rightarrow 0$ *in probability; and (b)* $\sup_{\|\theta-\theta_0\|\geq\epsilon} M(\theta) < M(\theta_0)$, *then* $\hat{\theta}_n \rightarrow \theta_0$ *in probability. Explain how to define* M_n *and* M *to recover as* $\hat{\theta}_n$ *the MLE, and rewrite conditions (a) and (b) in that particular case. Note: for the MLE, condition (a) is a uniform law of large numbers, which is much easier to establish when* Θ *is compact. This explains why much of MLE theory is stated assuming the parameter space is compact.*

14.3 *Discuss whether the EM algorithm remains valid when* Y *is a deterministic function of* X. *Then derive the EM update for the following simple model:* $X = (X_1, \ldots, X_n)$, $X_i \sim \mathcal{N}(\mu, \sigma^2)$, $Y_i = \min(X_i, c)$, *for a known constant* c. *(Note: you may express the EM update in terms of the gamma function,* $\Gamma(s,x) = \int_x^{+\infty} t^{s-1}e^{-t}\,\mathrm{d}t$, *which may be evaluated accurately in several numerical libraries.)*

14.4 *Prove Proposition 14.3.*

14.5 *Derive the EM update for a multivariate version of the model in Example 14.1, i.e.* $\mathcal{X} = \mathbb{R}^{d_x}$, $\mathcal{Y} = \mathbb{R}^{d_y}$,

$$X_t = X_{t-1} + U_t, \quad U_t \sim \mathcal{N}(0, \Sigma_X)$$
$$Y_t = X_t + V_t, \quad V_t \sim \mathcal{N}(0, \Sigma_Y)$$

and we wish to estimate matrices Σ_X, Σ_Y *from data* $y_{0:T}$. *(Hint: to apply Proposition 14.3, you may parameterise the model in terms of vectorised versions of the inverses of these matrices. Alternatively, you may recover the EM update directly, without using Proposition 14.3.)*

14.6 *Consider the following variant of the model of Example 14.1:*

$$X_t = X_{t-1} + U_t, \quad U_t \sim \mathcal{N}(0, \sigma_X^2)$$
$$Y_t = \lambda X_t + V_t, \quad V_t \sim \mathcal{N}(0, \sigma_Y^2)$$

and again $X_0 \sim \mathcal{N}(0, \sigma_X^2)$. *Show that this model is not identifiable for parameter* $(\lambda, \sigma_X^2, \sigma_Y^2)$. *(To do so, derive first the marginal distribution of* $Y_{0:T}$.*) Show that it is still possible to implement the EM algorithm (derive the corresponding update). Discuss the potential behaviour of the EM algorithm in this case.*

14.7 *Consider a finite state-space model, as in Chap. 6, that is* $\mathcal{X}$ *is finite. Assume furthermore that the transition matrix of process* $\{X_t\}$ *is fixed and known, and* *that the Y_t are conditionally independent, with a density taken from an exponential family:*

$$f_t(y_t|k) = \exp\{\lambda_k^T s(y_t) - \kappa(\lambda_k) - \xi(y_t)\}$$

and we wish to estimate $\theta = (\lambda_1, \ldots, \lambda_K)$ from data $y_{0:T}$. Show that the EM update takes the following form in this case: the k-th component of θ_n is the solution of:

$$\nabla\kappa(\lambda_k) = \frac{\sum_{t=0}^T s(y_t)\mathbb{P}^{\theta_{n-1}}(X_t = k|Y_{0:T} = y_{0:T})}{\sum_{t=0}^T \mathbb{P}^{\theta_{n-1}}(X_t = k|Y_{0:T} = y_{0:T})}.$$

Explain how these probabilities above may be computed at each EM iteration. (Again see Chap. 6 for an inspiration.) Apply this formula in the case where the Y_t are conditionally Poisson, $Y_t|X_t = k \sim \mathcal{P}(\lambda_k)$.

14.8 *Consider the same generic model as in Exercise 14.7, except now the (fixed) transition matrix of process $\{X_t\}$ is also unknown. First explain why the model is not identifiable in this case, and why a constraint (such as e.g. $\lambda_1 < \ldots < \lambda_K$ in the Poisson example) is needed to make the model identifiable. Then derive the expression for the EM update. Should this EM update involve the identifiability constraint in any way? Discuss.*

14.9 *Consider a state-space model such that $X_t|X_{t-1}$ is Poisson-distributed with rate some function of X_{t-1}. Explain why the CDF interpolation method presented in Sect. 14.3 does not make the likelihood estimate a smooth function of θ in this case. What about a model where $Y_t|X_t$ would be Poisson-distributed?*

14.10 *Derive the update formulae of the EM algorithm for the model discussed in the numerical experiment of Sect. 14.6.*

14.11 *Consider the basic stochastic volatility model discussed in Example 14.2. Express the mean, variance, and lag-one covariance of process $\log(Y_t^2)$ as a function of the parameter. (Assume stationarity, i.e. $\rho \in (-1, 1)$.) Derive moment estimators of θ from these expressions. (Note: if $V_t \sim \mathcal{N}(0, 1)$, then $\mathbb{E}[\log(V_t^2)] \approx -1.27$ and $\mathrm{Var}[\log(V_t^2)] \approx 4.93$.)*

Bibliographical Notes

This chapter is partly based on the more in-depth review of Kantas et al. (2015) on parameter estimation in state-space models. A good entry into the theory of maximum likelihood estimation of state-space models is Chapter 12 of Cappé et al. (2005); see also references therein. For a more general presentation of asymptotic Statistics, see eg. the book of van der Vaart (1998). The EM algorithm was given its name and presented in a general form in the classical paper of Dempster et al.

(1977). An authoritative source on the EM algorithm is the book of McLachlan and Krishnan (2008).

The importance sampling approach described in Sect. 14.3 is due to Hürzeler and Künsch (2001) and is related to the general MCMC-MLE approach introduced in Geyer and Thompson (1992). The interpolation approach described in the same section was developed in Pitt (2002) and Malik and Pitt (2011).

A proof of convergence for on-line MLE (Sect. 14.9) for hidden Markov models is given in Le Gland and Mevel (1997). As mentioned previously, one may consider alternatives to the MLE for a variety of reasons; see e.g. Andrieu et al. (2012) who propose an on-line estimation method based on composite likelihood.

Maximum likelihood estimation of state-space models with intractable dynamics (i.e. such that the density of $X_t|X_{t-1}$ is intractable) is particularly tricky. We have covered only one approach that may applied to this kind of problem: apply some gradient-free optimisation scheme (such as the simplex algorithm) to the noisy log-likelihood estimates obtained from a bootstrap filter. We now mention three alternatives. First, Dahlin and Lindsten (2014) use Bayesian optimisation to maximise the noisy log-likelihood. Bayesian optimisation assigns a Gaussian process prior to the objective function, and treats each evaluation as noisy observations of that function. This approach is appealing when the objective function is expensive to compute, which is precisely the case here.

Second, Ionides et al. (2011) show that the gradient of the log-likelihood (at a given θ) may be approximated as follows: replace the fixed parameter θ by a random walk process $\{\xi_t\}$, where $\xi_0 \sim \mathcal{N}(\theta, \tau I_d)$, and $\xi_t = \xi_{t-1} + \epsilon_t$, $\epsilon_t \sim \mathcal{N}(0, \sigma^2 I_d)$. Then the gradient at θ is obtained as the limit, as $\tau^2 \to 0$, and $\sigma^2/\tau^2 \to 0$, of some transformation of the expectation and variance of ξ_T given $Y_{0:T}$. This leads to the following approach (called iterated filtering by the authors): use gradient ascent, where at each iteration a bootstrap filter is run for the augmented state-space model, with decreasing values for τ^2, and σ^2. The difficulties with this approach are (a) that one needs to increase N at each iteration to properly approximate the gradient, since the variance of the artificial noise is decreasing; and (b) in our experience, this method is not always easy to tune for good performance (particularly the starting value for θ).

Third, one may consider computing an alternative estimator; in particular a Bayesian estimator that may be approximated by any of the methods discussed in the forthcoming chapters (Chaps. 16 and 18). For models with intractable dynamics, such an approach is typically not much more expensive than the aforementioned approaches. It also provides confidence regions at no extra cost. Finally, Bayesian estimates usually have the same asymptotic properties as the MLE.

Bibliography

Andrieu, C., Doucet, A., & Tadić, V. B. (2012). One-line parameter estimation in general state-space models using a pseudo-likelihood approach. *IFAC Proceedings Volumes, 45*(16), 500–505. 16th IFAC Symposium on System Identification.

Cappé, O., Moulines, E., & Rydén, T. (2005). *Inference in hidden Markov models. Springer series in statistics*. New York: Springer.

Cérou, F., Del Moral, P., & Guyader, A. (2011). A nonasymptotic theorem for unnormalized Feynman-Kac particle models. *Annales de l'Institut Henri Poincaré - Probabilités et Statistiques, 47*(3), 629–649.

Dahlin, J., & Lindsten, F. (2014). Particle filter-based Gaussian process optimisation for parameter inference. *Proceedings of the 19th IFAC World Congress, 47*(3), 8675–8680.

Dempster, A. P., Laird, N. M., & Rubin, D. B. (1977). Maximum likelihood from incomplete data via the EM algorithm. *The Journal of the Royal Statistical Society, Series B (Statistical Methodology), 39*(1), 1–38. With discussion.

Geyer, C. J., & Thompson, E. A. (1992). Constrained Monte Carlo maximum likelihood for dependent data. *The Journal of the Royal Statistical Society, Series B (Statistical Methodology), 54*(3), 657–699. With discussion and a reply by the authors.

Hürzeler, M., & Künsch, H. R. (2001). Approximating and maximising the likelihood for a general state-space model. In *Sequential Monte Carlo methods in practice. Statistics for engineering and information science* (pp. 159–175). New York: Springer.

Ionides, E. L., Bhadra, A., Atchadé, Y., & King, A. (2011). Iterated filtering. *Annals of Statistics, 39*(3), 1776–1802.

Kantas, N., Doucet, A., Singh, S. S., Maciejowski, J., & Chopin, N. (2015). On particle methods for parameter estimation in state-space models. *Statistical Science, 30*(3), 328–351.

Le Gland, F., & Mevel, L. (1997). Recursive estimation in hidden Markov models. In *Proceedings of the 36th IEEE Conference on Decision and Control* (Vol. 4, pp. 3468–3473).

Malik, S., & Pitt, M. K. (2011). Particle filters for continuous likelihood evaluation and maximisation. *Journal of Econometrics, 165*(2),190–209.

McLachlan, G. J., & Krishnan, T. (2008). *The EM algorithm and extensions. Wiley series in probability and statistics* (2nd ed.). Hoboken, NJ: Wiley.

Peters, G. W., Hosack, G. R., & Hayes, K. R. (2010). Ecological non-linear state space model selection via adaptive particle Markov chain Monte Carlo (AdPMCMC). *arXiv e-prints 1005.2238*.

Pitt, M. K. (2002). Smooth particle filters for likelihood evaluation and maximisation. *Warwick Economic Research Papers 651*.

Temereanca, S., Brown, E. N., & Simons, D. J. (2008). Rapid changes in thalamic firing synchrony during repetitive whisker stimulation. *Journal of Neuroscience, 28*(44), 11153–11164.

van der Vaart, A. W. (1998). *Asymptotic statistics. Cambridge series in statistical and probabilistic mathematics* (Vol. 3). Cambridge: Cambridge University Press.

Chapter 15
Markov Chain Monte Carlo

Summary A complementary to SMC framework for effective sampling in high-dimensions is that of Markov chain Monte Carlo (MCMC). In this framework there is a single target distribution but samples are drawn sequentially as a Markov chain that is invariant with respect to the target. For the purposes of this book we are primarily interested in the MCMC kernels that are used to generate the successive samples because we can use them as transition kernels in Feynman-Kac models. The combination of SMC with MCMC kernels, which we develop in subsequent chapters, creates opportunities for optimising parameters in the MCMC kernels using information about the target extracted from the particle population. This chapter lays down the fundamental notions, the most important of which for the purposes of this book being the notion of invariance, and a few typical MCMC kernels.

15.1 Markov Chain Simulation

In this chapter we revisit the fundamental problem, first considered in Chap. 3, of sampling from a given distribution $\mathbb{Q}(\mathrm{d}\theta)$, for $\theta \in \Theta = \mathbb{R}^{d_\theta}$. To focus on the most essential aspects we assume throughout that $\mathbb{Q}(\mathrm{d}\theta) = q(\theta)\nu(\mathrm{d}\theta)$ for some common dominating measure $\nu(\cdot)$.

The approach we summarise here uses Markov chains to draw samples from $\mathbb{Q}(\mathrm{d}\theta)$, say $\Theta_0, \Theta_1, \ldots, \Theta_N$, that although neither independent not identically distributed can be used for Monte Carlo calculations as described in Chap. 8. The samples are generated successively according to a transition kernel $M(\theta_0, \mathrm{d}\theta_1)$. The Markov chains used in sampling are constructed such that they obey laws of large numbers, i.e., as $N \to +\infty$,

$$\left\{\frac{1}{N}\sum_{n=1}^{N}\varphi(\Theta_n) - \mathbb{Q}(\varphi)\right\} \to 0 \quad \text{a.s.},$$

© Springer Nature Switzerland AG 2020

N. Chopin, O. Papaspiliopoulos, *An Introduction to Sequential Monte Carlo*, Springer Series in Statistics, https://doi.org/10.1007/978-3-030-47845-2_15

and under additional assumptions central limit theorems

$$N^{1/2}\left\{\frac{1}{N}\sum_{n=1}^{N}\varphi(\Theta_n)-\mathbb{Q}(\varphi)\right\}\Rightarrow\mathcal{N}\left(0,\tau_\varphi(\mathbb{Q}(\varphi^2)-\mathbb{Q}(\varphi)^2)\right),$$

where τ_φ is known as the integrated autocorrelation time,

$$\tau_\varphi=1+2\sum_{n=1}^{\infty}\mathrm{Cor}\left(\varphi(\Theta_0),\varphi(\Theta_n)\right),\quad \Theta_0\sim\mathbb{Q}.$$

(Symbol Cor denotes the correlation between two random variables.) Compared to the basic Monte Carlo as described in Chap. 8, the variance of the averages is inflated by the factor τ_φ due to the autocorrelation of the variables. A fundamental property of Markov chains used for sampling is that they are constructed so that the distribution of Θ_N converges to $\mathbb{Q}$ as $N\to\infty$. This convergence can be defined in different ways; one is that the total variation distance between the law of the Markov chain after N steps and $\mathbb{Q}$ goes to zero as $N\to\infty$; see Sect. 11.4.1 for a formal definition of this distance. The output Markov chain simulation can be analysed using time series techniques.

The Markov chain draws are not independent since each is used to generate the next. They are not identically distributed since only for large N the law of Θ_N is close to $\mathbb{Q}$. This is due to the initialisation of the Markov chain, i.e., the way Θ_0 is chosen. If Θ_0 could be drawn from $\mathbb{Q}$ then this initialisation bias, which is known as burn-in bias, would disappear. It can be argued that the burn-in bias is a lower order effect compared to the additional variance induced by autocorrelation in the series, the errors due to the former are $\mathcal{O}(1/N)$ and to the latter $\mathcal{O}(1/\sqrt{N})$. In practice for high-dimensional problems the burn-in bias might be a real concern, especially how to initialise the Markov chain in an area that it is not extremely far out in the tails of $\mathbb{Q}$.

As was the case with importance sampling in Chap. 8, the effective sample size in a Markov chain sample $\Theta_0,\ldots,\Theta_N$ is not the nominal number of draws due to their autocorrelation. In many instances of MCMC the draws are positively correlated and this makes a Markov chain sample worth less than an i.i.d. sample for estimating expectations $\mathbb{Q}(\varphi)$. We can define a notion of relaxation time for a Markov chain as the number of iterations that the Markov chain needs to decorrelate, i.e., to obtain draws that have arbitrarily little dependence. The effective sample size of an MCMC output could be defined as N/relaxation time. Additionally, for any test function φ, we have the bound $\tau_\varphi\leq 2\{\text{relaxation time}\}$, hence the relaxation time provides an upper bound on the inflation factor in the asymptotic variance of averages computed using the Markov chain. We can define the complexity of MCMC as the product of the relaxation time and the computational cost per iteration.

It is not easy at all to compute the relaxation time in specific applications of MCMC, even for simple algorithms and target distributions. However, useful asymptotic results exist for the relaxation time of MCMC algorithms when the dimension d_θ is large and when the target has certain properties. Simulation studies have empirically verified that these results are practically relevant well beyond the assumptions the asymptotic theory is based upon.

The topic of Markov chain Monte Carlo is vast and this chapter only aims at providing the very essentials.

15.2 Invariant Probability Kernels

Recall the notion of probability kernels of Sect. 4.1. In this chapter we are interested in kernels from a space $(\Theta, \mathcal{B}(\Theta))$ to itself, $M(\theta_0, \mathrm{d}\theta_1)$. The defining property of MCMC kernels is invariance.

Definition 15.1 We say that a probability kernel $M(\theta_0, \mathrm{d}\theta_1)$ is invariant with respect to a probability distribution $\mathbb{Q}(\mathrm{d}\theta)$ if

$$\mathbb{Q}(A) = \int \mathbb{Q}(\mathrm{d}\theta) M(\theta, A), \quad \forall A \in \mathcal{B}(\Theta).$$

Invariance is a fixed point equation and states that if $\Theta_0 \sim \mathbb{Q}$ then $\Theta_1 \sim \mathbb{Q}$. Reversibility is a stronger condition that implies invariance, and simply asks that the kernel is equal to its time reversal, i.e., $\overleftarrow{M}(\theta_1, \mathrm{d}\theta_0) = M(\theta_1, \mathrm{d}\theta_0)$. MCMC kernels are constructed to satisfy invariance with respect to the target distribution.

15.3 Metropolis-Hastings Kernels

Metropolis-Hastings kernels are constructed to be reversible with respect to the target distribution. They use a probability kernel to generate proposals, say $\widetilde{M}(\theta_0, \mathrm{d}\widetilde{\theta})$. These are retained as the next state of the chain with probability $\alpha(\theta_0, \widetilde{\theta})$, and if rejected the chain's next state is the same as its current. Any such Markov chain has a probability transition kernel which is a mixture of a point mass (which corresponds to the event of rejecting a proposed move), and the proposal transition kernel:

$$M(\theta_0, \mathrm{d}\theta_1) = \widetilde{M}(\theta_0, \mathrm{d}\theta_1)\alpha(\theta_0, \theta_1) + \delta_{\theta_0}(\mathrm{d}\theta_1) \int \widetilde{M}(\theta_0, \mathrm{d}\widetilde{\theta})[1 - \alpha(\theta_0, \widetilde{\theta})].$$

There are many valid ways to choose $\alpha(\theta_0, \theta_1)$ so that $M(\theta_0, \mathrm{d}\theta_1)$ is reversible with respect to a given $\mathbb{Q}(\mathrm{d}\theta_1)$. The so-called Metropolis-Hastings acceptance probability is

$$\alpha(\theta_0, \theta_1) = \min\{1, r(\theta_0, \theta_1)\}, \quad r(\theta_0, \theta_1) = \frac{\mathbb{Q}(\mathrm{d}\theta_1)\widetilde{M}(\theta_1, \mathrm{d}\theta_0)}{\mathbb{Q}(\mathrm{d}\theta_0)\widetilde{M}(\theta_0, \mathrm{d}\theta_1)} \tag{15.1}$$

and it is optimal in the sense of producing, for the give proposal kernel $\widetilde{M}(\theta_0, \mathrm{d}\widetilde{\theta})$, the Markov chain with the smallest possible τ_φ, for all square integrable functions φ. Algorithm 15.1 describes the corresponding kernel.

Algorithm 15.1: Metropolis-Hastings kernel

Input: Θ

$\widetilde{\Theta} \sim \widetilde{M}(\Theta, \mathrm{d}\widetilde{\theta})$

$U \sim \mathcal{U}([0, 1])$

if $\log U \leq \log r(\Theta, \widetilde{\Theta})$ ▷ see (15.1)

then

  return $\widetilde{\Theta}$

else

  return Θ

The simplest choice is to propose values around the current state, e.g.,

$$\widetilde{M}(\theta_0, \mathrm{d}\widetilde{\theta}) \equiv \mathcal{N}(\theta_0, \delta\Sigma) .$$

The resultant MCMC algorithm is known as random walk Metropolis. The acceptance ratio is simply $r(\theta_0, \theta_1) = q(\theta_1)/q(\theta_0)$. The covariance matrix Σ is known as a pre-conditioner and aims to capture the geometry of $q(\theta)$, that is the correlations between the different variables under $\mathbb{Q}$ and the relative variances thereof. The parameter δ is a step size that controls the distance between current and proposed values. When $q(\theta)$ is smooth, small δ will produce proposals that will be accepted with high probability but the chain will be exploring $\mathbb{Q}(\mathrm{d}\theta)$ very slowly, and the autocorrelation of the samples will be large; large δ returns proposals far away from the current state, but then $q(\widetilde{\theta})$ will be very different from $q(\theta)$ and the proposals will be rejected most of the time, hence the autocorrelation of the samples will be large again due to several identical successive samples.

A well-developed theory exist for optimal scaling which applies to high-dimensional problems (d_θ large) and under specific assumptions on $q(\theta)$, but has been found to be providing good guidelines when the assumptions do not hold. The rule of thumb is to tune the step size so as to achieve an acceptance probability somewhere between 0.2-0.4. In fact, a still more automatic rule of thumb is to take $\delta = 2.38^2/d_\theta$. Note that the steps we attempt become smaller as the dimension of the target increases. This theory implies that the relaxation time of random walk Metropolis for standard problems is $\mathcal{O}(d_\theta)$. When combined with the cost of computing the acceptance probability and generating the proposals, the resultant complexity of random walk Metropolis in standard problems is $\mathcal{O}(d_\theta^3)$. This already shows the advantages of Markov chain simulation relative to importance sampling, whose complexity as shown in Chap. 8 can be exponential in d_θ.

The choice of Σ is more complicated matter and the harder the larger d_θ is. Roughly speaking there are four generic approaches for its choice. The first is to simply set $\Sigma = I_{d_\theta}$, especially after doing coordinate-wise transformations to ensure that each coordinate has support on the whole of the real line (e.g. logarithmic transformation for positive random variables) and the different coordinates are measured in comparable scales (e.g. scaling random variables by an estimate of their standard deviation). The second is to set Σ to an estimate of $\text{Cov}(\Theta)$ obtained by an initial exploration of $\mathbb{Q}$, for example one obtained by setting the preconditioner to the identity matrix. A third is to set the preconditioner to some analytic approximation of $\text{Cov}(\Theta)$; we provide a concrete such construction for latent Gaussian models, such as state-space models with Gaussian hidden process, later on. A fourth, which can be used both for Σ and δ, is to use adaptive MCMC. This is an approach to Markov chain simulation, akin to stochastic optimisation, where the training and sampling stages are coupled and the two parameters evolve during sampling using the information about $\mathbb{Q}$ which has been obtained thus far. The convergence theory of adaptive MCMC is considerably harder compared to that for time-homogeneous Markov chains, although it has now reached a mature stage.

Better exploration of $\mathbb{Q}(\mathrm{d}\theta)$ can be achieved by using geometric information such as gradients of $q(\theta)$. If $q(\theta)$ is continuously differentiable, we can define gradient-based Metropolis-Hastings algorithms. A well-studied gradient-based algorithm is the Metropolis-adjusted Langevin algorithm (MALA), for which

$$\widetilde{M}(\theta_0, \mathrm{d}\widetilde{\theta}) \equiv \mathcal{N}(\theta_0 + (\delta/2)\Sigma\nabla \log q(\theta_0), \delta\Sigma).$$

An asymptotic optimal scaling theory analogous to that for the random walk Metropolis shows that the relaxation time of MALA is only $\mathcal{O}(d_\theta^{1/3})$, which makes the overall complexity in standard problems $\mathcal{O}(d_\theta^{7/3})$. A rule of thumb is to choose δ so that to achieve an acceptance probability between 0.4 and 0.6.

There are target distributions for which tailored Metropolis-Hastings algorithms achieve what is known as dimension-independent scaling, that is the δ does not have to decrease as the dimension increases to achieve a constant acceptance probability. In other words, the relaxation time of these algorithms is $\mathcal{O}(1)$, and their overall complexity $\mathcal{O}(d^2)$ in standard problems. A concrete such setting is the class of latent Gaussian models, for which $\mathbb{Q}(\mathrm{d}\theta) = q(\theta)\nu(\mathrm{d}\theta)$, where $\nu(\mathrm{d}\theta) \equiv \mathcal{N}(\mu, C)$. Such $\mathbb{Q}$ naturally arises as a posterior distribution under a Gaussian prior ν and likelihood $q(\theta)$, and it is very common in Statistics, Machine Learning and Inverse Problems. In the context of state-space models this distribution arises as the smoothing distribution of states when the signal is linear Gaussian. We highlight two MCMC algorithms appropriate for latent Gaussian models. One is an auto-regressive proposal which has recently become known as preconditioned

Crank-Nicolson (pCN),

$$\widetilde{M}(\theta_0, \mathrm{d}\widetilde{\theta}) \equiv \mathcal{N}\left(\frac{2}{2+\delta}\theta_0, \left[1-\left(\frac{2}{2+\delta}\right)^2\right]C\right).$$

To link with the previous discussion on the choice of Σ in random walk Metropolis algorithms, in pCN the preconditioner $\Sigma = C$ is used. The second MCMC algorithm for latent Gaussian models we highlight uses a geometric proposal that combines both the gradients of q and the Gaussian prior ν, and takes the form

$$\widetilde{M}(\theta_0, \mathrm{d}\widetilde{\theta}) \equiv \mathcal{N}\left(\frac{2}{\delta}A\left(\theta_0 + \frac{\delta}{2}\nabla q(\theta_0)\right), \frac{2}{\delta}A^2 + A\right)$$

for

$$A = \frac{\delta}{2}\left(C + \frac{\delta}{2}I\right)^{-1} C.$$

This is known as the marginal gradient-based sampler.

15.4 The Gibbs Sampler

The main ingredient of the Gibbs sampler is a decomposition of $\mathbb{Q}$ into conditional distributions. Since the presentation in this chapter is only meant to provide a high-level understanding we concentrate on the following setting, which albeit not fully general, is the one that it is encountered in most applications. The random variable $\Theta \sim \mathbb{Q}$ is decomposed into K components, $\Theta(k), k = 1 : K$, and the dominating measure $\nu(\mathrm{d}\theta)$ factorised accordingly as $\nu(\mathrm{d}\theta) = \nu_1(\mathrm{d}\theta(1)) \times \cdots \times \nu_K(\mathrm{d}\theta(K))$. The target density $q(\theta)$ has well-defined full conditional densities $q_k(\theta(k) \mid \theta(-k))$ where $\theta(-k)$ denotes all components but k. The Gibbs sampler consists of iterative sampling from these conditionals; this can be done in a systematic or random scan; the former is described as Algorithm 15.2.

Algorithm 15.2: (Systematic scan) Gibbs sampler kernel

Input: Θ

for $k = 1 : K$ **do**

  Sample

  $\widetilde{\Theta}(k) \sim q_k\left(\cdot \mid \widetilde{\Theta}(1), \ldots, \widetilde{\Theta}(k-1), \Theta(k+1), \ldots, \Theta(K)\right)$

return $\widetilde{\Theta}$

The algorithmic choices are the number and composition of blocks ("blocking scheme"); the order at which the defined blocks are updated ("scanning scheme"); the marginalisation if possible of certain components and the application of Gibbs sampling on the resultant marginal distribution ("collapsing"); the expansion of the state-space by defining $\mathbb{Q}(d\theta, du)$ which admits $\mathbb{Q}(d\theta)$ as a marginal and the application of Gibbs sampling on the expanded distribution ("auxiliary variables"); the definition of a distribution $\mathbb{P}(d\phi)$ and a transformation $\phi \to T(\phi) = \theta$, such that if $\Phi \sim \mathbb{P}$ then $\Theta \sim \mathbb{Q}$, and the application of Gibbs sampling on $\mathbb{P}$ ("reparameterisation").

All these choices affect the relaxation time of the Markov chain in non-obvious ways which is understood only partially. A high-level message is that high-dependence among the updated components will lead to slow convergence. A fairly developed methodology for reparameterisation is available that suggests transformations that can reduce the dependence and improve convergence. There are several interesting families of problems for which the Gibbs sampler (maybe using tricks like reparameterisation or collapsing) has relaxation time independent of d_θ and complexity $\mathcal{O}(d_\theta)$.

In many practical applications some, or even all, of the full conditionals $q\,(\theta(k)\,|\,\theta(-k))$ cannot be simulated directly. We can use a Metropolis-Hastings step to sample the conditional, and we can use any of the Metropolis-Hastings algorithms we have discussed for this task. This hybrid is known as Metropolis-within-Gibbs or Hastings-within-Gibbs.

15.5 Numerical Experiments: Smoothing in Stochastic Volatility Models

We revisit the stochastic volatility model introduced in Sect. 2.4.3 and used in the numerical experiments of Sect. 10.5.2 (using also the same data and parameter values). The aim is to sample from the smoothing distribution $\mathbb{P}\,(dx_{0:T}\,|\,Y_{0:T} = y_{0:T})$ using MCMC. Adjusting to the notation of this chapter, the state vector is $\theta := x_{0:T}$ and $\mathbb{Q}$ is the smoothing distribution. We consider two schemes. The first one is a single-site Gibbs sampler that uses $(T+1)$ blocks, each containing the volatility on a single time point. Each conditional is sampled using rejection sampling exploiting the inequality $\exp(-x) \geq 1 + x$ to build a proposal. The second scheme is a geometric Metropolis-Hastings algorithm, in particular the marginal algorithm of Titsias and Papaspiliopoulos (2018) that exploits the fact that the stochastic volatility model is a latent Gaussian one.

Figure 15.1 compares the marginal distributions obtained from both samplers, using QQ-plots. This kind of comparison is useful to detect any error in the implementation. Figure 15.2 plots the ACF (auto-correlation function) of selected components of $X_{0:T}$, again for both samplers. This type of plot is a standard way to assess the mixing of a MCMC algorithm. Here we see that the gradient-based

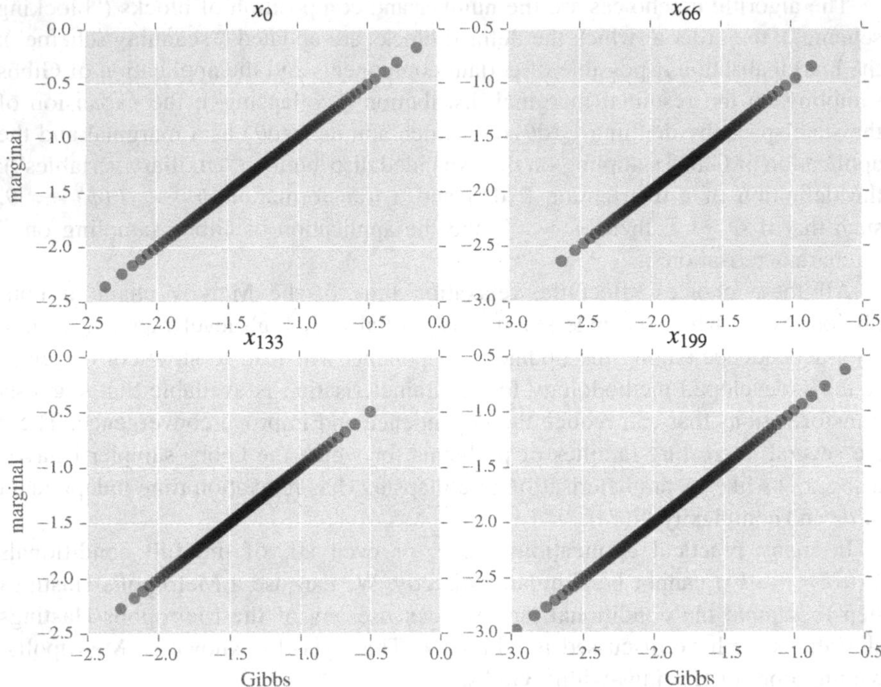

Fig. 15.1 QQ-plots comparing the empirical distributions of the one-at-a-time Gibbs sampler and the marginal sampler, for several components X_t; same settings as Fig. 15.2. Each dot corresponds to a $(k/100)$-percentile, with $k = 1, \ldots, 99$

scheme mixes better than the first. This is hardly surprising, given that the second scheme takes in account the correlation between the components. In Chap. 16 we employ another diagnostic for the performance of MCMC, based on estimating the squared jumping distance, see Sect. 16.5.

Exercises

15.1 ***The simplest Markov chain for sampling.*** *Consider the standard Gaussian target,* $\mathcal{N}(0, 1)$*, and the Markov chain with transition distribution* $\Theta_n|\Theta_{n-1} \sim \mathcal{N}(\rho\Theta_{n-1}, 1 - \rho^2)$*, for* $|\rho| < 1$.

1. *Show that this Markov chain is reversible with respect to* $\mathcal{N}(0, 1)$.
2. *Work out the integrated autocorrelation time for the identity test function* $\varphi(x) = x$*. Can the asymptotic variance of a Monte Carlo estimator based on samples from this chain be lower than that based on i.i.d. draws from the target?*

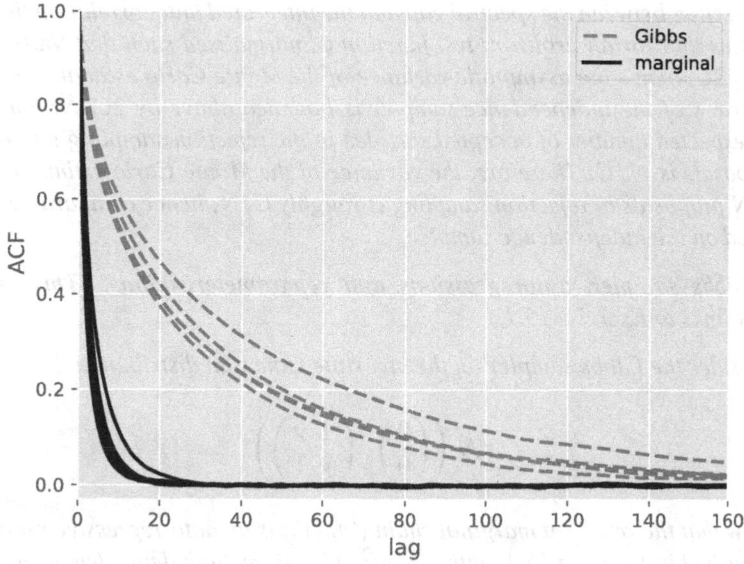

Fig. 15.2 ACF (auto-correlation function) for the one-at-a-time Gibbs sampler (grey lines) and the marginal sampler (black lines) and components $t = 0, 49, 99, 149, 199$ of $X_{0:t}$; chains were run for 10^6 iterations, and he first 10^5 simulations were discarded

15.2 *A transition kernel $M(\theta_0, d\theta_1)$ is invariant with respect to a distribution $\mathbb{Q}(d\theta)$ if it satisfies $\int \mathbb{Q}(d\theta_0)M(\theta_0, A) = \mathbb{Q}(A)$ for all measurable A. It is called reversible with respect to $\mathbb{Q}(d\theta)$ if it equals its time reversal $\overleftarrow{M}(\theta_0, d\theta_1)$. Using the basic definitions of backward kernels in Sect. 4.3, show that reversibility implies invariance.*

15.3 ***Rejection sampling and MCMC.*** *Recall rejection sampling in Algorithm 8.1. To use the same notation as the current Chapter, let $\mathbb{Q}(d\theta)$ be the target, $\mathbb{M}(d\theta)$ be the proposal, $w(\theta) = \mathbb{Q}(d\theta)/\mathbb{M}(d\theta)$ is assumed to be bounded by a known constant $C = \sup_\theta w(\theta)$, and let $\alpha(\theta) = w(\theta)/C$.*

1. *Consider a Metropolis-Hastings algorithm with $\widetilde{M}(\theta_0, d\theta_1) = w(\theta_1)\mathbb{M}(d\theta_1)$. Show that this MCMC algorithm generates i.i.d. draws from $\mathbb{Q}(d\theta)$. When simulation from $\widetilde{M}(\theta_0, d\theta_1)$ is done using rejection sampling, then the MCMC and rejection sampling algorithms become equivalent.*
2. *Consider now a Metropolis-Hastings algorithm with $\widetilde{M}(\theta_0, d\theta_1) = \mathbb{M}(d\theta_1)$. Show that in this algorithm each proposal stands a higher chance to be accepted compared to the rejection sampling algorithm with the same proposal, and that successive draws are dependent. This type of Metropolis-Hastings algorithm where the proposal does not depend on the current value is known as an independence sampler.*
3. *Smith and Tierney (1996), extending the analysis in Liu (1996) showed that the spectral gap of the independence sampler is $1 - 1/C$. Exploring a known*

connection between the spectral gap and the integrated autocorrelation time, one obtains that for an arbitrary test function φ, normalised such that $\mathrm{Var}(\varphi) = 1$, $\tau_\phi \le 2C$, *hence the asymptotic variance of the Monte Carlo estimator based on N draws of the independence sampler is bounded above by $2C/N$. Show that the expected number of accepted samples in the rejection sampling based on N proposals is N/C. Therefore, the variance of the Monte Carlo estimator based on N proposals in rejection sampling is roughly C/N, hence comparable to that based on the independence sampler.*

15.4 Gibbs sampler, autoregressions and reparameterisations. *(This exercise directly links to Exercise 15.1.)*

1. *Consider the Gibbs sampler on the bivariate Gaussian distribution*

$$\mathcal{N}\left(\begin{pmatrix}0\\0\end{pmatrix}, \begin{pmatrix}1 & \gamma\\ \gamma & 1\end{pmatrix}\right).$$

   Show that the resultant marginal chain $\{\Theta_n(1)\}$ is an auto-regressive process, as described in Exercise 15.1, with $\rho = \gamma^2$. Therefore, high dependence among the two components in the target results in high autocorrelation times in the marginal chain.
2. *Consider now the basic state-space model $X_t|X_{t-1} \sim \mathcal{N}(\mu, \sigma_X^2)$ and $Y_t|X_t \sim \mathcal{N}(X_t, \sigma_Y^2)$. This is a model for which there are no temporal dynamics in the hidden state. In this analysis σ_X^2, σ_Y^2 are going to be assumed known and the aim will be to sample from the posterior distribution of $(\mu, X_{0:T})$; the improper prior $p(\mu) \propto 1$ will be used for μ.*
   a) *Consider first the Gibbs sampler with $\Theta(1) = \mu$ and $\Theta(2) = X_{0:T}$.*
      i. *Work out analytically the two conditionals used in the Gibbs sampler. (Note that the conditional distribution of $\Theta(2)$ is that of $(T+1)$ independent random variables. Also note that the conditional of $\Theta(1)$ depends on $X_{0:T}$ only through $\sum_t X_t$.)*
      ii. *Using the model definitions obtain the marginal (with respect to $X_{0:T}$) posterior distribution of μ and the marginal (with respect to μ) posterior distribution of $X_{0:T}$. From the latter work out the marginal posterior distribution of $\sum_t X_t$.*
      iii. *Using the results obtained in the previous step show that the marginal chain $\{\Theta_n(1)\}$ is an auto-regressive process, as described in Exercise 15.1, with $\rho = \sigma_Y^2/(\sigma_X^2 + \sigma_Y^2)$.*
      iv. *The result in the previous step is very interesting! First, it suggests a simplification in the analysis of the Gibbs sampler for this problem. We can equivalently study the simpler model with $T = 0$. Hence, if you found difficult doing the previous calculations for general T, do them now for $T = 0$, since they are elementary. Second, it suggests that the efficiency of the Gibbs sampler in this model is independent of T! This is an example*

of what has been discussed in this Chapter, that for certain problems the Gibbs sampler has complexity that scales linearly with the dimension of the problem. Third, it implies that the Gibbs sampler is very efficient when the data are very informative relative to the system precision, that is when $\sigma_Y \ll \sigma_X$.

b) *Taking advantage from the simplification we discovered in the previous steps, focus now on the case* $T = 0$. *Consider the alternative parameterisation of the same model* $\tilde{X} \sim \mathcal{N}(\mu, \sigma_X^2)$ *and* $Y \sim \mathcal{N}(\tilde{X} + \mu, \sigma_Y^2)$, *and the Gibbs sampler that updates the blocks* $\Theta(1) = \mu$, $\Theta(2) = \tilde{X}$. *Show that the marginal chain* $\{\Theta_n(1)\}$ *is an auto-regressive process, as described in Exercise 15.1, with* $\rho = \sigma_X^2/(\sigma_X^2 + \sigma_Y^2)$. *Hence, this sampler is efficient precisely when the one developed in the previous steps is not, and vice versa! Hence, we get two alternative Gibbs samplers, based on two parameterisations of the same state-space model, with complementary efficiencies.*

Python Corner

Package `particles` provide some basic implementation for MCMC sampling; particularly Gibbs and (adaptive and non-adaptive) random walk Metropolis; see the documentation of module `mcmc`. As you might expect, the implementation is something along the lines of:

```
for n in range(niters):
    # Do something
    theta[n] = ..  # result
```

This is not very efficient, first because it is a loop (over typically a very large number of iterations); and second because element indexing is slow in `NumPy`.

Our main motivation for including a `mcmc` module in `particles` is to provide an implementation of PMCMC algorithms (which will be covered in Chap. 16). These are MCMC algorithms which generate, at each iteration, a complete particle filter. Hence a single iteration is expensive. The loop overhead is less of an issue is this kind of scenario. (Again, loop overhead depends to some extent to how slim or how fat is the body of the loop.) In general however, you may want to consider implementing your MCMC sampler in a compiled language for better performance.

Of course, another option is to rely on existing software. In particular, software like JAGS (http://mcmc-jags.sourceforge.net), or STAN (http://mc-stan.org/) takes as an input a description of the considered model, derives a MCMC sampler that leaves invariant the posterior, and runs it for you. JAGS relies on Gibbs sampling (i.e. JAGS computes automatically all the full conditional distributions of the posterior), while STAN relies on an advanced version of HMC (Hamiltonian Monte Carlo). Both JAGS and STAN have Python (and R) interfaces.

Another practical point regarding MCMC is the difficulty to spot "bugs"; in particular, for Gibbs sampling, it is hard to detect any error in the expressions that are used to sample each conditional distribution. Fortunately, there is a nice way to address this issue: simply add an extra step that re-generate the data (from its distribution given all the parameters and latent variables). This changes the stationary distribution to the joint distribution of all the variables; in particular, the parameter θ should be distributed according to the prior. If the simulated chain does not recover the prior, then it is a clear sign that something is amiss. To investigate further, one may fix one component of θ, and re-run the algorithm. If one recovers the prior for the other components, then it is likely that the error lies in the update of the component that has been fixed.

Bibliographical Notes

The topic of MCMC is vast with major advances in methodology, theory and applications over the last 30 years. In this chapter we provided a sketch of what is mostly relevant for the remainder of this book. We refer to Papaspiliopoulos et al. (2021) for a thorough exposition of the topic and hundreds of specific references. In this bibliography we provide a discussion and some references for the topics we touched upon.

The relative importance of burn-in bias and Monte Carlo variance in MCMC was discussed in Sokal (1997), who also provides a rule of thumb for estimating τ_ϕ. Optimal scaling for Metropolis-Hastings algorithms, which is the choice of the step size δ, has been intensively studied since the mid-1990s, see for example Roberts and Rosenthal (2001) for an overview. The gradient-based sampler we discussed in Sect. 15.3 is due to Titsias and Papaspiliopoulos (2018) who called it the marginal algorithm and demonstrated enormous efficiency gains relative to pCN but also relative to other geometric (gradient-based) methods, such as the so-called Metropolis-adjusted Langevin algorithm (MALA), effectively introduced in Roberts and Tweedie (1996). The main idea of pCN dates back at least to Neal (1996), but the algorithm has received considerable new interest following up from Cotter et al. (2013). The strengths and weaknesses of adaptive MCMC have been investigated in the last 15 years, Roberts and Rosenthal (2009), Andrieu and Thoms (2008) provide tutorials on different possibilities and numerical evidence. Parameterisation and the convergence of the Gibbs sampler has been systematically studied in Papaspiliopoulos et al. (2003, 2007).

An interesting application of MCMC in the context of particle filtering is to sample from the optimal proposal in the guided filter. Indeed, we have seen in Sect. 10.3.2 that it is typically not possible to sample (independently) from this optimal proposal. The price to pay is that we introduce correlations between the particles. See Finke et al. (2018) for some supporting theory and a discussion of this approach, called sequential MCMC.

Bibliography

Andrieu, C., & Thoms, J. (2008). A tutorial on adaptive MCMC. *Statistics and Computing, 18*(4), 343–373.

Cotter, S. L., Roberts, G. O., Stuart, A. M., & White, D. (2013). MCMC methods for functions: Modifying old algorithms to make them faster. *Statistics and Computing, 28*(3), 424–446.

Finke, A., Doucet, A., & Johansen, A. M. (2018). Limit theorems for sequential MCMC methods. *ArXiv e-prints 1807.01057.*

Liu, J. S. (1996). Metropolized independent sampling with comparisons to rejection sampling and importance sampling. *Statistics and Computing, 6*(2), 113–119.

Neal, R. M. (1996). *Bayesian learning for neural networks*. Berlin/Heidelberg: Springer.

Papaspiliopoulos, O., Roberts, G. O., & Sköld, M. (2003). Non-centered parameterizations for hierarchical models and data augmentation. In *Bayesian statistics, 7 (Tenerife, 2002)* (pp. 307–326). New York: Oxford University Press. With a discussion by Alan E. Gelfand, Ole F. Christensen and Darren J. Wilkinson, and a reply by the authors.

Papaspiliopoulos, O., Roberts, G. O., & Sköld, M. (2007). A general framework for the parametrization of hierarchical models. *Statistical Science, 22*(1), 59–73.

Papaspiliopoulos, O., Roberts, G. O., & Tweedie, R. (2021). *The methodology, ergodicity and optimisation of MCMC*. (in preparation).

Roberts, G. O., & Rosenthal, J. S. (2001). Optimal scaling for various Metropolis-Hastings algorithms. *Statistical Science, 16*(4), 351–367.

Roberts, G. O., & Rosenthal, J. S. (2009). Examples of adaptive MCMC. *Journal of Computational and Graphical Statistics, 18*(2), 349–367.

Roberts, G. O., & Tweedie, R. L. (1996). Exponential convergence of Langevin distributions and their discrete approximations. *Bernoulli, 2*(4), 341–363.

Smith, R. L., & Tierney, L. (1996). Exact transition probabilities for the independence Metropolis sampler. *Preprint*. Dept. of Statistics, Univ. of North Carolina

Sokal, A. (1997). Monte Carlo methods in statistical mechanics: Foundations and new algorithms. In *Functional integration (Cargèse, 1996). NATO advanced science institutes series B: Physics* (Vol. 361, pp. 131–192). New York: Plenum.

Titsias, M. K., & Papaspiliopoulos, O. (2018). Auxiliary gradient-based sampling algorithms. *Journal of the Royal Statistical Society: Series B (Statistical Methodology), 80*(4), 749–767.

Chapter 16
Bayesian Estimation of State-Space Models and Particle MCMC

Summary We consider again the problem of estimating the parameter θ of a state-space model, this time from a Bayesian viewpoint. That is, we view θ as the realisation of a random variable Θ, with prior distribution $\nu(\mathrm{d}\theta)$, and we wish to compute the distribution of Θ conditional on data $y_{0:T}$. This conditional distribution is called the posterior distribution, and represents the remaining uncertainty regarding the value of θ once data has been observed.

Posterior distributions are often intractable, and are commonly approximated by MCMC sampling. We discuss the various ways MCMC may be used to sample from the posterior distribution of a state-space model. We give particular emphasis to PMCMC (particle MCMC), a general methodology that involves running a particle filter at each MCMC iteration. Compared to vanilla MCMC (e.g. Gibbs sampling), the PMCMC methodology has the advantages of being more generally applicable, and showing better performance.

Notation/Terminology Recall that since Chap. 14 we have reinstated θ in the kernels, emission densities, etc., hence write $p_t^\theta(x_t|x_{t-1})$, $f_t^\theta(y_t|x_t)$, etc., a notation we first introduced in Chap. 2. In this chapter we will also deal with densities marginalised over a prior density $\nu(\mathrm{d}\theta)$, so whereas $p_t^\theta(y_{0:t})$ denotes the density of the observed data conditionally on the parameters with the states integrated out in a state-space model, $p_t(y_{0:t})$ will denote averaged quantity $\int_{\theta\in\Theta} p_t^\theta(y_{0:t})\nu(\mathrm{d}\theta)$.

16.1 Preliminaries

16.1.1 Posterior Distribution

As in Chap. 14, we consider a generic state-space model, depending on parameter $\theta \in \Theta$. We treat θ as the realisation of a random variable Θ, with prior distribution

© Springer Nature Switzerland AG 2020

N. Chopin, O. Papaspiliopoulos, *An Introduction to Sequential Monte Carlo*, Springer Series in Statistics, https://doi.org/10.1007/978-3-030-47845-2_16

$\nu(\mathrm{d}\theta)$. In many cases of interest, θ is a vector in $\mathbb{R}^{d_\theta}$, and $\nu(\mathrm{d}\theta) = \nu(\theta)\mathrm{d}\theta$, where $\nu(\theta)$ is a density with respect to Lebesgue measure.

The Bayesian viewpoint is particularly natural in the context of state-space models. We are already representing our uncertainty with respect to the state variables X_t through their conditional distribution given the data (the observed y_t's). To take into account our uncertainty with respect to θ, we consider the joint conditional distribution:

$$\mathbb{P}_T(\mathrm{d}\theta, \mathrm{d}x_{0:T}|Y_{0:T} = y_{0:T}) = \frac{1}{p_T(y_{0:T})}\nu(\mathrm{d}\theta)\mathbb{P}_T^\theta(\mathrm{d}x_{0:T})\prod_{t=0}^{T} f_t^\theta(y_t|x_t) . \tag{16.1}$$

Note that $p_T(y_{0:T})$ denotes here the marginal likelihood (i.e. the marginal density of the data, after θ has been integrated out):

$$p_T(y_{0:T}) = \int_\Theta \nu(\mathrm{d}\theta)p_T^\theta(y_{0:T}) = \int_{\Theta\times\mathcal{X}^{T+1}} \nu(\mathrm{d}\theta)\mathbb{P}_T^\theta(dx_{0:T})\prod_{t=0}^{T} f_t^\theta(y_t|x_t) .$$

If we marginalise out the states in (16.1), we obtain the marginal posterior distribution of θ:

$$\mathbb{P}_T(\mathrm{d}\theta|Y_{0:T} = y_{0:T}) = \nu(\mathrm{d}\theta)\frac{p_T^\theta(y_{0:T})}{p_T(y_{0:T})} .$$

16.1.2 Prior Specification

The prior distribution $\nu(\mathrm{d}\theta)$ should represent the user's beliefs regarding the potential values of θ. In practice, unfortunately, translating such beliefs into a probability distribution is not straightforward. We give below a few guidelines, and refer to the bibliography for further discussion.

Parameters of state-space models often have a 'physical' interpretation: in target tracking, the standard deviation σ_Y of the observation noise relates to the accuracy of the tracking device; in the theta-logistic model (Sect. 2.4.5), τ_0 is the growth rate of the considered population; and so on. It is usually easy to derive a range of scientifically plausible values for such parameters (from expert knowledge or from past experiments); the prior distribution should then be chosen so as to put most of its mass over that range.

One may also choose a prior distribution so as to deal with problematic features of the likelihood, as illustrated by the following example.

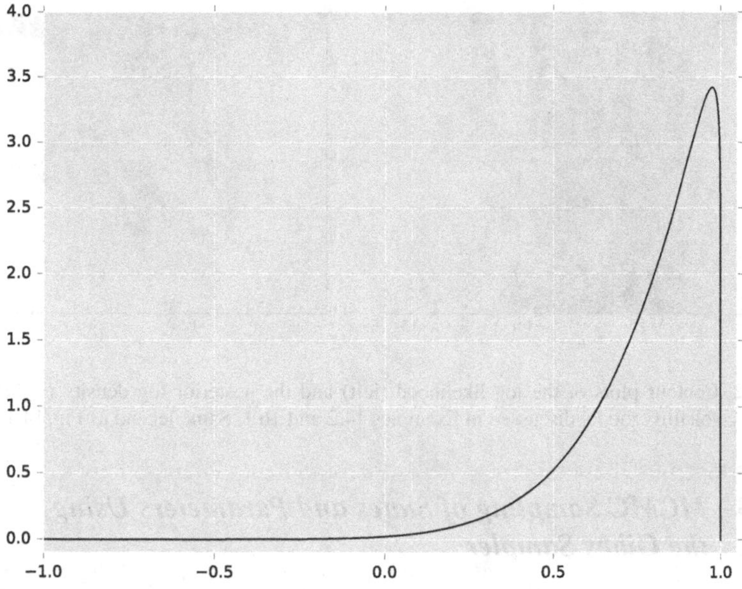

Fig. 16.1 Density of a $\mathcal{B}eta(8.9, 1.1)$ distribution, re-scaled to the interval $[-1, 1]$, used as a prior for parameter ρ in Example 16.1

Example 16.1 In Example 14.2, the log-likelihood function of the considered stochastic volatility model was bi-modal, and flat with respect to μ when $\rho \to 1$. Now consider the following prior distribution for ρ: $(1+\rho)/2 \sim \mathcal{B}eta(8.9, 1.1)$ (and some Gaussian prior for μ with a large scale). The prior density of ρ puts most of it mass on $[0.5, 1]$, but it vanishes at $\rho = 1$, thus preventing ρ to get too close to 1; see Fig. 16.1. The posterior log-density of the model is better behaved that the log-likelihood, see Fig. 16.2. The spurious mode such that $\rho \ll 0$ has disappeared, and the posterior log-density is, by construction, not flat in μ.

Note that the chosen prior also makes sense from a field expert point of view: stochastic volatility models are used to model volatility clustering, that is, the fact that absolute log-returns tend to be strongly (positively) correlated. Hence it is reasonable to choose a prior that favours large values for ρ. On the other hand, preventing $\rho = 1$ ensures the model is stationary, which may lead to better predictions.

Another motivation for choosing a particular prior may be computational convenience; for instance making the corresponding Gibbs sampler easily implementable, as illustrated in next section.

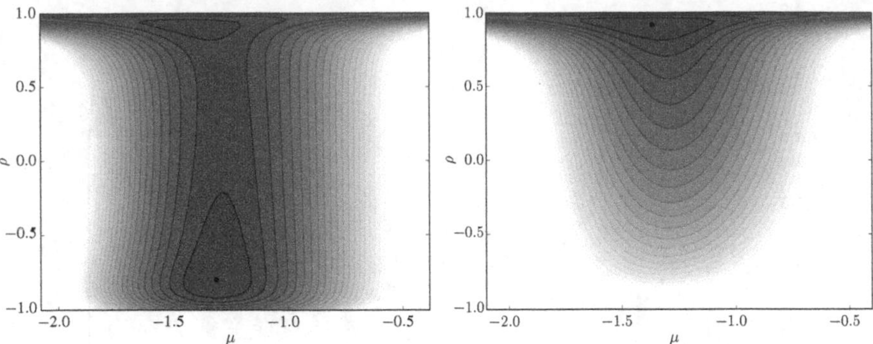

Fig. 16.2 Contour plots of the log likelihood (left) and the posterior log density (right) of the stochastic volatility model discussed in Examples 14.2 and 16.1. Same legend as Fig. 14.1

16.1.3 MCMC Sampling of States and Parameters Using the Gibbs Sampler

A common approach to sample from the joint posterior (16.1) is to implement a Gibbs sampler. Given the structure of the considered model, it may be possible to sample Θ given the states $X_{0:T}$, and to sample the states given Θ. If not, one may further split Θ and $X_{0:T}$ into smaller components, until one obtains conditional distributions that are easy to sample from.

The following example illustrates how such conditional distributions may be derived.

Example 16.2 We consider again the basic stochastic volatility model, where $X_t|X_{t-1} = x_{t-1} \sim \mathcal{N}(\mu + \rho(x_{t-1} - \mu), \sigma^2)$, and $Y_t|X_t = x_t \sim \mathcal{N}(0, e^{x_t})$. For the sake of simplicity, we take $X_0 \sim \mathcal{N}(\mu, \sigma^2)$, and let $X_{-1} = \mu$; The case where $X_0 \sim \mathcal{N}(\mu, \sigma^2/(1-\rho^2))$ is covered in Exercise 16.1. In addition, we assume that the chosen prior density factorises as follows (abusing notations): $\nu(\theta) = \nu(\mu)\nu(\sigma^2)\nu(\rho)$.

A quick way to compute the conditional posterior density of each component of the parameter is to write down the joint probability density of $(\Theta, X_{0:T}, Y_{0:T})$:

$$\nu(\mu)\nu(\sigma^2)\nu(\rho)\left(\frac{1}{2\pi}\right)^{T+1}\left(\frac{1}{\sigma^2}\right)^{(T+1)/2}$$
$$\times \exp\left\{-\frac{1}{2\sigma^2}\sum_{t=0}^{T}(x_t - \mu - \rho(x_{t-1} - \mu))^2 - \frac{1}{2}\sum_{t=0}^{T}(x_t + y_t^2 e^{-x_t})\right\}$$

(continued)

Example 16.2 (continued)
and then drop any factor that does not depend on the considered component. For instance, for μ, we obtain:

$$\propto \nu(\mu) \exp\left\{-\frac{1}{2\sigma^2}\sum_{t=0}^{T}(x_t - \rho x_{t-1} - (1-\rho)\mu)^2\right\}$$

which is clearly Gaussian in μ, provided we take $\nu(\mu)$ to be Gaussian as well: for $\mu \sim \mathcal{N}(m, s^2)$, $\mu|\sigma, \rho, x_{0:T}, y_{0:T} \sim \mathcal{N}(m_p, s_p^2)$ with

$$\frac{1}{s_p^2} = \frac{1}{s^2} + \frac{(T+1)(1-\rho)^2}{\sigma^2},$$

$$\frac{m_p}{s_p^2} = \frac{m}{s^2} + \frac{(1-\rho)\sum_{t=0}^{T}(x_t - \rho x_{t-1})}{\sigma^2}.$$

Similarly, for σ^2, we obtain:

$$\propto \nu(\sigma^2)\left(\frac{1}{\sigma^2}\right)^{(T+1)/2} \exp\left\{-\frac{1}{2\sigma^2}\sum_{t=0}^{T}(x_t - \rho x_{t-1} - (1-\rho)\mu)^2\right\}$$

so if we choose an inverse gamma prior density, i.e. $\nu(\sigma^2) \propto (1/\sigma^2)^{-a-1}\exp(-b/\sigma^2)$, we obtain as a conditional density an inverse gamma as well, with parameters $a_p = a + (T+1)/2$, $b_p = b + (1/2)\sum_{t=0}^{T}(x_t - \mu - \rho(x_{t-1} - \mu))^2$.

The expression of the full conditional density of ρ is discussed in Exercise 16.1. The conditional density of each component x_t does not belong to a standard parametric family, but it may be sampled using rejection, as discussed in Sect. 15.5.

A first drawback of Gibbs sampling is that its implementation is highly model-dependent. In the example above, changing only the distribution of X_0 requires deriving new conditional distributions (see Exercise 16.1). The same may be said for the prior distribution. As a side remark, it is often the case that the prior is chosen so as to ensure that the conditional distributions are easy to simulate from (as implicitly done in the example above). Of course, for some models, these conditional distributions may be of no known form whatever the prior.

Example 16.3 In the theta-logistic model (Sect. 2.4.5), $X_t = X_{t-1} + \tau_0 - \tau_1 \exp(\tau_2 X_{t-1}) + U_t$, $U_t \sim \mathcal{N}(0, \sigma_X^2)$, and the full conditional density of τ_2 is of the form:

$$\propto \nu(\tau_2) \exp\left\{-\frac{1}{2\sigma_X^2} \sum_{t=1}^{T} (x_t - x_{t-1} - \tau_0 + \tau_1 \exp(\tau_2 x_{t-1}))^2\right\}$$

which does not seem straightforward to simulate from (whatever the prior). The same thing may be said for the conditional distribution of each state x_t; for $0 < t < T$, the log-density of which is, up to constant (in x_t) terms,

$$-\frac{1}{2\sigma_X^2}\left[(x_t - x_{t-1} - \tau_0 + \tau_1 \exp(\tau_2 x_{t-1}))^2 + (x_{t+1} - x_t - \tau_0 + \tau_1 \exp(\tau_2 x_t))^2\right].$$

Implementing (and calibrating for good performance) a Metropolis-Hastings step for τ_2 and each state variable seems non-trivial.

A second drawback of Gibbs sampling is that the simulated chain often mixes poorly, because of the strong dependencies between the components of $(\Theta, X_{0:T})$. In Sect. 15.5, we illustrated the fact that the states may be strongly correlated (with respect to the posterior): in particular, in the stochastic volatility example above, when ρ is close to one. The problem is compounded when parameter estimation comes into play: typically Θ and $X_{0:T}$ are strongly dependent, as different values for Θ lead to very different behaviours for the states.

A third drawback of Gibbs sampling (or more generally standard MCMC) is that it is not applicable to state-space models with intractable dynamics, that is models such that the distribution of $X_t|X_{t-1}$ does not admit a tractable density. For such models the simulation of Θ and $X_{0:T}$ from their conditional distributions is typically entirely intractable.

The rest of the chapter is devoted to PMCMC (Particle MCMC), a class of MCMC algorithms that do not suffer from these drawbacks. However, as the theoretical justification of PMCMC is more involved, we discuss first simpler algorithms which are based on similar ideas.

16.2 Pseudo-Marginal Samplers

Consider a generic Bayesian model, with prior distribution $\nu(d\theta)$, and likelihood function $p^\theta(y)$, and assume furthermore that this likelihood is intractable; that is, it is not possible (or is too expensive) to compute $p^\theta(y)$ at any θ. This prevents us from deriving a Metropolis-Hastings sampler that would leave invariant the corresponding posterior distribution:

$$\mathbb{P}(\mathrm{d}\theta|Y=y) = \nu(\mathrm{d}\theta)\frac{p^{\theta}(y)}{p(y)}.$$

Indeed, the Metropolis-Hastings ratio would take the form (assuming densities for all the involved distributions):

$$r(\theta, \widetilde{\theta}) = \frac{\nu(\widetilde{\theta})p^{\widetilde{\theta}}(y)\widetilde{m}(\theta|\widetilde{\theta})}{\nu(\theta)p^{\theta}(y)\widetilde{m}(\widetilde{\theta}|\theta)}$$

which is therefore not tractable.

A pseudo-marginal MCMC sampler may be described, superficially, as a Metropolis-Hastings sampler where the intractable likelihood is replaced by an unbiased estimate. That is, we assume we know how to sample an auxiliary variable $Z \sim \mathbb{M}^{\theta}(\mathrm{d}z)$, and how to compute a *non-negative* function $L(\theta, z)$, such that, for any $\theta \in \Theta$,

$$\mathbb{E}[L(\theta, Z)] = \int \mathbb{M}^{\theta}(\mathrm{d}z)L(\theta, z) = p^{\theta}(y). \tag{16.2}$$

Then, from the current pair (Θ, Z), we generate a new pair $(\widetilde{\Theta}, \widetilde{Z})$, and use the ratio:

$$r_{\mathrm{PM}}(\theta, z, \widetilde{\theta}, \widetilde{z}) = \frac{\nu(\widetilde{\theta})L(\widetilde{\theta}, \widetilde{z})\widetilde{m}(\theta|\widetilde{\theta})}{\nu(\theta)L(\theta, z)\widetilde{m}(\widetilde{\theta}|\theta)} \tag{16.3}$$

to decide on the acceptance of the proposed values.

Algorithm 16.1 describes one step of a pseudo-marginal sampler.

Algorithm 16.1: Generic pseudo-marginal kernel

Input: Θ, Z

$\widetilde{\Theta} \sim \widetilde{M}(\Theta, \mathrm{d}\theta)$

$\widetilde{Z} \sim \mathbb{M}^{\widetilde{\Theta}}(\mathrm{d}z)$

$U \sim \mathcal{U}([0, 1])$

$v \leftarrow \log r_{\mathrm{PM}}(\Theta, Z, \widetilde{\Theta}, \widetilde{Z})$ ▷ $r_{\mathrm{PM}}(\theta, z, \widetilde{\theta}, \widetilde{z}) := \frac{\nu(\widetilde{\theta})L(\widetilde{\theta},\widetilde{z})\widetilde{m}(\theta|\widetilde{\theta})}{\nu(\theta)L(\theta,z)\widetilde{m}(\widetilde{\theta}|\theta)}$

if $\log U \leq v$ **then**
  return $\widetilde{\Theta}$, $\widetilde{Z}$
else
  return Θ, Z

At first sight, pseudo-marginal sampling seems very ad hoc: each step of the Algorithm approximates in some way a Metropolis-Hastings step with respect to the posterior. It is not clear how the approximation error accumulates over time. However, pseudo-marginal samplers happen to be exact, in a sense made precise by the following proposition.

Proposition 16.1 *Algorithm 16.1 defines the Markov kernel of the Metropolis-Hastings sampler with the following characteristics: the sampling space is $\Theta \times \mathcal{Z}$, the proposal kernel is $M(\theta, \mathrm{d}\widetilde{\theta})\mathbb{M}^{\theta}(\mathrm{d}z)$, and the invariant probability distribution is:*

$$\pi(\mathrm{d}\theta, \mathrm{d}z) = \nu(\mathrm{d}\theta)\mathbb{M}^{\theta}(\mathrm{d}z)\frac{L(\theta, z)}{p(y)}. \tag{16.4}$$

This invariant distribution admits, as a marginal distribution for variable Θ, the posterior distribution of the model:

$$\int_{z \in \mathcal{Z}} \pi(\mathrm{d}\theta, \mathrm{d}z) = \nu(\mathrm{d}\theta)\frac{p^{\theta}(y)}{p(y)}. \tag{16.5}$$

Proof To go from (16.4) to (16.5), simply integrate out Z, using (16.2). Note that if we now integrate out Θ, we obtain one, and this shows that (16.4) is indeed a probability measure. Now consider the Metropolis-Hastings sampler (Algorithm 15.1) mentioned in the proposition above. Its acceptance ratio should then be (assuming that all involved distributions admit densities with respect to appropriate dominating measures, i.e. $m^{\theta}(z)$ is the density of $\mathbb{M}^{\theta}(\mathrm{d}z)$ and so on):

$$\frac{\pi(\widetilde{\theta}, \widetilde{z})\widetilde{m}(\theta|\widetilde{\theta})m^{\theta}(z)}{\pi(\theta, z)\widetilde{m}(\widetilde{\theta}|\theta)m^{\widetilde{\theta}}(\widetilde{z})} = \frac{\nu(\widetilde{\theta})\cancel{m^{\widetilde{\theta}}(\widetilde{z})}L(\widetilde{\theta}, \widetilde{z})\widetilde{m}(\theta|\widetilde{\theta})\cancel{m^{\theta}(z)}}{\nu(\theta)\cancel{m^{\theta}(z)}L(\theta, z)\widetilde{m}(\widetilde{\theta}|\theta)\cancel{m^{\widetilde{\theta}}(\widetilde{z})}} = r_{\mathrm{PM}}(\theta, z, \widetilde{\theta}, \widetilde{z})$$

the ratio we defined in (16.3). Hence such a Metropolis-Hastings kernel corresponds precisely to Algorithm 16.1. □

In other words, pseudo-marginal samplers are not approximate MCMC algorithms. Rather, they are standard Metropolis-Hastings algorithms with respect to an "extended" distribution that includes the auxiliary variable Z. And since that extended distribution admits the true posterior as its marginal, a pseudo-marginal algorithm samples exactly from the posterior. This property of being a standard MCMC sampler with respect to an extended distribution will apply to all the other algorithms discussed in this chapter.

Any unbiased estimate of the likelihood may be used within a pseudo-marginal sampler. One may wonder how the choice of the unbiased estimate affects the performance of the algorithm. The following example provides some insight.

Example 16.4 Let $c \in (0, 1]$, $Z \sim \mathcal{B}er(c)$, and $L(\theta, z) = (Z/c)p^\theta(y)$; thus the estimate is zero with probability $1 - c$. By construction, the acceptance probability of any given step is upper-bounded by c. It is then possible to obtain arbitrarily bad mixing chains by taking c arbitrarily small.

Of course, the above example is artificial, but it suggests that the variability of the unbiased estimate may degrade the acceptance rate, and therefore the mixing properties of the corresponding pseudo-marginal algorithm. We shall return to this point when discussing PMCMC algorithms. We now turn our attention to a specific pseudo-marginal sampler, called GIMH.

16.3 GIMH (Grouped Independence Metropolis-Hastings)

We consider again a generic model with an intractable likelihood $p^\theta(y)$. However, we now assume that the likelihood may be expressed as an integral with respect to a latent variable X:

$$p^\theta(y) = \int_{\mathcal{X}} \mathbb{P}^\theta(\mathrm{d}x) f^\theta(y|x) \,.$$

Furthermore, we assume we know a distribution $\mathbb{M}^\theta(\mathrm{d}x)$, which is easy to simulate from, and such that $\mathbb{M}^\theta(\mathrm{d}x) \gg \mathbb{P}^\theta(\mathrm{d}x) f^\theta(y|x)$. Thus there exists a ($\theta$-dependent) density $G^\theta(x)$ such that

$$\mathbb{P}^\theta(\mathrm{d}x) f^\theta(y|x) = \mathbb{M}^\theta(\mathrm{d}x) G^\theta(x) \,,$$

and we assume we are able to compute G^θ point-wise.

In that case, a simple way to obtain an unbiased estimate of the likelihood is to use importance sampling: take $Z = X^{1:N}$, where the X^n are i.i.d. variables generated from $\mathbb{M}^\theta$ and let

$$L(\theta, x^{1:N}) = \frac{1}{N} \sum_{n=1}^{N} G^\theta(x^n) \,.$$

This particular pseudo-marginal sampler is called GIMH (grouped independence Metropolis-Hastings), and is described as Algorithm 16.2.

Algorithm 16.2: GIMH (one step)

Input: Θ, $X^{1:N}$

$\widetilde{\Theta} \sim \widetilde{M}(\Theta, \mathrm{d}\widetilde{\theta})$

$\widetilde{X}^n \sim \mathbb{M}^{\widetilde{\Theta}}(\mathrm{d}x)$ for $n = 1, \ldots, N$

$U \sim \mathcal{U}([0, 1])$

$v \leftarrow \log r_{\mathrm{GIMH}}(\Theta, X^{1:N}, \widetilde{\Theta}, \widetilde{X}^{1:N})$ *where*

$$r_{\mathrm{GIMH}}(\theta, x^{1:N}, \widetilde{\theta}, \widetilde{x}^{1:N}) := \frac{\nu(\widetilde{\theta})\{\sum_{n=1}^{N} G^{\widetilde{\theta}}(\widetilde{x}^n)\}\widetilde{m}(\theta|\widetilde{\theta})}{\nu(\theta)\{\sum_{n=1}^{N} G^{\theta}(x^n)\}\widetilde{m}(\widetilde{\theta}|\theta)}$$

if $\log U \leq v$ **then**
  return $\widetilde{\Theta}$, $\widetilde{X}^{1:N}$
else
  return Θ, $X^{1:N}$

As per Proposition 16.1, this algorithm leaves invariant distribution

$$\pi(\mathrm{d}\theta, \mathrm{d}x^{1:N}) = \nu(\mathrm{d}\theta) \prod_{n=1}^{N} \mathbb{M}^{\theta}(\mathrm{d}x^n) \left[\frac{\sum_{n=1}^{N} G^{\theta}(x^n)}{Np(y)}\right] . \tag{16.6}$$

This invariant distribution turns out to have several interesting properties, which suggest various possible extensions to the basic GIMH algorithm.

To describe these properties, we introduce an extra variable, K, which selects randomly one of the N variables X^n, and consider the following extended distribution:

$$\pi(\mathrm{d}\theta, \mathrm{d}x^{1:N}, k) := \nu(\mathrm{d}\theta) \prod_{n=1}^{N} \mathbb{M}^{\theta}(\mathrm{d}x^n) \left(\frac{G^{\theta}(x^k)}{Np(y)}\right) . \tag{16.7}$$

Notation/Terminology Measure-theoretic notations, e.g $\pi(\mathrm{d}k)$, are very uncommon for discrete random variables. Thus, we simply write k, instead of $\mathrm{d}k$, whenever a joint distribution involves a discrete variable K; and we do not write $\mathrm{d}k$ on the right-hand side. (This $\mathrm{d}k$ would be the counting measure.) The definition has the same interpretation as usual: e.g. the integral

(continued)

$\int_A \pi(d\theta, dx^{1:N}, k)$ is the joint probability that $(\Theta, X^{1:N}) \in A$ and $K = k$ (for a measurable set $A \subset \Theta \times \mathcal{X}^N$).

Proposition 16.2 *Consider random variables $\Theta, X^{1:N}, K$, taking values respectively in Θ, $\mathcal{X}^N$, and $1 : N$, and being jointly distributed according to* (16.7). *Then:*

1. *Variables $(\Theta, X^{1:N})$ are jointly distributed according to* (16.6).
2. *Variable K is distributed marginally according to a $\mathcal{U}(1 : N)$ distribution, and, conditionally on $\Theta = \theta$, $X^{1:N} = x^{1:N}$, takes value n with probability $G^\theta(x^n)/\sum_{m=1}^N G^\theta(x^m)$.*
3. *(Θ, X^K) is distributed according to the joint posterior distribution of the model, i.e.*

$$\mathbb{P}(d\theta, dx|Y = y) = \nu(d\theta)\mathbb{P}^\theta(dx)\frac{f^\theta(y|x)}{p(y)}. \tag{16.8}$$

4. *Conditional on (Θ, X^K), the random variables $(X^n)_{n\neq K}$ form a collection of i.i.d. variables distributed according to $\mathbb{M}^\theta(dx)$.*

Proof To go from (16.7) to (16.6), simply sum over k. To obtain the conditional distribution of K, divide (16.7) by (16.6). To obtain, the marginal distribution of K, integrate out $X^{1:N}$, then Θ in (16.7). To establish the last two properties, apply the change of variables that corresponds to the transformation $(\Theta, X^{1:N}, K) \to (\Theta, X^\star, X^{-K}, K)$, where $X^\star = X^K$, $X^{-K} = (X^n)_{n\neq K}$. The Jacobian of this transformation is clearly one. Thus we obtain the following joint distribution for $(\Theta, X^\star, X^{-K}, K)$:

$$\left\{\nu(d\theta)\mathbb{M}^\theta(dx^\star)\frac{G^\theta(x^\star)}{p(y)}\right\}\prod_{n\neq k}\mathbb{M}^\theta(dx^n)\left(\frac{1}{N}\right)$$

and given the definition of G^θ, we recognise the factor inside the curly brackets as the joint posterior (16.8). From the same expression we obtain that K is uniformly distributed over the range $1 : N$. □

The properties above offer several extensions to the basic GIMH algorithm. For instance, we have described GIMH as a pseudo-marginal sampler that leaves invariant the posterior of Θ, $\mathbb{P}(d\theta|Y = y)$. But, by Property 3, we see now that it is possible to obtain samples from the joint posterior $\mathbb{P}(d\theta, dx|Y = y)$. For this, we simply add an extra step to the GIMH algorithm, where we sample K from its conditional distribution; then (Θ, X^K) is distributed according to the joint posterior.

We can also use the properties above to derive an alternative, Gibbs sampler-like MCMC kernel, which has the same invariant distribution as GIMH; see Algorithm 16.3.

Algorithm 16.3: Gibbs sampler-like variant of GIMH

Input: Θ, X

$X^1 \leftarrow X$

$X^n \sim \mathbb{M}^{\Theta}(\mathrm{d}x)$ for $n = 2, \ldots, N$

$W^n \leftarrow G^{\Theta}(X^n)/\sum_{m=1}^{N} G^{\Theta}(X^m)$ for $n = 1, \ldots, N$

$K \sim \mathcal{M}(W^{1:N})$

$\widetilde{X} \leftarrow X^K$

$\widetilde{\Theta} \sim \mathbb{P}(\mathrm{d}\theta | X = \widetilde{X}, Y = y)$ ▷ `conditional distribution of Θ with respect to joint distribution` (16.8)

return $\widetilde{\Theta}$, $\widetilde{X}$

A few comments are in order. In the first instruction, we arbitrarily label the input X as X^1. Alternatively, we could have described the algorithm as an MCMC kernel that updates variables Θ, X, and K; then input X would have been assigned to X^K, and we would have re-generated the other X^n's, for $n \neq K$. However, it is easy to see that the labelling of the X^n's has no impact on the random distribution of the output (in the same way as multinomial resampling does not depend on the particle labels). Thus, we may as well discard the current value of K, and label the input X as X^1.

The last instruction samples from the conditional distribution of Θ (with respect to the target distribution). Depending on the model, this conditional distribution may or may not be easy to simulate from. For models such that this operation is feasible, Algorithm 16.3 offers a complete replacement to Algorithm 16.2. Their relative pros and cons seem more or less the same as Metropolis-Hastings versus Gibbs sampler: the latter does not require specifying a proposal distribution, while the former may mix better, if we manage to construct a good proposal.

Technically, Algorithm 16.3 is not exactly a Gibbs sampler. It does consist of a sequence steps that sample from conditional distributions. However, some of these conditional distributions are with respect to (16.6), while the others are with respect to the distribution that corresponds to the change of variables described in Properties 3 and 4 above: $(\Theta, X^{1:N}, K) \rightarrow (\Theta, X^{\star}, X^{-K}, K)$, where $X^{\star} = X^K$, $X^{-K} = (X_n)_{n \neq K}$. Moreover, although this algorithm does not involve an accept/reject step (like a Metropolis-Hastings sampler), it has a non-zero probability of not updating X; this happens when the realisation of variable K is one.

PMCMC algorithms, which we discuss next, are essentially advanced versions of the previous two algorithms, where the N instrumental variables $X^{1:N}$ are replaced by the random variables generated during the course of a particle algorithm.

16.4 Particle Markov Chain Monte Carlo

16.4.1 Unbiased Estimates from SMC

We start by showing that an SMC algorithm provides unbiased estimates of the normalising constants L_t of the associated Feynman-Kac model. To fix ideas, we restrict to the case where resampling is triggered at every step (i.e. Algorithm 10.1 in Chap. 10).

Proposition 16.3 *The random variables generated by Algorithm 10.1 are such that, at any time $T \geq 0$,*

$$L_T^N = \left(\frac{1}{N}\sum_{n=1}^N G_0(X_0^n)\right)\prod_{t=1}^T\left(\frac{1}{N}\sum_{n=1}^N G_t(X_{t-1}^{A_t^n}, X_t^n)\right) \tag{16.9}$$

is an unbiased estimator of L_T, the normalising constant at time T of the associated Feynman-Kac model.

Before stating the proof, we recall that Algorithm 10.1 is assumed to rely on an unbiased resampling scheme. This implies that

$$\mathbb{E}\left[\frac{1}{N}\sum_{n=1}^N \varphi(X_t^{A^n}) \,\middle|\, X_t^{1:N}\right] = \sum_{n=1}^N W_t^n \varphi(X_t^n) \tag{16.10}$$

for any function φ. In fact, this is the defining property of unbiased resampling scheme; see Chap. 9.

Proof For the sake of space, we take $T = 1$; the general case follows exactly along the same lines. Let $\mathcal{F}_0 = \sigma(X_0^{1:N})$, $M_1(x_0, G_1) = \int_{\mathcal{X}} M_1(x_0, \mathrm{d}x_1)G_1(x_0, x_1)$, and $W_0^n = G_0(X_0^n)/\sum_{m=1}^N G_0(X_0^m)$. Then

$$\begin{aligned}
\mathbb{E}\left[L_1^N \,\middle|\, \mathcal{F}_0\right] &= \left(\frac{1}{N}\sum_{n=1}^N G_0(X_0^n)\right)\mathbb{E}\left[\frac{1}{N}\sum_{n=1}^N M_1(X_0^{A_1^n}, G_1)\,\middle|\,\mathcal{F}_0\right]\\
&= \left(\frac{1}{N}\sum_{n=1}^N G_0(X_0^n)\right)\left(\sum_{n=1}^N W_0^n M_1(X_0^n, G_1)\right)\\
&= \frac{1}{N}\sum_{n=1}^N G_0(X_0^n)M_1(X_0^n, G_1)
\end{aligned}$$

where we have integrated out X_1^n in the first equality, and applied (16.10) in the second equality. Since $X_0^n \sim \mathbb{M}_0(\mathrm{d}x_0)$, by the tower property of conditional expectation

$$\mathbb{E}\left[L_1^N\right] = \int_{\mathcal{X}} \mathbb{M}_0(\mathrm{d}x_0) M_1(x_0, G_1) G_0(x_0) = L_1 . \qquad \square$$

The property above also holds when adaptive resampling is used; see Exercise 16.4, or when SQMC is used instead of SMC, as established by the following proposition.

Proposition 16.4 *Proposition 16.3 also holds for SQMC (Algorithm 13.3).*

Proof Algorithm 13.3 relies on RQMC (randomised QMC) sequences; i.e. sequences of random variables that are marginally distributed according to an appropriate uniform distribution. This implies that, conditional on $\mathcal{F}_0 = \sigma(X_0^{1:N})$, the pairs $(X_0^{A_1^n}, X_1^n)$ are marginally distributed according to

$$\sum_{n=1}^{N} W_0^n \delta_{X_0^n}(\mathrm{d}x_0) M_1(X_0^n, \mathrm{d}x_1)$$

hence we still have that:

$$\mathbb{E}\left[\frac{1}{N}\sum_{n=1}^{N} G_1(X_0^{A_1^n}, X_1^n) \,\middle|\, \mathcal{F}_0\right] = \sum_{n=1}^{N} W_0^n M_1(X_0^n, G_1)$$

and the rest of the calculations follows as in the previous proof. $\square$

16.4.2 PMMH: Particle Marginal Metropolis-Hastings

We return to the main topic of this chapter: we have a state-space model, parameterised by $\theta \in \Theta$, and we want to sample from its posterior (given data $y_{0:T}$). Pseudo-marginal sampling, together with Propositions 16.3 and 16.4 give us a general recipe: run a pseudo-marginal sampler, where the intractable likelihood $p_T^\theta(y_{0:T})$ of the model is replaced by an unbiased estimate obtained from a particle filter.

To go further, we introduce some notation. We associate our parametric state-space model with a parametric Feynman-Kac model, with kernels $M_t^\theta(x_{t-1}, \mathrm{d}x_t)$ and functions G_t^θ that depend on θ. This Feynman-Kac model must be such that the

Feynman-Kac measure $\mathbb{Q}_T^\theta(dx_{0:T})$ matches the posterior distribution of the states (conditional on $\Theta = \theta$):

$$\begin{aligned}\mathbb{Q}_T^\theta(\mathrm{d}x_{0:T}) &= \frac{1}{L_T^\theta}\mathbb{M}_0^\theta(\mathrm{d}x_0)\prod_{t=1}^T M_t^\theta(x_{t-1},\mathrm{d}x_t)G_0^\theta(x_0)\prod_{t=1}^T G_t^\theta(x_{t-1},x_t)\\ &= \frac{1}{p_T^\theta(y_{0:T})}\mathbb{P}_T^\theta(\mathrm{d}x_{0:T})\prod_{t=0}^T f_t^\theta(y_t|x_t)\\ &= \mathbb{P}_T(\mathrm{d}x_{0:T}|\Theta=\theta, Y_{0:T}=y_{0:T});\end{aligned}$$

in particular the normalising constant L_T^θ equals the likelihood $p^\theta(y_{0:T})$, for all $\theta \in \Theta$.

To specify this parametric Feynman-Kac model, we use the same recipes as in Chap. 10: i.e. the bootstrap formalism, the guided formalism, or the auxiliary formalism. Perhaps it is simpler to focus on the guided formalism, where $M_t^\theta(x_{t-1},\mathrm{d}x_t) \gg P_t^\theta(x_{t-1},\mathrm{d}x_t)f_t^\theta(y_t|x_t)$, and

$$G_t^\theta(x_{t-1},x_t) = \frac{P_t^\theta(x_{t-1},\mathrm{d}x_t)f_t^\theta(y_t|x_t)}{M_t^\theta(x_{t-1},\mathrm{d}x_t)}$$

for $t \geq 1$ (and a similar expression for G_0^θ). The bootstrap formalism is recovered as a special case (take $M_t^\theta(x_{t-1},\mathrm{d}x_t) = P_t^\theta(x_{t-1},\mathrm{d}x_t)$); for the auxiliary formalism, see Exercise 16.3.

Since $L_T^\theta = p_T^\theta(y_{0:T})$, we may, for any θ, run an SMC (or SQMC) algorithm associated with the considered Feynman-Kac model, and compute

$$L_T^N(\theta, X_{0:T}^{1:N}, A_{1:T}^{1:N}) = \left(\frac{1}{N}\sum_{n=1}^N G_0^\theta(X_0^n)\right)\prod_{t=1}^T\left(\frac{1}{N}\sum_{n=1}^N G_t^\theta(X_{t-1}^{A_t^n}, X_t^n)\right) \tag{16.11}$$

as an unbiased estimate of $p_T^\theta(y_{0:T})$. The quantity above is the same thing as (16.9), except that the new notation makes explicit the dependence on θ and the particles. (Again, we consider a particle filter where the resampling step is triggered at every time step.)

Algorithm 16.4 describes PMMH (particle marginal Metropolis-Hastings), which is the standard name of pseudo-marginal algorithms based on particle filters.

Algorithm 16.4: PMMH (one step)

Input: Θ, $X_{0:T}^{1:N}$, $A_{1:T}^{1:N}$

$\widetilde{\Theta} \sim \widetilde{M}(\Theta, \mathrm{d}\theta)$

Run associated SMC (or SQMC) algorithm to generate variables $(\widetilde{X}_{0:T}^{1:N}, \widetilde{A}_{1:T}^{1:N})$ given parameter value $\widetilde{\Theta}$.

$U \sim \mathcal{U}([0, 1])$

$v \leftarrow \log r_{\text{PMMH}}(\Theta, X_{0:T}^{1:N}, A_{1:T}^{1:N}, \widetilde{\Theta}, \widetilde{X}_{0:T}^{1:N}, \widetilde{A}_{1:T}^{1:N})$ where

$$r_{\text{PMMH}}(\theta, x_{0:T}^{1:N}, a_{1:T}^{1:N}, \widetilde{\theta}, \widetilde{x}_{0:T}^{1:N}, \widetilde{a}_{1:T}^{1:N}) := \frac{\nu(\widetilde{\theta}) L_T^N(\widetilde{\theta}, \widetilde{x}_{0:T}^{1:N}, \widetilde{a}_{1:T}^{1:N}) \widetilde{m}(\theta|\widetilde{\theta})}{\nu(\theta) L_T^N(\theta, x_{0:T}^{1:N}, a_{1:T}^{1:N}) \widetilde{m}(\widetilde{\theta}|\theta)}$$

if $\log U \leq v$ **then**
  return $\widetilde{\Theta}$, $\widetilde{X}_{0:T}^{1:N}$, $\widetilde{A}_{1:T}^{1:N}$
else
  return Θ, $X_{0:T}^{1:N}$, $A_{1:T}^{1:N}$

Since PMMH is a pseudo-marginal sampler, it leaves invariant a certain distribution on an extended space, such that the marginal of Θ matches the posterior distribution of the parameter. As for GIMH, we may derive extensions of, or even alternatives to, PMMH by looking more closely at this invariant distribution. This is the point of the next section.

16.4.3 Particle Gibbs Sampler

We now leave aside SQMC and focus on standard (Monte Carlo) particle filters. Furthermore, we restrict our attention to the case where multinomial resampling is used. In that case, the distribution of the random variables generated during the course of the filtering algorithm have the following expression:

$$\psi_T^\theta(\mathrm{d}x_{0:T}^{1:N}, a_{1:T}^{1:N}) = \left(\prod_{n=1}^N \mathbb{M}_0^\theta(\mathrm{d}x_0^n)\right) \prod_{t=1}^T \left(\prod_{n=1}^N W_{t-1}^{a_t^n} M_t(X_{t-1}^{a_t^n}, \mathrm{d}x_t^n)\right) .$$

where we have used the short-hand $W_t^n = G_t^\theta(X_{t-1}^{A_t^n}, X_t^n)/\sum_{m=1}^N G_t^\theta(X_{t-1}^{A_t^m}, X_t^m)$. Note that factor $W_{t-1}^{a_t^n}$ means that variable A_t^n takes value m with probability W_{t-1}^m.

We obtain the invariant distribution of PMMH by applying Proposition 16.1 (replacing variable z by $(x_{0:T}^{1:N}, a_{1:T}^{1:N})$ and so on):

$$\pi_T(\mathrm{d}\theta, \mathrm{d}x_{0:T}^{1:N}, a_{1:T}^{1:N}) = \nu(\mathrm{d}\theta)\psi_T^\theta(\mathrm{d}x_{0:T}^{1:N}, a_{1:T}^{1:N})\frac{L_T^N(\theta, x_{0:T}^{1:N}, a_{1:T}^{1:N})}{p(y_{0:T})}.$$

This invariant distribution has properties that are similar to those of GIMH:

Proposition 16.5 *Consider random variables Θ, $X_{0:T}^{1:N}$, $A_{1:T}^{1:N}$, and K, taking values respectively in Θ, $\mathcal{X}^{N(T+1)}$, $(1:N)^T$ and $1:N$, that are jointly distributed as follows:*

$$\pi_T(\mathrm{d}\theta, \mathrm{d}x_{0:T}^{1:N}, a_{1:T}^{1:N}, k) := \pi_T(\mathrm{d}\theta, \mathrm{d}x_{0:T}^{1:N}, a_{1:T}^{1:N})\left(\frac{G_T^\theta(x_{T-1}^{A_T^k}, x_T^k)}{\sum_{m=1}^N G_T^\theta(x_{T-1}^{a_T^m}, x_T^m)}\right). \tag{16.12}$$

In addition, define the star trajectory as follows: $B_T := K$, $B_t := A_{t+1}^{B_{t+1}}$ for $t < T$ (recursively), and finally $X_t^\star := X_t^{B_t}$ for $t = 0, \ldots, T$. Then:

1. *Variables Θ and $X_{0:T}^\star$ are jointly distributed according to the posterior distribution of the model, $\mathbb{P}_T(\mathrm{d}\theta, \mathrm{d}x_{0:T}|Y_{0:T} = y_{0:T})$.*
2. *Variable K is marginally distributed according to a $\mathcal{U}(1:N)$ distribution, and, conditionally on all the other variables, is distributed according to $\mathcal{M}(W_T^{1:N})$, with $W_T^n = G_T^\theta(X_{T-1}^{A_T^n}, X_T^n)/\sum_{m=1}^N G_T^\theta(X_{T-1}^{A_T^m}, X_T^m)$.*
3. *Conditional on Θ, $B_{0:T}$ and $X_{0:T}^\star$, the remaining variables follow a CSMC (conditional SMC) distribution, that is, the distribution of a particle system $(X_{0:T}^{1:N}, A_{1:T}^{1:N})$, conditional on $X_t^{B_t} = X_t^\star$ (for $t = 0, \ldots, T$), and $A_{t+1}^{B_{t+1}} = B_t$ (for $t = 0, \ldots, T-1$). This distribution may be decomposed as follows: at time 0, $X_0^n \sim \mathbb{M}_0(\mathrm{d}x_0)$ for $n \neq B_0$; at time 1, $A_1^n \sim \mathcal{M}(W_0^{1:N})$, $X_1^n \sim M_1(X_0^{A_1^n}, \mathrm{d}x_1)$ for $n \neq B_1$ where $W_0^n = G_0(X_0^n)/\sum_{m=1}^N G_0(X_0^m)$; and so on for $t > 0$.*

Proof Again take $T = 1$ for the sake of space (the general case follows exactly along the same lines), and plug (16.11) in (16.12) to obtain:

$$\pi_1(\mathrm{d}\theta, \mathrm{d}x_{0:1}^{1:N}, a_1^{1:N}, k) = \nu(\mathrm{d}\theta) \prod_{n=1}^{N} \mathbb{M}_0^\theta(\mathrm{d}x_0) \prod_{n=1}^{N} W_0^{a_1^n} M_1^\theta(x_0^{a_1^n}, \mathrm{d}x_1^n)$$

$$\times \left(\frac{1}{N^2} \sum_{n=0}^{N} G_0^\theta(x_0^n) \right) \frac{G_1^\theta(x_0^{a_1^k}, x_1^k)}{p(y_{0:1})}.$$

Now extract factor k from the second product, factor a_1^k from the first product, and re-arrange as follows:

$$\pi_1(\mathrm{d}\theta, \mathrm{d}x_{0:1}^{1:N}, a_1^{1:N}, k) = \prod_{n \neq a_1^k} \mathbb{M}_0^\theta(\mathrm{d}x_0) \prod_{n \neq k} W_0^{a_1^n} M_1^\theta(x_0^{a_1^n}, \mathrm{d}x_1^n)$$

$$\times \frac{1}{N^2} \nu(\mathrm{d}\theta) \mathbb{M}_0^\theta(\mathrm{d}x_0^{a_1^k}) M_1^\theta(x_0^{a_1^k}, \mathrm{d}x_1^k) \frac{G_0^\theta(x_0^{a_1^k}) G_1^\theta(x_0^{a_1^k}, x_1^k)}{p(y_{0:1})}.$$

Letting $B_1 := K$, $B_0 := A_1^K$, $X_1^\star := X_1^K$, $X_0^\star := X_0^{B_0}$, the second line of the expression above gives us the marginal distribution of $(\Theta, X_0^\star, X_1^\star)$, which is simply the joint posterior of the model; and the marginal distribution of K and $B_0 = A_1^K$, which is the uniform distribution over $1 : N$. The first line defines the CSMC distribution, that is, the distribution of the remaining variables of the particle system, given Θ, $X_{0:1}^\star$, and $B_{0:1}$. This distribution may be described sequentially as follows (from left to right): at time 0, the particles X_0^n, for $n \neq B_0$, are distributed according to $\mathbb{M}_0^\theta(\mathrm{d}x_0)$; at time 1 (conditional on the variables generated at time 0), for $n \neq k$, the A_1^n's follow a $\mathcal{M}(W_0^{1:N})$ distribution with $W_0^n \propto G_0(X_0^n)$, and the X_1^n's are distributed according to $M_1(X_0^{A_1^n}, \mathrm{d}x_1)$. □

In Property 3, we described the CSMC distribution as the distribution of the particle system conditional on one 'frozen' trajectory, labelled $B_0, \ldots, B_T$. However, the particle system is exchangeable: its distribution does not change if we permute the particle labels. Hence we may as well assign labels $1, 1, \ldots$ to the frozen trajectory. Algorithm 16.5 shows how one may sample from the CSMC distribution using this property.

Algorithm 16.5: CSMC (Conditional SMC)

Input: Star trajectory $X^{\star}_{0:T}$, parameter Θ.

Output: Output of a particle filter: $X^{1:N}_{0:T}$, $A^{1:N}_{1:T}$.

$X^1_0 \leftarrow X^{\star}_0$

$X^n_0 \sim \mathbb{M}^{\Theta}_0(\mathrm{d}x_0)$ for $n = 2, \ldots, N$

$w^n_0 \leftarrow G^{\Theta}_0(X^n_0)$ and $W^n_0 \leftarrow w^n_0 / \sum^N_{m=1} w^m_0$ for $n = 1, \ldots, N$

for $t = 1$ **to** T **do**

$\quad A^1_t \leftarrow 1$

$\quad A^{2:N}_t \sim \mathcal{M}(W^{1:N}_t)$ ▷ `multinomial resampling (Algorithm 9.3)`

$\quad X^1_t \leftarrow X^{\star}_t$

$\quad X^n_t \sim M^{\Theta}_t(X^{A^n_t}_{t-1}, \mathrm{d}x_t)$ for $n = 2, \ldots, N$

$\quad w^n_t \leftarrow G^{\Theta}_t(X^{A^n_t}_{t-1}, X^n_t)$ and $W^n_t \leftarrow w^n_t / \sum^N_{m=1} w^m_t$ for $n = 1, \ldots, N$

As for GIMH, we may use the properties above to either extend PMMH, or derive an alternative MCMC kernel.

Consider the first option. We have initially described PMMH is a pseudo-marginal sampler that leaves invariant $\mathbb{P}_T(\mathrm{d}\theta | Y_{0:T} = y_{0:T})$. From Property 1, we see that we may also obtain samples from the joint posterior distribution $\mathbb{P}_T(\mathrm{d}\theta, \mathrm{d}x_{0:T} | Y_{0:T} = y_{0:T})$, by simply adding an extra step, which generates the random index K, and "pulls" the corresponding trajectory. This extra step is described in Algorithm 16.6.

Algorithm 16.6: Sampling the star trajectory (optional extra step in PMMH)

Input: Θ, $X^{1:N}_{0:T}$, $A^{1:N}_{1:T}$

Output: $X^{\star}_{0:T}$

$B_T \sim \mathcal{M}(W^{1:N}_T)$, where $W^n_T \propto G^{\theta}_T(X^{A^n_T}_{T-1}, X^n_T)$

$X^{\star}_T \leftarrow X^{B_T}_T$

for $t = T - 1$ **to** 0 **do**

$\quad B_t \leftarrow A^{B_{t+1}}_{t+1}$

$\quad X^{\star}_t \leftarrow X^{B_t}_t$

Now, consider the second option: by combining Algorithms 16.5 and 16.6, we define a Markov kernel that leaves invariant the conditional distribution of $X_{0:T}$, given Θ and the data. If, in addition, we know how to sample Θ given $X_{0:T}$ and $Y_{0:T}$, we may construct a Markov kernel that leaves invariant the joint posterior distribution; see Algorithm 16.7. Such a Markov kernel is called a Particle Gibbs kernel, and may be used as a complete replacement to the PMMH kernel.

Algorithm 16.7: Particle Gibbs kernel

Input: Θ, $X_{0:T}$

Generate particle system $(X_{0:T}^{1:N}, A_{1:T}^{1:N})$ from star trajectory $X_{0:T}$ (using Algorithm 16.5)

Generate a new star trajectory $\widetilde{X}_{0:T}$ from $(X_{0:T}^{1:N}, A_{1:T}^{1:N})$ (using Algorithm 16.6)

Generate $\widetilde{\Theta} \sim \mathbb{P}(\mathrm{d}\theta | X_{0:T} = \widetilde{X}_{0:T}, Y_{0:T} = y_{0:T})$

return $\widetilde{\Theta}$, $\widetilde{X}_{0:T}$

In simple cases, one may be able to sample from $\Theta | X_{0:T}, Y_{0:T}$ exactly. More commonly, one may able to decompose it into conditional distributions for each component that are easy to sample from (e.g. the stochastic volatility model of Sect. 16.1.3). Then we may implement a Gibbs step that leaves invariant this conditional distribution. More generally, in the usual Hastings-within-Gibbs fashion, the sampling of Θ might be achieved using any Metropolis-Hastings kernel that targets the conditional.

For some models, this conditional distribution does not even admit a tractable density (e.g. models with intractable dynamics). In such cases, Particle Gibbs is not feasible. For models where both PMMH and Particle Gibbs may be implemented, an important question is which algorithm one should implement in practice. We shall discuss this point in our numerical examples in Sect. 16.5.

16.4.4 The Backward Sampling Step

The Particle Gibbs algorithm suffers from one particular deficiency, which is related to the coalescence of particle trajectories. Let's look again at the genealogical tree in Fig. 12.2. If we assume that one of these trajectories is the star trajectory, and the other trajectories have been generated by the CSMC step, then, clearly, any trajectory selected by Algorithm 16.6 will be identical to the star trajectory up to time 40 (where coalescence occurs). Thus, early state variables are unlikely to change.

One way to avoid this problem is to increase N so that trajectories are more likely to differ. Recent theory suggests that one should take $N = \mathcal{O}(T^\gamma)$, $\gamma > 1$ to obtain stable performance for Particle Gibbs as T grows (see bibliography). An alternative approach is to try to change as much as possible the selected trajectory, by sampling the variables B_t that determine the genealogy of the star trajectory. The following proposition outlines how to do this.

Proposition 16.6 *Let Θ, $X_{0:T}^{1:N}$, $A_{1:T}^{1:N}$, K, and $B_{0:T}$ be distributed as in Proposition 16.5. Assume furthermore that the Markov kernels $M_t(x_{t-1}, \mathrm{d}x_t)$ admit probability density $m_t(x_t|x_{t-1})$ with respect to a common (i.e. independent of x_{t-1}) dominating probability measure. Then:*

- *Given Θ and $X_{0:T}^{1:N}$, $B_T \sim \mathcal{M}(W_T^{1:N})$ where $W_T^n \propto G_T^\Theta(X_{T-1}^{A_T^n}, X_T^n)$.*
- *For $t < T$, given $B_{t+1:T}$, $X_{0:T}^{1:N}$ and $A_{1:t-1}^{1:N}$, $B_t \sim \mathcal{M}(\widetilde{W}_t^{1:N})$ where*

$$\widetilde{W}_t^n \propto G_t^\Theta(X_{t-1}^{A_t^n}, X_t^n) m_{t+1}(X_{t+1}^{B_{t+1}}|X_t^n).$$

When the functions G_t^Θ depend only their second argument, x_t, the proposition above gives a backward decomposition of the distribution of $B_{0:T}$ given Θ and $X_{0:T}^{1:N}$. However, this is not true in general, as the probabilities above depend on the A_t^n's.

A simple way to understand this proposition is to consider the following theoretical Gibbs step: first sample $K = B_T$ given all the variables; then for $t = T, \ldots, 1$ (backward in time), sample the block of ancestor variables $A_t^{1:N}$, given Θ, the particles $X_{0:T}^{1:N}$ and the other blocks $A_s^{1:N}$, $s \neq t$. Clearly this Gibbs step leaves invariant the distribution described in the above proposition. Now, when we sample block $A_t^{1:N}$, we re-compute variable B_{t-1}, which is a deterministic function of B_t and $A_t^{1:N}$. This is equivalent to sampling the B_t's as described in the above proposition, see Exercise 16.6.

Thanks to this proposition, we may implement the so-called 'backward sampling step', which generates a new star trajectory, while actively trying to make it differ from the previous star trajectory; see Algorithm 16.8.

Algorithm 16.8: Backward sampling step (alternative to Algorithm 16.6 for generating the star trajectory)

Input: Θ, $X_{0:T}^{1:N}$, $A_{1:T}^{1:N}$

Output: $X_{0:T}^\star$

$B_T \sim \mathcal{M}(W_T^{1:N})$, where $W_T^n \propto G_T^\Theta(X_{T-1}^{A_T^n}, X_T^n)$

$X_T^\star \leftarrow X_T^{B_T}$

for $t = T - 1$ **to** 0 **do**

  $B_t \sim \mathcal{M}(\widetilde{W}_t^{1:N})$ where $\widetilde{W}_t^n \propto G_t^\Theta(X_{t-1}^{A_t^n}, X_t^n) m_{t+1}(X_{t+1}^{B_{t+1}}|X_t^n)$

  $X_t^\star \leftarrow X_t^{B_t}$

The backward sampling step requires that the Markov kernels $M_t(x_{t-1}, dx_t)$ (a) admit a density with respect to a common dominating measure; (b) which may be computed point-wise. A simple example where requirement (a) is not met is when $M_t(x_{t-1}, dx_t)$ involves a Dirac mass; then the backward sampling step amounts to sampling from a degenerate distribution (Exercise 16.5).

When implementable, the backward sampling step often improves significantly the performance of Particle Gibbs; see the numerical experiment section. Empirical evidence suggests that the performance does not degrade any more, for a fixed N, when T goes to infinity. However, this has not been established theoretically. What is known is that the version of Particle Gibbs with a backward sampling step leads to smaller asymptotic variance of estimates compared to that of the version without (in other terms, the former dominates the latter in what is known as Peskun ordering).

16.5 Numerical Experiments

PMCMC algorithms are formally appealing. In practice, they are not so easy to use, as they require the user to make several non-trivial choices, such as: which algorithm to use? (PMMH, Particle Gibbs, or some combination of the two?) How to set N? Which particle algorithm should be implemented? How to assess their performance? And so on.

The numerical experiments in this section are designed to give some guidance on these points.

16.5.1 Calibration of PMMH on a Toy Example

We consider the following toy example: $X_0 \sim \mathcal{N}(0, \sigma_X^2)$, and

$$X_t = \rho X_{t-1} + U_t\,, \qquad U_t \sim \mathcal{N}(0, \sigma_X^2)\,,$$

$$Y_t = X_t + V_t\,, \qquad V_t \sim \mathcal{N}(0, \sigma_Y^2)\,,$$

the components of $\theta = (\rho, \sigma_X^2, \sigma_Y^2)$ are assigned independent prior distributions: $\rho \sim \mathcal{U}([-1, 1])$, $\sigma_X^2, \sigma_Y^2 \sim \mathcal{IG}(2, 2)$. The dataset is simulated from the model, using $T + 1 = 100$, $\rho = 0.9$, $\sigma_X = 1$, $\sigma_Y = 0.2$.

We implement a PMMH algorithm based on (a) a Gaussian random walk proposal, with covariance τI_3; and (b) the bootstrap filter associated to this model, for $N = 100, \ldots, 1500$. We also implement an "ideal" Metropolis-Hastings sampler, based on the same proposal, and on the exact likelihood (computed using the Kalman filter). At each execution of these algorithms, the number of iterations is set to 10^5, the burn-in period to 10^4, and the starting point is sampled from the prior.

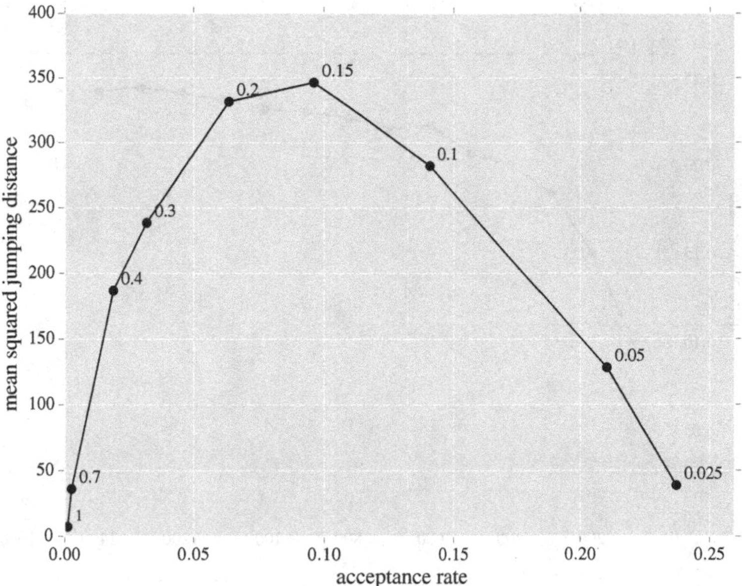

Fig. 16.3 Mean squared jumping distance versus acceptance rate, for PMMH with $N = 100$, and different random walk scales τ; the value of τ is printed next to each dot

Our point is to discuss how to set both N and τ to obtain good performance. The following basic strategy may seem sensible at first sight: set N to an arbitrary value, and calibrate the scale of the random walk so as to obtain an acceptance rate close to 0.234; see discussion in Sect. 15.3. This approach often fails spectacularly. To see this, we run our PMMH sampler, with $N = 100$, and for different values of τ, and plot in Fig. 16.3 the mean squared jumping distance versus the acceptance rate. (The mean squared jumping distance is the average of the Euclidean distance between successive states of the considered Markov chain; here successive values of parameter θ.) Clearly, targeting a 23.4% acceptance rate would make us choose a very small value for τ, which would lead to poor mixing, hence small mean squared jumping distance. In fact, the chains corresponding to these high acceptance rate have not even reached stationarity after 10^5 iterations. In this particular case, the highest efficiency (according to the mean square jumping criterion) is obtained when the acceptance rate is close to 10%.

This basic strategy (based on the acceptance rate) works well for standard Metropolis-Hastings samplers; however it is very misleading for PMMH. The following experiments give insight on why this is the case.

We now set τ so that the acceptance rate of the ideal sampler is in the 20–30% range. We run this ideal sampler, and the PMMH sampler for $N = 100, \ldots, 1000$, using the same value for τ. Figure 16.4 compares the acceptance rate of these algorithms. We see that replacing the true likelihood by a noisy likelihood degrades the acceptance rate; however, this degradation is less and less severe as N grows.

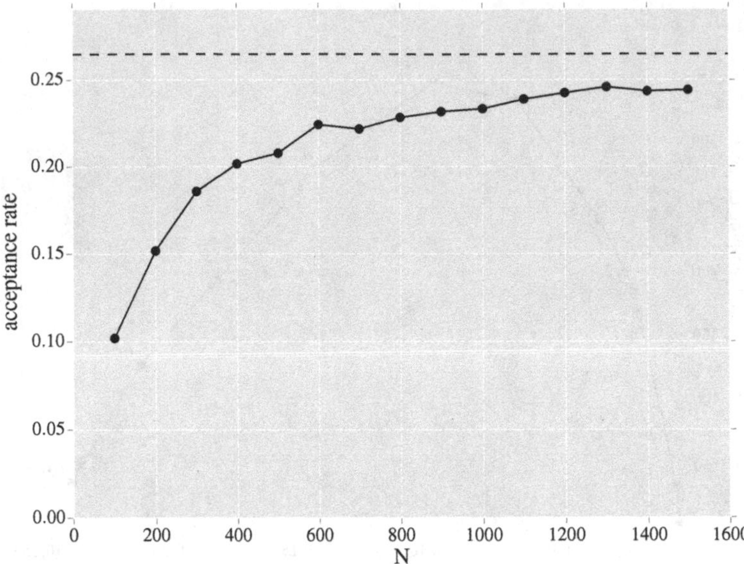

Fig. 16.4 Acceptance rate versus N, for the random walk PMMH algorithm described in the text (linear Gaussian toy example); dashed line gives the acceptance rate of the ideal sampler

Figure 16.5 plots the ACFs of two components of θ: the same phenomenon is observed: the ACFs of the PMMH sampler gets closer and closer to the ACFs of the ideal sampler as N increases.

We also plot in Fig. 16.6, for each PMMH sampler, (a) the mean squared jumping distance (MSJD) versus (b) the variance of the log-likelihood estimates. For (b), we selected randomly ten points generated from the ideal sampler; for each point, we computed the empirical variance (over 100 runs) of the log-likelihood estimate generated by the bootstrap filter, with N set to the same value as the one in the corresponding PMMH algorithm.

These experiments suggest the following recipe for choosing N: take N high enough so that the variance of the log-likelihood estimates is $\ll 1$. Figure 16.6 in particular shows that increasing further N (and therefore the CPU cost of the algorithm) would not significantly improve the performance. Once N is chosen, the random walk scale τ should be calibrated so as to maximise directly some efficiency measure, such as the mean squared jumping distance, or the integrated autocorrelation time. The acceptance rate is no longer a reliable measure of efficiency, and should not be used as such.

Our recipe for choosing N is also related to recent theoretical results (Doucet et al. 2015; Sherlock et al. 2015) on the optimal trade-off between efficiency and CPU cost. For instance, Doucet et al. (2015) consider the following criterion: the integrated autocorrelation time divided by N. The division by N is to account for

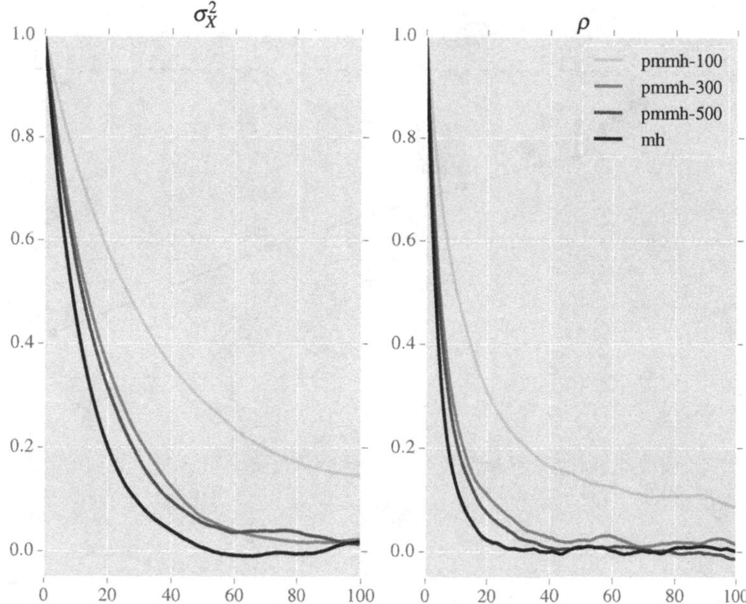

Fig. 16.5 ACFs (auto-correlation function) of two components of θ of the ideal sampler and selected PMMH samplers (based on the bootstrap filter for $N = 100, 300, 500$), in the linear Gaussian toy example

CPU time. They show that this criterion is maximised when the variance of the log-likelihood estimate equals approximately 1.

These results rely on several assumptions such as: (a) the CPU cost of PMMH is proportional to N; (b) the noise of the log-likelihood estimate is distributed according to a $\mathcal{N}(-\sigma^2/2, \sigma^2)$ distribution, where σ does not depend on θ.

These assumptions are of course simplistic. Regarding (a), the CPU cost of our implementation of PMMH behaves more like an affine function, i.e. CPU $= a+bN$, where a is non-negligible in the $N = 10^2 - 10^3$ range. This might be specific to our implementation (see comments in Python corner), but still, this suggests that, in practice, it may be worth 'overshooting the target', i.e. take N high enough so that the variance is $\ll 1$ (rather than about 1).

Regarding (b), it is not too hard to design examples where this assumption is strongly violated. Consider the same model, but with $\mathcal{G}(0.5, 0.5)$ priors for both σ_X^2 and σ_Y^2 (instead of inverse-gamma priors). The left panel of Fig. 16.7 compares the marginal posterior distributions of these two parameters, as estimated from the ideal sampler, and PMMH with $N = 100$. Note the discrepancy, in particular when $\sigma_Y^2 \to 0$. The right panel shows the trace of σ_Y^2 obtained with PMMH: the chain gets 'sticky' when σ_Y^2 gets close to zero.

What happens here is that (a) the true posterior density behaves like function $x \to x^{-1/2}$ as $\sigma_Y^2 \to 0$ (see Exercise 16.7); while (b) the variance of the log-

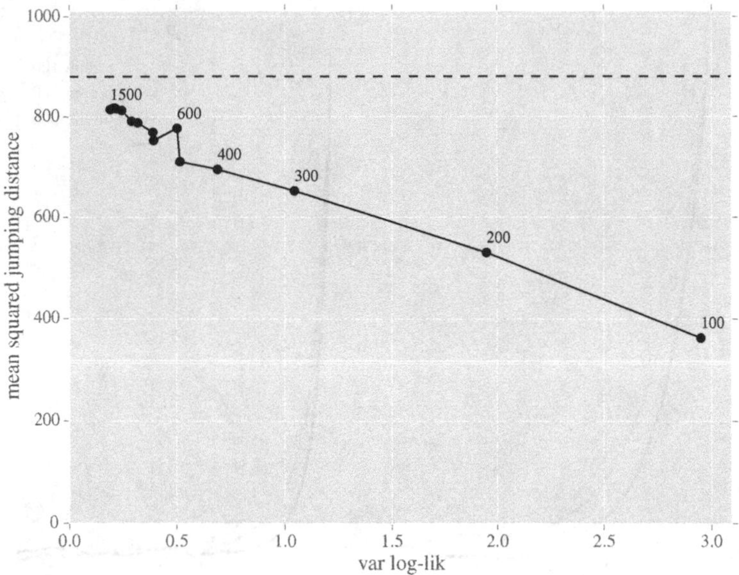

Fig. 16.6 Mean squared jumping distance versus log-likelihood variance, for the PMMH algorithms described in the toy example ($N = 100, \ldots, 1500$). The latter quantity is in fact the average over 10 values of θ sampled from the posterior by the ideal sampler of the empirical variance of the log-likelihood estimate generated by the bootstrap filter for a given N. The dashed line gives the mean squared jumping distance of the ideal sampler. The value of N is printed next to selected dots

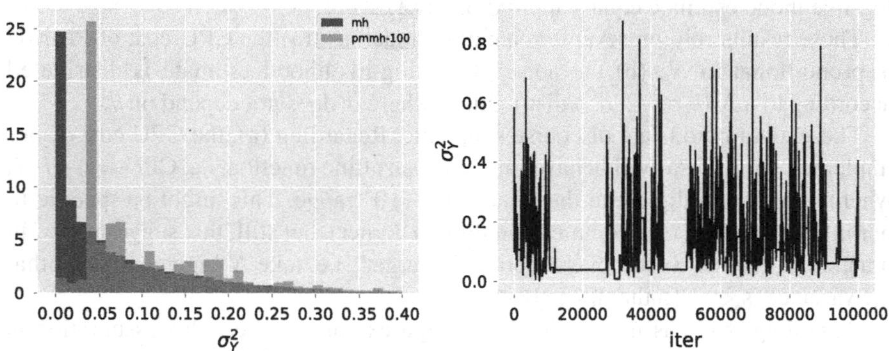

Fig. 16.7 Linear Gaussian model with alternative prior. Left: Marginal posterior distribution of σ_Y^2, as estimated by the ideal sampler (black), and PMMH with $N = 100$ (grey). Right: MCMC trace of σ_Y^2 for the PMMH algorithm

likelihood estimate blows up as $\sigma_Y^2 \to 0$. Indeed, we have already explained in Chap. 10 the following deficiency of the bootstrap filter: when the observation noise is small, the density of $Y_t|X_t$ gets very peaky, and very few particles get a non-negligible weight at any given iteration. Because of this difficulty, PMMH does not even seem to converge in this particular case.

In this specific case, we may avoid this problem by taking a prior for the observation variance that vanishes at zero. However, it is not clear how to avoid this type of difficulty in general; i.e. the fact that the variance of the log-likelihood estimate may get very high in certain regions of the posterior mass, and that as a result that the corresponding PMMH sampler may fail to explore that region properly. If anything, this gives an extra justification to choose N conservatively.

Note finally that, rather than increasing N, we may improve the performance of PMMH by switching to a more efficient particle algorithm. In this particular example, if we use the SQMC version of the guided filter with an optimal proposal (where particles are simulated according to $X_t|X_{t-1}, Y_t$), one obtains performance similar to that of the ideal sampler with N as small as $N = 10$.

16.5.2 Theta-Logistic Model in Ecology

We return to the Theta-logistic model (Example 16.3 and Sect. 2.4.5): X_t stands for the logarithm of a population size, which evolves according to:

$$X_t = X_{t-1} + \tau_0 - \tau_1 \exp(\tau_2 X_{t-1}) + U_t, \quad U_t \sim \mathcal{N}(0, \sigma_X^2),$$

and $Y_t = X_t + V_t$, $V_t \sim \mathcal{N}(0, \sigma_Y^2)$. (Take $X_0 \sim \mathcal{N}(0, 1)$ for simplicity.) The parameter is $\theta = (\tau_0, \tau_1, \tau_2, \sigma_X^2, \sigma_Y^2)$, and we consider a prior distribution where (independently): $\tau_i \sim \mathcal{N}_+(0, 1)$ (a normal distribution truncated to $\mathbb{R}^+$), $\sigma_X^2, \sigma_Y^2 \sim \mathcal{IG}(2, 1)$ (inverse-gamma).

As discussed in Example 16.3, implementing a standard, one-at-a-time Gibbs sampler for this model is not very appealing; in particular it is not clear how to sample from the conditional distribution of a given state, X_t, given Θ and the other states.

However, updating Θ given $X_{0:T}$ and $Y_{0:T}$ turns out to be reasonably straightforward. The blocks (τ_0, τ_1) and (σ_X^2, σ_Y^2) admit full conditional distributions which are easy to sample from (see Exercise 16.2 for some indications on how to derive these conditional distributions). The full conditional distribution of τ_2 is non-standard, but it may be updated by a random walk Metropolis-Hastings step (with scale s_τ) that leaves that conditional distribution invariant.

This makes it possible to run Particle Gibbs, with or without a backward step at each iteration. We run these two algorithms for 10^5 iterations, and discard the first 10^4 iterations as burn-in. We set $s_\tau = 0.2$, $N = 50$ and use the nutria dataset from Peters et al. (2010) (sample size is $T + 1 = 120$).

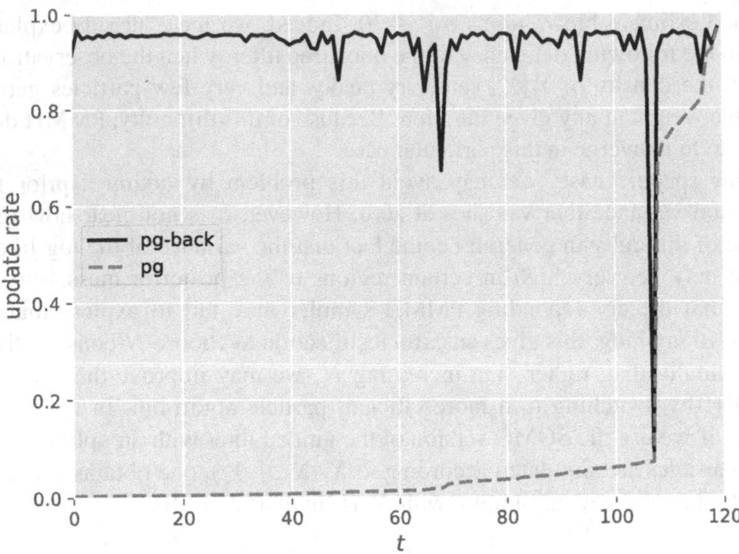

Fig. 16.8 Update rates for each component of $X_{0:T}$ of the two Particle Gibbs samplers (one with the backward step, one without implemented for the theta-logistic example

Figure 16.8 plots the update rates (i.e. proportion of iterations that change the value of the considered component) of the components of $X_{0:T}$ for both algorithms. The update rate stays close to one when backward sampling is used, and drops to zero for the first states when it is not. Consequently, the ACFs of X_0 and certain parameter components are much better when the backward step is applied; see Fig. 16.9.

The good performance of Particle Gibbs with the backward step is remarkable, given the small value of N; in fact, the performance is similar even when N is reduced to 10 (results not shown).

We now turn our attention to PMMH. Calibrating this algorithm for reasonable performance turns out to be particularly tricky for this model. As a first attempt, we used a bootstrap filter with $N = 50$ to estimate the likelihood, and a Gaussian random walk proposal with covariance τI_5; however we did not manage to find a value of τ for which the algorithm would even reach stationarity (within 10^5 iterations). As a second attempt, we replaced the bootstrap filter by the SQMC version of a (locally optimal) guided filter with $N = 100$; in addition, we used the vanishing adaptation scheme to adaptively scale the covariance matrix of the proposal (i.e. at each MCMC step this matrix is updated based on the current position of the chain). This approach gives much more reasonable results, but remains strongly outperformed by Particle Gibbs (in terms of ACFs for instance), and, on some occasions (over repeated runs), failed to converge as well. (Results not shown.)

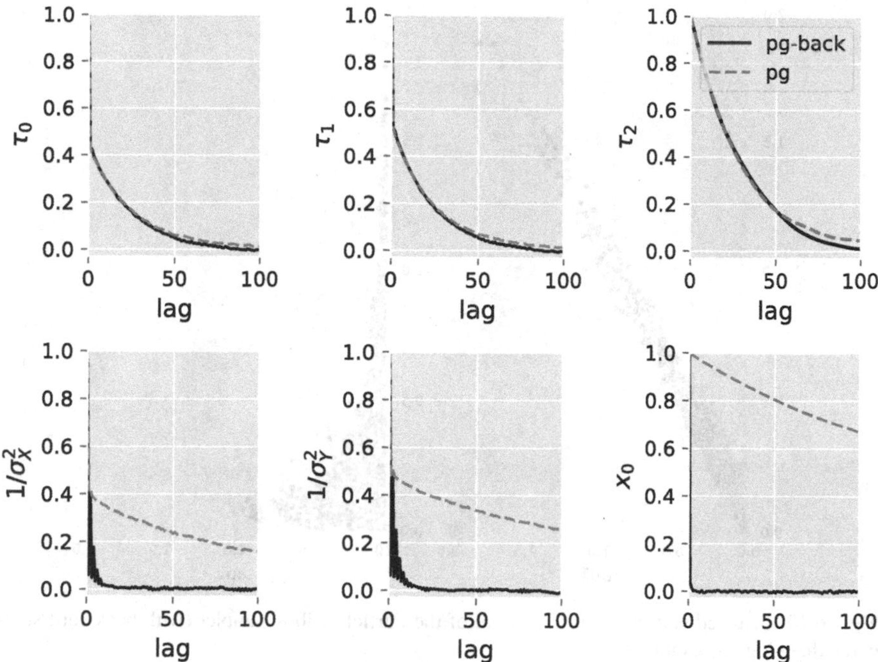

Fig. 16.9 ACFs (auto-correlation functions) for components of Θ and X_0, from the two versions of Particle Gibbs discussed in the text (with or without the backward step) in the theta-logistic example

There are several reasons why this particular example is so challenging for PMMH. First, the posterior induces strong, non-linear dependencies between certain components of θ; see Fig. 16.10. Second, the model is such that the log-likelihood is difficult to estimate accurately. To see this, we generated a large number of parameters from the prior, and, for each θ, we computed the empirical variance (over 10 runs) of the likelihood estimate obtained from the considered particle algorithm (first, bootstrap with $N = 50$; then SQMC guided with $N = 100$). In the former case, only 4% of the estimates had a variance lower than 1. In the latter case, this proportion increased to 62%. Increasing N leads to higher proportions, but only to a small extent. What happens here is that the model fits reasonably well to the data only for a small region of θ-values; outside of this region, even a SQMC guided filter performs poorly in estimating the likelihood.

To be fair, simple models like this one are rarely as challenging to PMMH as in this particular case. It remains noteworthy that Particle Gibbs performs so well here, without any need for calibration. In general, when Particle Gibbs is implementable, we often observe that it outperforms PMMH (but not necessarily to this extent).

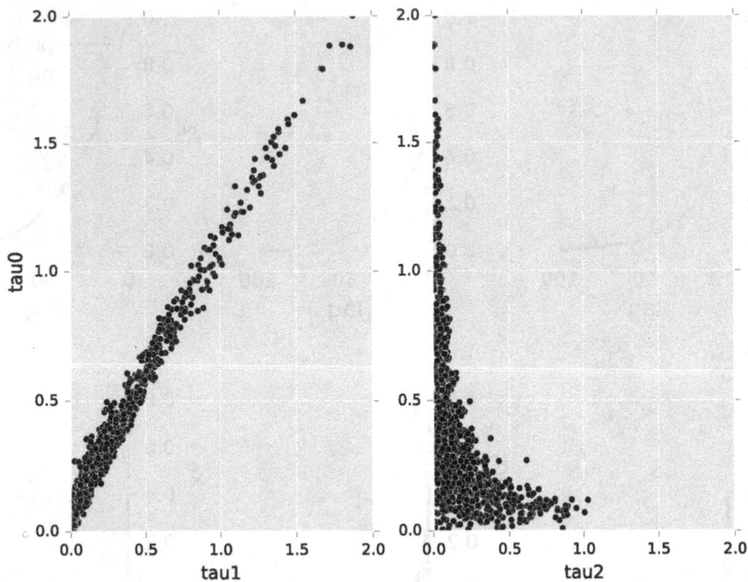

Fig. 16.10 Selected pair plots for the output of the Particle Gibbs sampler (with backward step) in the theta-logistic example

16.6 Practical Recommendations

Given the results of the numerical experiments, we propose the following practical guidelines.

Relative to Particle Gibbs, PMMH is easier to implement, and is applicable more generally. However, it is a bit tricky to calibrate. The first thing to do is to explore all possible options to reduce the variance of the likelihood estimates generated by the particle algorithm at each iteration; e.g. set the resampling scheme to something else than multinomial, design good proposal distributions (that may be used in a guided or auxiliary particle filter), or implement the SQMC variant presented in Chap. 13. (In our experience, the combination of the last two approaches is particularly effective.)

When this is done, one may perform pilot runs to set the value of N so that the variance of the likelihood estimate is typically below one (for representative values of the parameter). Finally, one may calibrate the parameters of the proposal kernels (e.g. the covariance matrix of a random walk proposal) by looking at the autocorrelation times (or the mean squared jumping distance) of the components of the parameter θ. On the other hand, we have seen that calibrating simultaneously N and the parameters of the proposal kernel may be very sub-optimal, especially if one attempts at obtaining a 0.234 acceptance rate.

Relative to PMMH, Particle Gibbs requires to design an extra step to sample from Θ given $X_{0:T}$ and $Y_{0:T}$; this is more work for the user, and may not always be feasible. However, when it is, it usually performs better than PMMH. The backward sampling step should be implemented whenever possible (i.e. when the distribution of X_t given X_{t-1} admits a tractable density), because it strongly improves performance. When the backward step is not feasible, another strategy to improve performance is to implement an alternative (to multinomial) resampling scheme. This is a bit tricky: one needs to implement a different algorithm to sample $N - 1$ ancestor variables at each time step in the CSMC algorithm. See the bibliography for papers explaining how to do this.

Whatever variant of Particle Gibbs is implemented, we recommend to monitor the update rates of the states (Fig. 16.8) in order to choose a value of N that ensures good mixing.

Of course, the discussion above is overly simplistic, and is meant only to help the reader avoid the most obvious pitfalls. The bibliography of this chapter lists several very recent papers that discuss more thoroughly the tuning of PMCMC algorithms.

Exercises

16.1 *In Example 16.2, derive the conditional distribution of ρ, assuming as a prior (a) a Gaussian truncated to $[-1, 1]$; (b) a $\mathcal{U}([-1, 1])$. (You may deduce (b) from (a). For (a), you may first do the calculations for a non-truncated Gaussian prior, and then discuss what happens when the prior is truncated.) Discuss how one may sample from this conditional distribution in case the prior is the Beta prior discussed in Example 16.1. For instance, if we use a Metropolis-Hastings step, what would be a good proposal?*

Re-do the calculations (for μ, σ^2, and ρ) for the same model, but with $X_0 \sim \mathcal{N}(\mu, \sigma^2/(1-\rho^2)$. In particular, what happens for ρ when one of the two priors above is chosen?

16.2 *For the theta-logistic model introduced in Sect. 2.4.5 and discussed in Example 16.3, derive the full conditional distribution of all parameters except τ_2. Show that is it possible to derive a Gibbs sampler that updates jointly parameters τ_0 and τ_1. (As in the previous exercise, you may first compute the conditional distribution of (τ_0, τ_1) when the prior for this pair is a (non-truncated) Gaussian; to check you results, you may have a look at standard results on the conjugacy of Gaussian priors for Gaussian regression models).*

16.3 *Following Exercise 10.8, explain why* (16.11) *remains an unbiased estimate of the likelihood in the APF case (i.e. when an auxiliary PF algorithm is used), provided G_t^θ is modified as follows:*

$$G_t^\theta(x_{t-1}, x_t) = \frac{P_t^\theta(x_{t-1}, dx_t) f_t^\theta(y_t|x_t)}{M_t^\theta(x_{t-1}, dx_t)} \times \frac{\eta_t^\theta(x_t)}{\eta_{t-1}^\theta(x_{t-1})}$$

for $t \geq 1$, taking $\eta_T^\theta(x_T) = 1$, and defining $G_0^\theta(x_0)$ similarly. (The notation η_t^θ reflects that the auxiliary functions typically depend on θ.)

16.4 *Show that the proof of Proposition 16.3 still holds if the resampling step is triggered according to a criterion which is a function of $X_0^{1:N}$ (e.g. the ESS).*

16.5 *Derive the distribution of $B_{0:T}$ when (a) the distribution of $X_t|X_{t-1} = x_{t-1}$ is a Dirac mass at x_{t-1}; (b) the distribution of certain components of X_t given $X_{t-1} = x_{t-1}$ are a Dirac mass at a function of x_{t-1}. Discuss the usefulness of the backward sampling step in these cases.*

16.6 *Prove in two ways Proposition 16.6. First, fill the blanks in the outline of the proof given below the proposition. In particular, derive the conditional distribution of $A_t^{1:N}$, and deduce the conditional distribution of $B_t|B_{t+1:T}$ and other variables described in that proposition. Second, write down for $T = 1$ the joint distribution of Θ, $X_{0:T}^{1:N}$, $A_{1:T}^{1:N}$, and deduce from this expression, first, the distribution of $B_1 = K$ given all the other variables; and second, the distribution of B_0 given B_1 and the other variables. Explain why this proof extend to any $T \geq 1$.*

16.7 *For the toy example of Sect. 16.5.1, explain why the likelihood converges to a positive value as $\sigma_Y^2 \to 0$, and why the corresponding posterior (based on a $\mathcal{G}(0.5, 0.5)$ prior for σ_Y^2) diverges as $\sigma_Y^2 \to 0$. as stated at the end of Sect. 16.5.1.*

Python Corner

Suppose we want to run PMMH for the stochastic volatility model defined in the Python corner of Chap. 5. This may be done as follows in `particles`:

```
from particles import mcmc
from particles import state_space_models as ssm

_, y = ssm.StochVol().simulate(100)  # simulated data
my_alg = mcmc.PMMH(niter=10**3, ssm_cls=ssm.StochVol,
                   prior=my_prior, data=y, Nx=100)
my_alg.run()
```

You can see that the number of MCMC iterations is 10^3 (parameter `niter`) and that the number of particles is $N = 100$ (parameter `Nx`). Argument `ssm_cls` is set to *class* `StochVol` (not one of its instance). We said before that a `StateSpaceModel` class such as `StochVol` represents a *parametric* class of models; hence it makes sense to pass it as a parameter to `PMMH`.

Argument `prior` requires a `StructDist` object (from module `distributions`), which specifies the distribution of each component of parameter θ through a dictionary. For instance, for a stochastic volatility model, we could define the following prior distribution:

```
from particles import distributions as dists

prior_dict = {'mu': dists.Normal(scale=2.),
              'rho': dists.Uniform(a=-1., b=1.),
              'sigma':dists.Gamma()}
my_prior = dists.StructDist(prior_dict)
```

(The components are independent here; in case you wonder how to specify conditional prior distributions, for one component given others, see the documentation of `StructDist`.)

Like other probability distributions defined in module `distributions`, `StructDist` objects implement methods `rvs` and `logpdf`; in the case of `StructDist`, these methods take as input or output structured arrays; that is `NumPy` arrays with named fields. For instance:

```
theta = my_prior.rvs(size=100)   # generate 100 theta's from the prior
avg = np.mean(theta['mu'])  # theta['mu'] behaves like a (N,) array
```

Structured arrays are akin to the data frame objects of library `Pandas`. Note however that, for our purposes, structured arrays are more versatile, in particular because they do not impose each named field to be univariate. For more information on structured arrays, check the documentation at https://docs.scipy.org/doc/numpy-1.15.0/user/basics.rec.html. Package `particles` uses structured arrays to store the output of PMCMC (and more generally MCMC) algorithms, and to store the N particles of an SMC sampler, see the Python corner of Chap. 17.

To go back to the snippet above, you might wonder what type of PMMH algorithm this runs exactly. By default, the starting point is sampled from the prior (may be changed by setting argument `theta0`), an adaptive random walk proposal is used (set option `adaptive` to False for the non-adaptive version), and the particle filter run at each iteration is a bootstrap filter (this may be changed by setting argument `fk_cls`).

We mentioned that the actual cost per iteration of PMMH is something like $a + bN$, rather than simply proportional to N. We illustrate this in Fig. 16.11, which plots the average cost per iteration of PMMH, over 100 iterations, as a function of N_x, for the PMMH algorithm discussed in Sect. 16.5.1. One factor seems to be that `NumPy` is not optimised for small arrays; see e.g. package `tinyarrays` (https://pypi.org/project/tinyarray/) which tries to address this issue. However, note that, by construction, any implementation is bound to have some overhead for small N. Hence our general recommendation to choose N conservatively (i.e. larger than possibly needed).

Module `mcmc` also includes an implementation of Particle Gibbs, as an *abstract class*. To define a particular Particle Gibbs sampler, one must sub-class `ParticleGibbs` and define method `update_theta`, which samples parameter

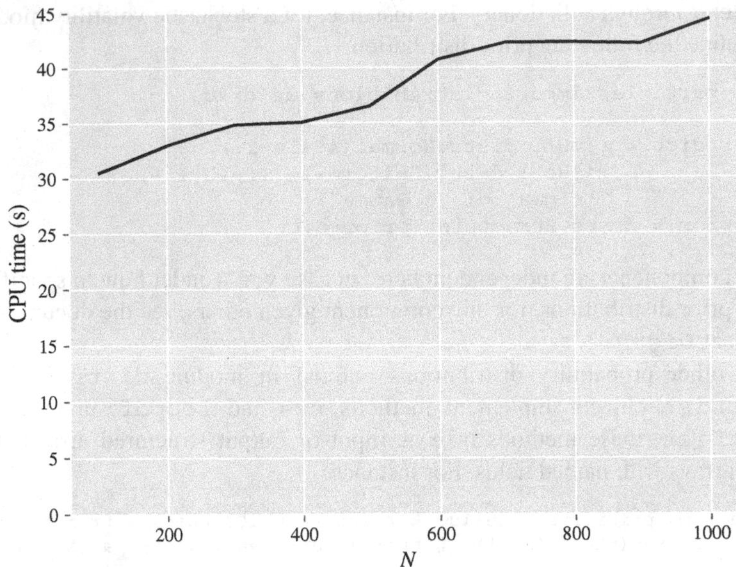

Fig. 16.11 CPU time (in seconds) versus N for 10^3 iterations of the PMMH algorithm of Sect. 16.5.1

θ conditional on a state trajectory and the data. Here is an example for a stochastic volatility model where σ and ρ would be fixed, and we only update μ.

```
class PGStochVol(mcmc.ParticleGibbs):
    def update_theta(self, theta, x):
        new_theta = theta.copy()
        sigma, rho = 0.2, 0.95  # fixed values
        xlag = np.array(x[1:] + [0.,])
        dx = (x - rho * xlag) / (1. - rho)
        s = sigma / (1. - rho)**2
        new_theta['mu'] = self.prior.laws['mu'].posterior(dx, sigma=s).rvs()
        return new_theta

alg = PGStochVol(ssm_cls=StochVol, data=y, prior=my_prior,
                 Nx=200, niter=1000)
alg.run()
```

The `__init__` method of `ParticleGibbs` accepts several extra options such as `backward_step` (to add the backward step described in Sect. 16.4.4), or `regenerate_data` (to add an extra step to re-generate the data, as discussed in the Python corner of Chap. 15). Regarding the latter, although the justification for this regeneration trick remains valid (if we add an extra step that generates the data given θ and the chosen state trajectory, then we effectively sample from the prior), in our experience the corresponding algorithm may mix poorly. Presumably, when we sample from the prior, we might generate parameter values for which Particle Gibbs (in particular the CSMC step) is much less efficient. Thus we recommend to use very informative prior when this extra step is used.

Bibliographical Notes

Prior specification (discussed briefly in Sect. 16.1.2) is a vast subject; a good introduction is Chapter 3 of Robert (2007).

The GIMH algorithm was introduced by Beaumont (2003). A good reference for pseudo-marginal algorithms and their theoretical properties is Andrieu and Roberts (2009). On the central issue of constructing positive unbiased estimators and their connection to exact Monte Carlo see the exact simulation framework of Beskos et al. (2006), the Bernoulli factory framework of Łatuszyński et al. (2011), and also Papaspiliopoulos (2011) and Jacob and Thiery (2015).

The PMCMC methodology was introduced in the influential paper of Andrieu et al. (2010). Interesting applications of PMCMC include ecology (Peters et al. 2010), systems biology (Golightly and Wilkinson 2011), finance (Pitt et al. 2012), electricity forecasting (Launay et al. 2013), natural language processing (Dubbin and Blunsom 2012), social networks (Everitt 2012), hydrology (Vrugt et al. 2013), actuarial science (Fung et al. 2017) among others.

Regarding the theoretical properties of PMMH algorithms and their optimal tuning, see Doucet et al. (2015) and Sherlock et al. (2015). These papers in particular give a theoretical justification for the recommendation to choose N so that variance of the log-likelihood is close to one.

The following papers establish various properties of Particle Gibbs, such as uniform ergodicity, reduction of asymptotic variance through the backward sampling step, complexity relative to the sample size T, and so on: Chopin and Singh (2015), Lindsten et al. (2015), Del Moral et al. (2016), and Andrieu et al. (2018). The first paper also discusses how to extend Particle Gibbs to resampling schemes that are not multinomial, a point we touched upon in Sect. 17.2.

See also Lindsten et al. (2014) for an alternative variant of the backward step in Particle Gibbs, and Deligiannidis et al. (2018) for a way to correlate successive likelihood estimates in PMMH, in order to improve mixing.

Bibliography

Andrieu, C., Doucet, A., & Holenstein, R. (2010). Particle Markov chain Monte Carlo methods. *Journal of the Royal Statistical Society: Series B (Statistical Methodology), 72*(3), 269–342.

Andrieu, C., Lee, A., & Vihola, M. (2018). Uniform ergodicity of the iterated conditional SMC and geometric ergodicity of particle Gibbs samplers. *Bernoulli, 24*(2), 842–872.

Andrieu, C., & Roberts, G. O. (2009). The pseudo-marginal approach for efficient Monte Carlo computations. *Annals of Statistics, 37*(2), 697–725.

Beaumont, M. A. (2003). Estimation of population growth or decline in genetically monitored populations. *Genetics, 164*(3), 1139–1160.

Beskos, A., Papaspiliopoulos, O., Roberts, G. O., & Fearnhead, P. (2006). Exact and computationally efficient likelihood-based estimation for discretely observed diffusion processes. *Journal of the Royal Statistical Society: Series B (Statistical Methodology)*, 68(3), 333–382. With discussions and a reply by the authors.

Chopin, N., & Singh, S. S. (2015). On particle Gibbs sampling. *Bernoulli, 21*(3), 1855–1883.

Del Moral, P., Kohn, R., & Patras, F. (2016). On particle Gibbs samplers. *Annales de l'Institut Henri Poincaré Probabilités et Statistiques, 52*(4), 1687–1733.

Deligiannidis, G., Doucet, A., & Pitt, M. K. (2018). The correlated pseudomarginal method. *Journal of the Royal Statistical Society: Series B (Statistical Methodology), 80*(5), 839–870.

Doucet, A., Pitt, M. K., Deligiannidis, G., & Kohn, R. (2015). Efficient implementation of Markov chain Monte Carlo when using an unbiased likelihood estimator. *Biometrika, 102*(2), 295–313.

Dubbin, G., & Blunsom, P. (2012). Unsupervised Bayesian part of speech inference with particle Gibbs. In *Machine learning and knowledge discovery in databases* (pp. 760–773). Berlin/Heidelberg: Springer.

Everitt, R. G. (2012). Bayesian parameter estimation for latent Markov random fields and social networks. *Journal of Computational and Graphical Statistics, 21*(4), 940–960.

Fung, M. C., Peters, G. W., & Shevchenko, P. V. (2017). A unified approach to mortality, modelling using state-space framework: Characterisation, identification, estimation and forecasting. *Annals of Actuarial Science, 11*(2), 343–389.

Golightly, A., & Wilkinson, D. J. (2011). Bayesian parameter inference for stochastic biochemical network models using particle Markov chain Monte Carlo. *Interface Focus, 1*(6), 807–820.

Jacob, P. E., & Thiery, A. H. (2015). On nonnegative unbiased estimators. *Annals of Statistics, 43*(2), 769–784.

Łatuszyński, K., Kosmidis, I., Papaspiliopoulos, O., & Roberts, G. O. (2011). Simulating events of unknown probabilities via reverse time martingales. *Random Structures Algorithms, 38*(4), 441–452.

Launay, T., Philippe, A., & Lamarche, S. (2013). On particle filters applied to electricity load forecasting. *Journal de la SFdS, 154*(2), 1–36.

Lindsten, F., Douc, R., & Moulines, E. (2015). Uniform ergodicity of the particle Gibbs sampler. *Scandinavian Journal of Statistics, 42*(3), 775–797.

Lindsten, F., Jordan, M. I., & Schön, T. B. (2014). Particle Gibbs with ancestor sampling. *Journal of Machine Learning Research, 15*, 2145–2184.

Papaspiliopoulos, O. (2011). Monte Carlo probabilistic inference for diffusion processes: A methodological framework. In *Bayesian time series models* (pp. 82–103). Cambridge: Cambridge University Press.

Peters, G. W., Hosack, G. R., & Hayes, K. R. (2010). Ecological non-linear state space model selection via adaptive particle Markov chain Monte Carlo (AdPMCMC). *arXiv e-prints 1005.2238.*

Pitt, M. K., Silva, R. S., Giordani, P., & Kohn, R. (2012). On some properties of Markov chain Monte Carlo simulation methods based on the particle filter. *Journal of Econometrics, 171*(2), 134–151.

Robert, C. P. (2007). *The Bayesian choice: From decision-theoretic foundations to computational implementation*. Berlin/Heidelberg: Springer.

Sherlock, C., Thiery, A. H., Roberts, G. O., & Rosenthal, J. S. (2015). On the efficiency of pseudo-marginal random walk Metropolis algorithms. *Annals of Statistics, 43*(1), 238–275.

Vrugt, J. A., ter Braak, C. J., Diks, C. G., & Schoups, G. (2013). Hydrologic data assimilation using particle Markov chain Monte Carlo simulation: Theory, concepts and applications. *Advances in Water Resources, 51*, 457–478.

Chapter 17
SMC Samplers

Summary This chapter covers SMC samplers, that is, particle algorithms that are able to track a sequence of probability distributions $\mathbb{P}_t(\mathrm{d}\theta)$, related by the recursion $\mathbb{P}_t(\mathrm{d}\theta) = \ell_t G_t(\theta)\mathbb{P}_{t-1}(\mathrm{d}\theta)$. Chapter 3 gave a few examples of applications of these algorithms; in some, one is genuinely interested in approximating each distribution in a given sequence, e.g., sequential Bayesian learning, in which case the sequence corresponds to the incorporation of more and more data; in others, one is interested in a single target distribution, but an artificial sequence is designed so as to be able to implement successfully an SMC sampler, and the sequence corresponds to increasing "temperatures". Despite the different applications and appearances, the two approaches are not that different. We can think the former as "data tempering" and the latter as "temperature tempering" and in both situations sampling early distributions is easier than later ones.

When used to approximate a single distribution, SMC samplers represent an alternative to MCMC, with the following advantages: first, they offer a simple way to estimate the normalising constant of the target; this quantity is of interest in many problems, in particular in Bayesian model choice. Second, they are easier to parallelise. Third, it is reasonably easy to develop adaptive SMC samplers; that is algorithms which calibrate their tuning parameters automatically, using the current sample of particles.

We start by describing a Feynman-Kac model that formalises a large class of SMC samplers, based on invariant kernels. We then provide practical recipes to obtain good performance in the following scenarios: Bayesian sequential learning, tempering, rare-event simulation, and likelihood-free inference. Finally, we explain how to further extend SMC samplers by using more general Markov kernels.

© Springer Nature Switzerland AG 2020

N. Chopin, O. Papaspiliopoulos, *An Introduction to Sequential Monte Carlo*, Springer Series in Statistics, https://doi.org/10.1007/978-3-030-47845-2_17

17.1 A Generic SMC Sampler Based on Invariant Kernels

For simplicity, we focus on a sequence of probability distributions $\mathbb{P}_t(\mathrm{d}\theta)$ defined with respect to a *fixed* probability space $(\Theta, \mathcal{B}(\Theta))$. These distributions have been chosen by the user (based for instance on one of the strategies described in the introduction), and they are of the form:

$$\mathbb{P}_t(\mathrm{d}\theta) = \frac{\gamma_t(\theta)}{L_t}\nu(\mathrm{d}\theta) \tag{17.1}$$

where $\nu(\mathrm{d}\theta)$ is a probability measure, and $L_t := \int_\Theta \gamma_t(\theta)\nu(\mathrm{d}\theta) > 0$.

We assume that we are able to compute the ratios $\gamma_t(\theta)/\gamma_{t-1}(\theta)$ reasonably quickly. This means we could implement sequential importance sampling; however this strategy is usually too naive for problems of interest.

We make a second important assumption: that we are able to construct, for each distribution $\mathbb{P}_t(\mathrm{d}\theta)$, a Markov kernel $M_{t+1}(\theta', \mathrm{d}\theta)$ that leaves invariant $\mathbb{P}_t(\mathrm{d}\theta)$. (Note the $t/t+1$.) To that end, we may use any MCMC algorithm (Chap. 15); e.g. Metropolis-Hastings or Gibbs.

We are now able to define a Feynman-Kac model such that its marginal distributions $\mathbb{Q}_t(\mathrm{d}\theta)$ match $\mathbb{P}_t(\mathrm{d}\theta)$. Indeed, we have (replacing θ by θ_t in the argument of $\mathbb{P}_t(\mathrm{d}\theta)$):

$$\begin{aligned}\mathbb{P}_t(\mathrm{d}\theta_t) &= \frac{L_{t-1}}{L_t}\frac{\gamma_t(\theta_t)}{\gamma_{t-1}(\theta_t)}\mathbb{P}_{t-1}(\mathrm{d}\theta_t) \\ &= \frac{1}{\ell_t}G_t(\theta_t)\int_{\theta_{t-1}\in\Theta}\mathbb{P}_{t-1}(\mathrm{d}\theta_{t-1})M_t(\theta_{t-1}, \mathrm{d}\theta_t)\end{aligned}$$

where $\ell_t := L_t/L_{t-1}$, $G_t(\theta_t) := \gamma_t(\theta_t)/\gamma_{t-1}(\theta_t)$, and the integral is with respect to variable θ_{t-1} only.

We recognise in the expression above the forward recursion of a Feynman-Kac model with Markov kernels M_t and potential functions G_t. The corresponding SMC sampler is described as Algorithm 17.1.

Input of an SMC Sampler

- A probability distribution $\nu(\mathrm{d}\theta)$ from which one may sample from.
- a sequence of distributions $\mathbb{P}_t(\mathrm{d}\theta)$, $t = 0, \ldots, T$ of the form given by (17.1) and such that function $\theta \to \gamma_0(\theta)$ and functions $\theta \to \gamma_t(\theta)/\gamma_{t-1}(\theta)$, $t \geq 1$ may be evaluated pointwise.

(continued)

- MCMC kernels $M_t(\theta_{t-1}, \mathrm{d}\theta_t)$ that leave invariant $\mathbb{P}_{t-1}(\mathrm{d}\theta_{t-1})$ for $t \geq 1$.
- The usual input of a particle filter: the number of particles N, the choice of an unbiased resampling scheme, and a threshold $\mathrm{ESS}_{\min}$.

Algorithm 17.1: Generic SMC sampler

```
Operations involving index n must be performed
  for n = 1, ..., N.
Θ_0^n ~ ν(dθ)
w_0^n ← γ_0(Θ_0^n)
W_0^n ← w_0^n / Σ_{m=1}^N w_0^m
for t = 1 to T do
    if ESS(W_{t-1}^{1:N}) < ESS_min then
        A_t^{1:N} ← resample(W_{t-1}^{1:N})
        ŵ_{t-1}^n ← 1
    else
        A_t^n ← n
        ŵ_{t-1}^n ← w_{t-1}^n
    Θ_t^n ~ M_t(Θ_{t-1}^{A_t^n}, dθ_t)
    w_t^n ← ŵ_{t-1}^n γ_t(Θ_t^n) / γ_{t-1}(Θ_t^n)
    W_t^n ← w_t^n / Σ_{m=1}^N w_t^m
```

Of course, to turn this generic algorithm into a practical one, we need to discuss how to choose the sequence $\mathbb{P}_t(\mathrm{d}\theta)$, and how to design invariant kernels. This is the point of next section.

17.2 Practical Guidelines to Make SMC Algorithms Adaptive

Throughout this chapter d denotes the dimension of θ.

17.2.1 Invariant Kernels

In the context of MCMC, calibrating tuning parameters to obtain good mixing is often tedious, and may require several pilot runs. By contrast, in SMC, we have at our disposal, at the beginning of iteration t, a weighted sample that approximates $\mathbb{P}_{t-1}(d\theta)$. This greatly facilitates the calibration of the chosen MCMC kernels.

Specifically, we recommend the following default strategy: set M_t to the Markov kernel that corresponds to k iterations of a Metropolis-Hastings kernel, based on a Gaussian random walk, with covariance matrix equals to $\lambda\hat{\Sigma}_{t-1}$, with, e.g., $\lambda = 2.38d^{-1/2}$, and $\hat{\Sigma}_{t-1}$ the empirical covariance matrix of the weighted sample $(W_{t-1}^{1:N}, \Theta_{t-1}^{1:N})$ obtained at the end of iteration $t-1$. This recommendation is based on the optimal scaling result for random walk Metropolis we mentioned in Sect. 15.3.

In our experience, this default strategy works well in a variety of problems. However, it may fail in the following cases:

- If the dimension d is high, the particle estimate $\hat{\Sigma}_{t-1}$ may be poor (or even degenerate). In that case, a better strategy is to set the proposal covariance to a diagonal matrix containing the particle estimates of the d marginal variances of the target. Another issue is that the mixing of random walk Metropolis degrades with the dimension. Thus, for high d, one may consider using gradient-based MCMC kernels, again calibrated from the current sample of particles.
- If the targets have several strongly separated modes, this default strategy will work only if the dimension d of Θ is small. How to design Markov kernels (within SMC) in order to deal with multi-modal targets is still an open problem.
- Of course, when the sampling space is discrete, a different approach is required. For instance, for $\Theta = \{0, 1\}^d$ (the set of binary words of length d), Schäfer and Chopin (2013) considered independent Metropolis kernels based on a proposal corresponding to a nested logistic regression model, with parameters fit to the current particle sample.

Another issue is how to choose the number k of MCMC iterations. One may set it adaptively, according to some criterion, e.g., the relative increase of the squared jumping distance (between the N initial particles, and the N values obtained after applying a number of MCMC iterations). Alternatively, one may use pilot runs; this is what we do in the numerical experiments of this chapter. At any rate, one should be ready to set this to a large value in high dimensions; in fact, optimal scaling theory suggests one should take $\mathcal{O}(d)$ iterations for a random walk proposal. Again, this point will be investigated in our numerical experiments.

17.2.2 IBIS

IBIS (Iterated Batch Importance Sampling) refers to an SMC sampler targeting a sequence of posterior distributions of a certain static model, with prior $\nu(\mathrm{d}\theta)$, and likelihood $\gamma_t(\theta) = p_t^\theta(y_{0:t})$ at time t. To implement IBIS, one needs to be able to compute

$$\frac{\gamma_t(\theta)}{\gamma_{t-1}(\theta)} = p_t^\theta(y_t|y_{0:t-1})$$

which is simply the likelihood factor.

IBIS is attractive when this quantity may be computed in $\mathcal{O}(1)$ time; for instance when the observations are i.i.d. or have Markovian dependence. Even in that favourable case, any iteration t such that the N particles are moved through an MCMC kernel that leaves invariant $\mathbb{P}_{t-1}$ will typically have complexity $\mathcal{O}(tN)$. (This is true for instance for any Metropolis kernel, as it requires computing the target density at $\mathcal{O}(N)$ proposed values). Hence a naive version of IBIS has overall complexity $\mathcal{O}(T^2)$ where $T+1$ is the sample size.

To avoid this quadratic complexity, one may move particles through an invariant kernel only at times t where the ESS drops below a certain level. Formally, this is equivalent to set adaptively the Markov kernels M_t to either the identity kernel, $M_t(\theta_{t-1}, \mathrm{d}\theta_t) = \delta_{\theta_{t-1}}(\mathrm{d}\theta_t)$, or to an MCMC kernel, according to the value of the ESS. Algorithm 17.2 describes the so-obtained algorithm.

Input of IBIS

- A Bayesian model such that one is able to
  - sample from the prior distribution $\nu(\mathrm{d}\theta)$;
  - compute pointwise the likelihood factor $\theta \to p_t^\theta(y_t|y_{0:t-1})$.
- MCMC kernels M_t that leave invariant $\mathbb{P}_{t-1}(\mathrm{d}\theta)$, the posterior distribution of the model, given data $y_{0:t-1}$.
- The usual input of an SMC algorithm: the number of particles N, the choice of an unbiased resampling scheme, and a threshold $\mathrm{ESS}_{\min}$.

Algorithm 17.2: IBIS

```
Operations involving index n must be performed
 for n = 1, ..., N.

Θ_0^n ~ ν(dθ)
w_0^n ← G_0(Θ_0^n)                          ▷ where G_0(θ) := p_0^θ(y_0)
W_0^n ← w_0^n / Σ_{m=1}^N w_0^m
for t = 1 to T do
    if ESS(W_{t-1}^{1:N}) < ESS_min then
        A_t^{1:N} ← resample(W_{t-1}^{1:N})
        ŵ_{t-1}^n ← 1
        Θ_t^n ~ M_t(Θ_{t-1}^{A_t^n}, dθ)
    else
        A_t^n ← n
        ŵ_{t-1}^n ← w_{t-1}^n
        Θ_t^n ← Θ_{t-1}^n
    w_t^n ← ŵ_{t-1}^n G_t(Θ_t^n)            ▷ where G_t(θ) := p_t^θ(y_t | y_{0:t-1})
    W_t^n ← w_t^n / Σ_{m=1}^N w_t^m
```

This adaptive strategy is motivated by Bayesian asymptotics. For regular models, the Bernstein-von Mises theorem states that the rescaled posterior distribution (the posterior of $\sqrt{t}(\theta - \theta_0)$) converges to a Gaussian with mean 0 and covariance the inverse of the Fisher information at θ_0, where θ_0 is the true value of the parameter. Less formally, this means that, for t large enough, the posterior distribution at time t behaves like a Gaussian distribution $\mathcal{N}_d(\theta_0, t^{-1}\Sigma_0)$. This implies in particular that thc sequence of target distributions $\mathbb{P}_t(\mathrm{d}\theta)$ evolves at a slower and slower rate.

The following proposition, whose proof is developed in Exercise 17.1, quantifies this phenomenon. Recall from Sect. 8.6 the definition of the chi-square pseudo-distance between the two probability measures and the property that it matches the variance of the weights of importance sampling from one to another.

Proposition 17.1 *Assume that* $\mathbb{P}_t(\mathrm{d}\theta)$ *and* $\mathbb{P}_{ct}(\mathrm{d}\theta)$ *stand respectively for the following Gaussian distributions:* $\mathcal{N}_d(\theta_0, t^{-1}\Sigma_0)$ *and* $\mathcal{N}_d(\theta_0, (ct)^{-1}\Sigma_0)$ *for integers*

t, $c > 1$. Then the chi-square pseudo-distance between the two distributions equal

$$\int \mathbb{P}_t(\mathrm{d}\theta)\left(\frac{\mathbb{P}_{ct}(\mathrm{d}\theta)}{\mathbb{P}_t(\mathrm{d}\theta)} - 1\right)^2 = \left(\frac{c^2}{2c-1}\right)^{d/2} - 1 .$$

In particular, it does not depend on t.

This result is both good and bad news. Good, because it means that, asymptotically, the time between two successive (ESS-based) resampling steps should increase at a geometric rate. A quick calculation shows that the resulting complexity for IBIS is then $\mathcal{O}_P(T)$. Bad, because the $d/2$ exponent reveals a curse of dimensionality: the decrease of the ESS between two time steps, t and ct, should increase exponentially with the dimension.

In practice, what is usually observed is that if the dimension of θ is too high, IBIS performs poorly because incorporating even a single data-point makes the ESS drop to a very low value. A possible way out of this problem is to hybrid IBIS with tempering: i.e. rather than reweight directly according to $p^\theta(y_t|y_{0:t-1})$, perform several steps with weight function $\{p^\theta(y_t|y_{0:t-1})\}^\lambda$, for a certain $0 < \lambda < 1$.

Finally, recall that SMC samplers may be used to approximate the normalising constants L_t. In sequential Bayesian estimation, L_t is the marginal likelihood of the data $y_{0:t}$. This means that IBIS may be used to perform sequential model choice, on top of sequential inference.

17.2.3 Tempering

Consider a generic target distribution, which we are able to express as:

$$\pi(\mathrm{d}\theta) = \frac{1}{L}\nu(\mathrm{d}\theta)\exp\{-V(\theta)\}$$

where $\nu(\mathrm{d}\theta)$ is a probability measure. We want either to sample from π or compute the normalising constant $L = \int_\Theta \nu(\mathrm{d}\theta)\exp\{-V(\theta)\} > 0$, or both, but it is difficult to do so directly, for some reason (e.g. π is multi-modal).

Tempering SMC amounts to construct an SMC sampler the tracks a sequence of tempered distributions defined as:

$$\mathbb{P}_t(\mathrm{d}\theta) = \frac{1}{L_t}\nu(\mathrm{d}\theta)\exp\{-\lambda_t V(\theta)\}$$

for $0 < \lambda_0 < \ldots < \lambda_T = 1$. The intermediate distributions are rarely of interest. Thus we are at liberty of choosing any sequence $\{\lambda_t\}$.

Setting this sequence manually often works poorly. A better approach is to set λ_t adaptively, so that the successive distributions $\mathbb{P}_t$ are "equidistant", e.g., in terms of their chi-square pseudo-distance. For instance, from the current set of particles Θ_t^n, one may set λ_t by solving for $\delta \in [0, 1 - \lambda_{t-1}]$ the following equation:

$$\frac{\left\{\sum_{n=1}^N \exp\{-\delta V(\Theta_t^n)\}\right\}^2}{\sum_{n=1}^N \exp\{-2\delta V(\Theta_t^n)\}} = \mathrm{ESS}_{\min}$$

and setting $\lambda_t \leftarrow \lambda_{t-1} + \delta$. This may be done by numerical root-finding. If the solution of this equation is above $1 - \lambda_{t-1}$, set $\lambda_t \leftarrow 1$.

The following proposition gives some intuition on how the sequence of λ_t's should evolve when they are chosen adaptively in this way.

Proposition 17.2 *Let $\nu(\mathrm{d}\theta)$ stand for a $\mathcal{N}_d(0, I_d)$ distribution, and $V(\theta) = \|\theta\|^2/2$. Then the chi-square pseudo-distance of $\mathbb{P}_t(\mathrm{d}\theta)$ relative to $\mathbb{P}_{t-1}(\mathrm{d}\theta)$ equals*

$$\mathbb{P}_{t-1}\left[\frac{\mathbb{P}_t(\mathrm{d}\theta)}{\mathbb{P}_{t-1}(\mathrm{d}\theta)} - 1\right]^2 = \left(\frac{(1+\delta_t)^2}{1+2\delta_t}\right)^{d/2} - 1$$

where $\delta_t := (\lambda_t - \lambda_{t-1})/(1 + \lambda_{t-1})$. To ensure that the successive $\mathbb{P}_t$'s remain "equidistant', i.e., that this quantity equals some constant $C > 0$ for all t's, one must set the λ_t's as follows:

$$\lambda_t = (1+\gamma)^{t+1} - 1 \tag{17.2}$$

where $\gamma > 0$ is a constant that depends only on d and C. If the λ_t's are set in this way, the smallest integer T such that $\lambda_T \geq 1$ is such that $T = \mathcal{O}(d^{1/2})$.

We draw two lessons from this proposition: first, the sequence of λ_t should increase geometrically. In particular, it should increase slowly at first, and then makes bigger and bigger steps. Incidentally, this explains also why setting the λ_t's manually is difficult.

Second, tempering seems to escape the curse of dimensionality, as the number of steps required to move from $\nu(\mathrm{d}\theta)$ to the target $\pi(\mathrm{d}\theta)$ is $\mathcal{O}(d^{1/2})$. Of course, the proposition above does not give the whole picture, as the performance of SMC tempering relies also on the MCMC kernels used at each iteration; and these kernels are also affected by d, as discussed in Chap. 15. The fact remains that, if provided with MCMC steps with good mixing properties, SMC combined with tempering may potentially work in high-dimensional problems.

To approximate the normalising constant L_T of the target distribution, one may either use the standard SMC estimate L_T^N, or an estimate based on the thermodynamic integration identity, see (3.2). In our numerical experiments, however, we found that the two estimates were very close numerically, so we do not elaborate further, and recommend using the standard estimate. (Exercise 17.2 gives some intuition on why the two estimates tend be to very close.)

Algorithm 17.3 recalls the general structure of SMC tempering.

Input of SMC Tempering

- A probability distribution $\nu(d\theta)$ which may be sampled from.
- A function $\theta \to V(\theta)$ which may be evaluated pointwise, and such that the probability distribution

$$\pi_\lambda(d\theta) \propto \nu(d\theta) \exp\{-\lambda V(\theta)\}$$

is properly defined for any $\lambda \in [0, 1]$ (meaning that $\int_\Theta \nu(d\theta) \exp\{-\lambda V(\theta)\} \in (0, \infty)$).
- A generic way to build, for any $\lambda \in [0, 1]$, a MCMC kernel $M^\lambda(\theta, d\theta')$ that leaves invariant $\pi_\lambda(d\theta)$.
- The usual input of an SMC algorithm: the number of particles N, the choice of an unbiased resampling scheme, and a threshold $\text{ESS}_{\min}$.

Algorithm 17.3: SMC tempering

Operations involving index n must be performed for $n = 1, \ldots, N$.

$\lambda_{-1} \leftarrow 0$
$t \leftarrow -1$
while $\lambda_t < 1$ **do**
  $t \leftarrow t + 1$
  if $t = 0$ **then**
    $\Theta_0^n \sim \nu(d\theta)$
  else
    $A_t^{1:N} \leftarrow \texttt{resample}(W_{t-1}^{1:N})$
    $\Theta_t^n \sim M^{\lambda_{t-1}}(\Theta_{t-1}^n, d\theta)$
  Solve in $\delta \in [0, 1 - \lambda_{t-1}]$ the equation:

$$\frac{\left\{\sum_{n=1}^N \exp\{-\delta V(\Theta_t^n)\}\right\}^2}{\sum_{n=1}^N \exp\{-2\delta V(\Theta_t^n)\}} = \text{ESS}_{\min}$$

  (Set $\delta \leftarrow 1 - \lambda_{t-1}$ if there is no solution in $[0, 1 - \lambda_{t-1}]$.)
  $\lambda_t \leftarrow \lambda_{t-1} + \delta$
  $w_t^n \leftarrow \exp\{-\delta V(\Theta_t^n)\}$
  $W_t^n \leftarrow w_t^n / \sum_{n=1}^N w_t^n$

17.2.4 Rare Events

SMC samplers for rare events target distributions constrained to a sequence of nested sets:

$$\mathbb{P}_t(\mathrm{d}\theta) = \frac{1}{L_t}\mathbb{1}_{\mathcal{A}_t}(\theta)\nu(\mathrm{d}\theta)\,.$$

Often, these sets are defined with respect to a score function S:

$$\mathcal{A}_t = \{\theta \in \Theta : S(\theta) \geq \lambda_t\}$$

with $\lambda_0 < \ldots < \lambda_T$.

In most applications, we are interested only in simulating from the final distribution $\mathbb{P}_T(\mathrm{d}\theta)$, or computing its normalising constant (the probability, under ν, that $\Theta \in \mathcal{A}_T$). Hence, we may choose the λ_t's adaptively in the same way as for tempering: set λ_t so that the ESS at each step equals a certain threshold $\mathrm{ESS}_{\min} = \alpha N$, for $\alpha \in (0, 1)$.

Since the potential functions of the associated Feynman-Kac model are indicator functions:

$$G_t(\theta_t) = \mathbb{1}_{\mathcal{A}_t}(\theta_t) = \mathbb{1}\{S(\theta_t) \geq \lambda_t\} \tag{17.3}$$

the adaptive strategy is particularly simple to carry out in this context: set λ_t to the α−quantile of the $S(\Theta_t^n)$'s, $n = 1, \ldots, N$.

Such an adaptive SMC sampler for rare-event simulation is very close in spirit to a class of algorithms known as the cross-entropy method in operations research. In fact, the basic cross-entropy algorithm simulates the particles independently at time t (from a parametric distribution fit on the previous particles), but extensions that simulate particles according to a Markov kernel fall also within the SMC framework we have just described.

17.2.5 SMC-ABC and Likelihood-Free Inference

Likelihood-free inference refers to the problem of estimating the parameter of a model defined through a black-box simulator: for any parameter θ, we are able to simulate data $Y \sim \mathbb{M}^\theta(\mathrm{d}y)$ from the model, but we are not able to compute the corresponding likelihood. Chapter 3 presented applications of likelihood-free inference in e.g. population genetics.

The ABC (Approximate Bayesian computation) approach to likelihood-free inference amounts to approximating the posterior distribution of the parameter by the marginal distribution $\pi(d\theta)$, which we obtain by integrating out Y in the joint distribution:

$$\pi(d\theta, dy) = \frac{1}{L}\nu(d\theta)\mathbb{M}^{\theta}(dy)\mathbb{1}\{d(y, y^{\star}) \leq \epsilon\} \tag{17.4}$$

where $\nu(d\theta)$ denotes the prior distribution, $y^{\star}$ the actual data, and $d(\cdot, \cdot)$ a certain distance. As $\epsilon \rightarrow 0$, $\pi(d\theta)$ should converge to the true posterior. (Often, the condition in the indicator function is replaced by $d(s(y), s(y^{\star})) \leq \epsilon$ where $s(y)$ is a low-dimensional summary of the data. This incurs a second level of approximation.)

The most basic ABC algorithm samples from (17.4) using rejection: sample Θ from the prior, Y from the data simulator (given $\Theta = \theta$), and accept when $d(Y, y^{\star}) \leq \epsilon$. Choosing ϵ is not trivial: taking ϵ smaller reduces the ABC bias, while increasing the CPU cost (since it reduces the acceptance rate).

A more elaborate approach is to implement an SMC-ABC sampler, that is, an SMC sampler that tracks the following sequence:

$$\mathbb{P}_t(d\theta, dy) = \frac{1}{L_t}\nu(d\theta)\mathbb{M}^{\theta}(dy)\mathbb{1}\{d(y, y^{\star}) \leq \epsilon_t\}$$

where $\epsilon_0 > \epsilon_1 > \ldots$. In addition, we can make the algorithm choose adaptively the ϵ_t's, as in previous section: set ϵ_t so that the proportion of alive particles equals a certain fixed value. In practice, one may stop the algorithm when the number of calls to the simulator has exceeded a certain budget.

Clearly, SMC-ABC samplers are close to the rare-event SMC samplers presented in the previous section. One difference however is that we are not able to implement a standard MCMC kernel on space $\Theta \times \mathcal{Y}$. We must take in account the constraint that the density of $Y|\Theta$ is not tractable.

Algorithm 17.4 describes the so-called MCMC-ABC kernel. This is a Metropolis-Hastings kernel that leaves invariant (17.4), a joint distribution for (Θ, Y), and such that the second component, Y, is simulated from the model. For component Θ, we use the same notations as in Chap. 15: a new value for Θ is proposed according to some proposal kernel $\widetilde{M}(\theta, d\widetilde{\theta})$. When used within an SMC sampler, one may use the usual recipe: set this proposal kernel to a random walk proposal calibrated on the current set of θ-particles.

Algorithm 17.4: MCMC-ABC kernel

Input: (Θ, Y)

$\widetilde{\Theta} \sim \widetilde{M}(\Theta, \mathrm{d}\widetilde{\theta})$

$\widetilde{Y} \sim \mathrm{M}^{\widetilde{\Theta}}(\mathrm{d}y)$

$U \sim \mathcal{U}([0,1])$

if $d(Y, y^\star) \leq \epsilon$ **and** $\log U \leq \log r(\Theta, \widetilde{\Theta})$ ▷ where $r(\theta, \widetilde{\theta}) = \frac{\nu(\widetilde{\theta})\widetilde{m}(\theta|\widetilde{\theta})}{\nu(\theta)\widetilde{m}(\widetilde{\theta}|\theta)}$

then

  return $(\widetilde{\Theta}, \widetilde{Y})$

else

  return (Θ, Y)

The MCMC-ABC kernel has a big drawback: its acceptance rate decreases as ϵ goes to zero (see also Exercise 17.3). Thus, the mixing of the MCMC steps is going to decrease significantly as time progresses. We could address this problem by increasing the number of MCMC steps at each iteration, but it is not clear how to do so in practice.

A more appealing approach is to use the one-hit ABC kernel, described in Algorithm 17.5. This kernel samples repetitively Y until it "hits the target", i.e. until $d(Y, y^\star) \leq \epsilon$. In this way, the CPU effort scales automatically with the difficulty of hitting the constraint.

This kernel leaves invariant the marginal distribution $\pi(\mathrm{d}\theta)$; see either Exercise 17.4 for some indications on how to establish this property, or Lee (2012), who introduced the more general r-hits kernels, of which the one-hit kernel is a particular example. By default, we recommend to use this type of kernel when implementing SMC-ABC.

Algorithm 17.5: One-hit ABC kernel

Input: Θ

$U \sim \mathcal{U}([0,1])$

if $\log U \geq \log r(\Theta, \widetilde{\Theta})$ **then**

  return Θ

$\widetilde{\Theta} \sim \widetilde{M}(\Theta, \mathrm{d}\widetilde{\theta})$

repeat

  $Y \sim \mathbb{M}^{\Theta}(\mathrm{d}y)$

  $\widetilde{Y} \sim \mathbb{M}^{\widetilde{\Theta}}(\mathrm{d}y)$

until $d(Y, y^{\star}) \leq \epsilon$ **or** $d(\widetilde{Y}, y^{\star}) \leq \epsilon$

if $d(Y, y^{\star}) \leq \epsilon$ **then**

  return $\widetilde{\Theta}$

else

  return Θ

17.2.6 Note Regarding the Normalising Constant Estimate

We have established in Sect. 16.4.1 that, for particle filters, L_t^N is an unbiased estimate of L_t. The proof assumes implicitly that the components of the considered Feynman-Kac model are fixed; e.g. G_t or M_t do not depend on particles $X_{t-1}^{1:N}$. This means that this unbiasedness property may not hold any more for adaptive SMC samplers (Exercise 17.5). This is easy to see in the context of rare events: take $\alpha = 1/2$, and assume that the λ_t's are set adaptively as described in Sect. 17.2.4. Then $L_t^N = (1/2)^{t+1}$, a deterministic constant, while $L_t = \nu\left(S(\Theta) > \lambda_t\right)$ itself is actually random, since λ_t is determined numerically from the particles simulated in the course of the algorithm. Hence the statement $\mathbb{E}[L_t^N] = L_t$ does not even make sense in that particular context.

Fortunately, although not necessarily unbiased, the estimate of the normalised constant provided by adaptive SMC samplers remains consistent and reasonably well-behaved; see the numerical experiment in Sect. 17.3.

17.2.7 Note Regarding the Starting Distribution

A simple way to improve the performance of most SMC samplers is to adapt the starting distribution $\nu(\mathrm{d}\theta)$. For instance, one may perform a pilot run, construct some approximation of the target $\mathbb{P}_T(\mathrm{d}\theta)$, and use it as the starting distribution $\nu(\mathrm{d}\theta)$ in the next run.

There are a few practical pitfalls however. First, one must be sure that the new starting distribution is not too narrow (e.g. miss a certain modal region in case some of the target distributions are multi-modal).

Second, when using IBIS for Bayesian inference, the user may still want to use a particular prior distribution, call it $\nu_{\text{actual}}(\mathrm{d}\theta)$. Thus, one may perform a pilot run to approximate the full posterior, use that approximation to construct a 'computer prior' $\nu(\mathrm{d}\theta)$, use it as the starting distribution in the next run, and finally recover the actual prior by performing an extra importance sampling step (reweighting particles according to $\nu_{\text{actual}}(\mathrm{d}\theta)/\nu(\mathrm{d}\theta)$).

This strategy is not appropriate in case the user wants to recover all the partial posteriors. In that case, an alternative strategy is to reinstate the actual prior after k observations have been incorporated; ideally $k = 1$, but in practice one may need to take k larger in case the importance sampling step for re-instating the prior is too inefficient when applied at time 1.

17.3 Numerical Experiments

17.3.1 Settings

We investigate the relative performance of SMC tempering and IBIS (data tempering) for approximating the posterior distribution and the marginal likelihood of a logistic regression model: the Y_t are independent variables, taking values in $\{-1, 1\}$, and the likelihood is:

$$p_t^\theta(y_{0:T}) = \prod_{t=0}^{T} F(y_t \theta' x_t) \tag{17.5}$$

where $x_t \in \mathbb{R}^d$ is a vector of (fixed) predictors (and not the realisation of a latent variable as in other chapters). The considered datasets are taken from the UCI repository (https://archive.ics.uci.edu/ml/) and are pre-processed so that predictors have zero mean and standard deviation 0.5, except for an intercept which is a predictor with constant value 1 for all t. The components of $\theta \in \mathbb{R}^d$ are assigned independent $\mathcal{N}(0, 5^2)$ priors.

As discussed in Chopin and Ridgway (2017), the posterior distribution of a logistic regression model is often very Gaussian-like, and thus not particularly

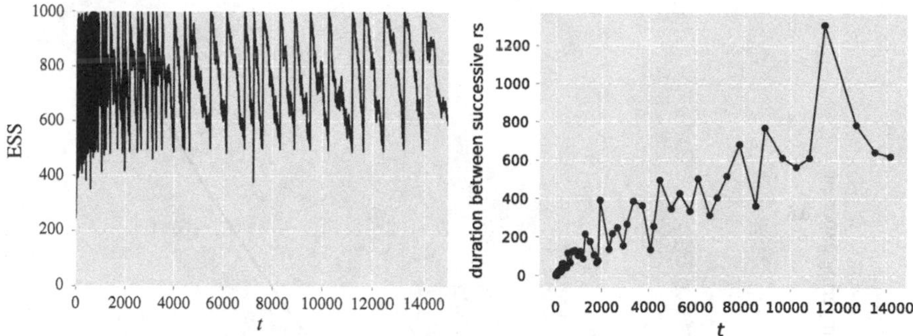

Fig. 17.1 Typical behaviour of the ESS for the IBIS algorithm applied to the EEG eye state data. Left: ESS versus time t; Right: duration until next resampling time, versus resampling time (each dot has coordinates $(t_i, t_{i+1} - t_i)$, where t_i denotes the successive times where resampling occurs)

challenging to simulate from. In fact, when $d \ll 100$, they find that best performance is often obtained by importance sampling, using as a proposal a Gaussian approximation computed in a preliminary step. Still, it is of interest to see how SMC samplers perform on Gaussian-like targets.

17.3.2 A Tall Dataset

We consider first the EEG eye state dataset ($T + 1 = 14,980$, $d = 15$). We run 50 times IBIS and adaptive tempering, with $N = 10^3$ particles. Resampling is triggered whenever the ESS drops below $N/2 = 500$; then particles are moved through k iterations of a random-walk Metropolis kernel, which is automatically calibrated using the default strategy described in Sect. 17.2.1; we set $k = 5$ for now, but will consider other values later on.

One of our motivations for considering a dataset with such a large sample size is to assess the long-time behaviour of IBIS. Figure 17.1 plots how the ESS evolves over time in a typical IBIS run. Recall that, in IBIS, iteration t targets the posterior of the $t + 1$ first datapoints, and we expect resampling to occur less and less. This is indeed the case here. In particular, we see in the right panel of Fig. 17.1 that the duration between two resampling steps seems to increase linearly with time.

Figure 17.2 plots how the tempering exponent λ_t increases in a typical run of SMC with adaptive tempering. Recall that in this algorithm the successive λ_t's are calibrated numerically, so that the ESS between two successive steps is $N/2 = 500$. As predicted in Sect. 17.2.3, the tempering exponents vary very slowly at the beginning, then increase exponentially.

We now compare the relative performance of IBIS and tempering. Interestingly, both algorithms have roughly the same CPU running time (when run for the same number of particles, N, and the same number of MCMC steps, k). Tempering

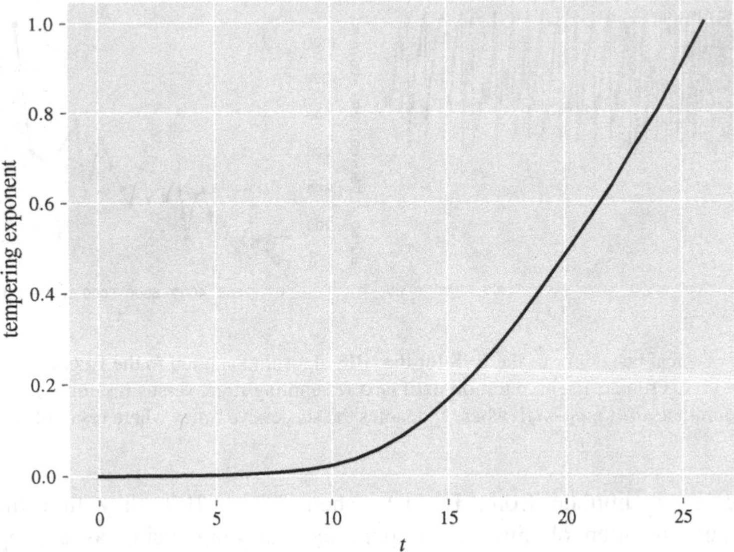

Fig. 17.2 Sequence of tempering exponents obtained in a single run of SMC with adaptive tempering, for the EEG dataset

performs a much smaller number of iterations than IBIS (25, as chosen adaptively by the algorithm, compared to $T + 1 = 14980$ iterations for IBIS). However, tempering iterations are much more expensive, as they require evaluating the full likelihood (whereas the importance sampling step of an IBIS iteration evaluates a single likelihood factor; the MCMC steps involves $t + 1$ factors but are performed only at certain times). More precisely, we observe that tempering performs (on average) 1.75 times more evaluations of likelihood factors (i.e. the factors in (17.5)).

Figure 17.3 compares, for different values of k (number of MCMC steps) the variability (over the 50 independent runs) of the following estimates: the posterior expectation of the first predictor, and the log of the marginal likelihood. Clearly, IBIS performs better than SMC tempering in this case. This is not terribly surprising, given that we are considering a "tall" dataset; hence the data tempering effect of IBIS may take full effect. More importantly, we see that increasing k is critical for good performance. Taking $k = 1$ leads to very poor estimates.

To better assess the impact of the number of MCMC steps, we plot in Fig. 17.4 the variance of the estimates of the posterior expectation of the d components of θ, times k, as a function of k (the number of MCMC steps). We multiply the variance by k in order to account for CPU time. We see that for IBIS the optimal value of k should be around 10. It is noteworthy that, even for $d = 15$, we already need about ten random walk iterations at each mutation step in order to obtain optimal performance.

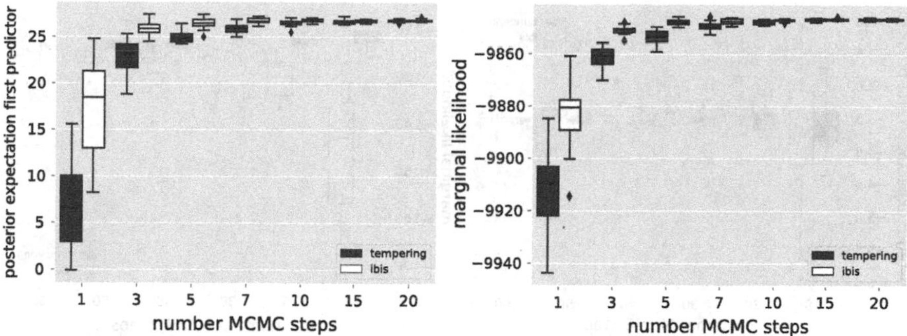

Fig. 17.3 Box-plots of estimates obtained from 50 runs of each of the two SMC samplers, as a function of the number of MCMC steps performed at each mutation step. Left: posterior expectation of coefficient of first predictor. Right: log of marginal likelihood

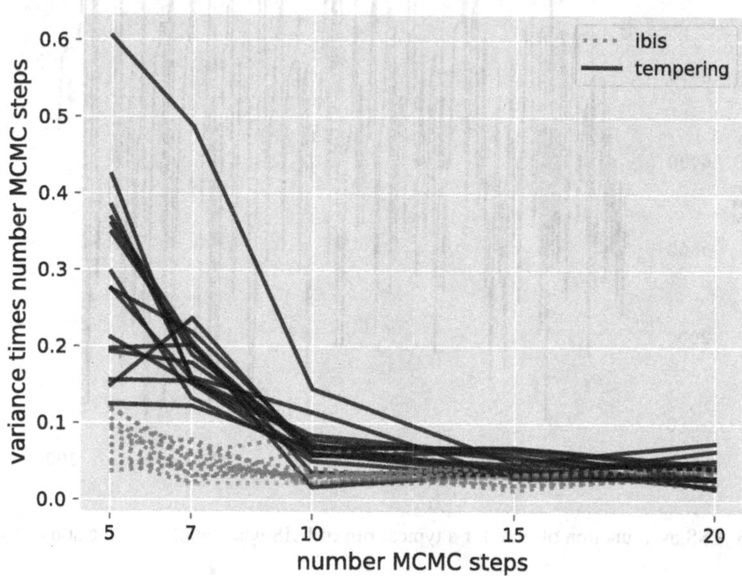

Fig. 17.4 Sample variance over the 50 runs of the algorithm multiplied by the number of MCMC steps used within each mutation step, one for each of the d estimates of the posterior expectation of the logistic regression coefficients for the EEG data, plotted versus number of MCMC steps. For readability, the results for $k = 1, 3$ are discarded

17.3.3 A Wide Dataset

We now consider a second dataset with a very different shape: the sonar dataset, where $T + 1 = 200$, $d = 60$. The corresponding posterior is more challenging, not only because d is large, but also because there is complete separation in the data, which suggests that the posterior is less Gaussian-like. Figure 17.5 compares

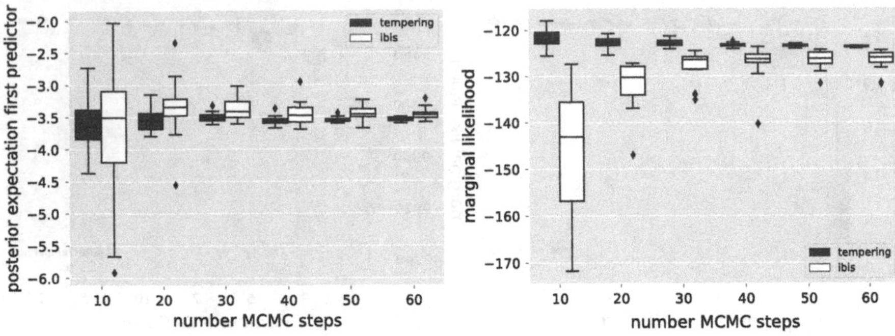

Fig. 17.5 Same caption as Fig. 17.3 for sonar dataset

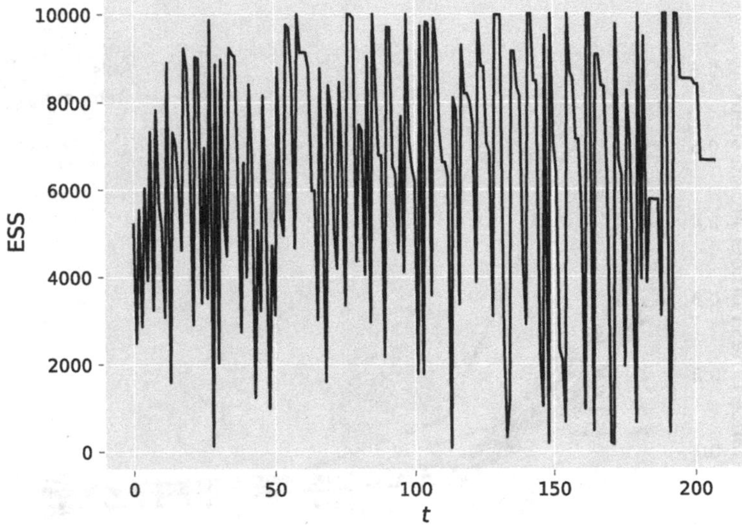

Fig. 17.6 ESS as a function of time for a typical run of IBIS when applied to the sonar dataset

the variability of estimates obtained from IBIS and SMC with tempering. (Again, the CPU times of both approaches are roughly the same, for fixed N and k.) This time, we take $N = 10^4$ particles, because for $N = 10^3$ the covariance matrix of the particles used to calibrate the random walk is often degenerate (while running IBIS).

Again, one needs to take k to a fairly large value to obtain optimal performance; e.g. $k = 50$. We also observe that IBIS is strongly outperformed by SMC with tempering, especially regarding the estimation of the marginal likelihood. This seems to be explained by Fig. 17.6, which plots the ESS versus t for a single run of IBIS. A few times, the ESS drops sharply to a very small value. This is an illustration of the curse of dimensionality discussed in 17.2.2: when d gets large, the discrepancy between two successive posterior distributions get large as well. (In addition, a few datapoints may be outliers in this specific dataset.)

17.3.4 Discussion

One may say that IBIS is better suited for tall datasets ($T \gg d$) and tempering is better suited for wide datasets ($d \approx T$ or even larger). However, tempering appears to be the more versatile approach of the two, in the sense that it still performs reasonably well in the tall case, whereas IBIS collapses in the wide case. On the other hand, IBIS has more to offer than tempering, since it makes it possible to approximate the sequence of partial posteriors, and thus to perform sequential inference, whereas tempering approximates only the full posterior.

In case one is interested in sequential inference for high-dimensional models, one should consider the hybrid approach outlined in Sect. 17.2.2; i.e. using intermediate tempered distribution to move from the posterior with t observations to the one with $t+1$ observations.

17.4 More General SMC Samplers Based on Backward Kernels

So far, we have considered Markov kernels M_t that are invariant (with respect to $\mathbb{P}_{t-1}(\mathrm{d}\theta)$). It is possible to derive SMC samplers based on more general kernels.

Assume that M_t admits a probability density $m_t(\theta_t|\theta_{t-1})$ (see Exercise 17.6 for the how to deal with the more general case where M_t does not). Consider the following potential function:

$$G_t(\theta_{t-1}, \theta_t) = \frac{\gamma_t(\theta_t) m_{t-1}^{\mathrm{back}}(\theta_{t-1}|\theta_t)}{\gamma_{t-1}(\theta_{t-1}) m_t(\theta_t|\theta_{t-1})} \tag{17.6}$$

where m_{t-1}^{back} is the density of a user-chosen backward kernel $M_{t-1}^{\mathrm{back}}(\theta_t, \mathrm{d}\theta_t)$. This function corresponds to an importance sampling step on the joint space of (Θ_{t-1}, Θ_t), from a proposal such that marginally $\Theta_{t-1} \sim \mathbb{P}_{t-1}(\mathrm{d}\theta_{t-1})$, to a target such that marginally $\Theta_t \sim \mathbb{P}_t(\mathrm{d}\theta_t)$. It is easy to check that the corresponding Feynman-Kac model is such that

$$\mathbb{Q}_t(\mathrm{d}\theta_{0:t}) = \mathbb{P}_t(\mathrm{d}\theta_t) \prod_{s=t-1}^{0} M_s^{\mathrm{back}}(\theta_{s+1}, \mathrm{d}\theta_s)$$

and thus $\mathbb{Q}_t(\mathrm{d}\theta_t) = \mathbb{P}_t(\mathrm{d}\theta_t)$.

It is not immediately clear how one should choose the backward kernel. The following result gives some insight.

Proposition 17.3 *For* (Θ_{t-1}, Θ_t), *a pair of random variables distributed according to* $\mathbb{P}_{t-1}(\mathrm{d}\theta_{t-1})M_t(\theta_{t-1}, \mathrm{d}\theta_t)$, *the variance of* $G_t(\Theta_{t-1}, \Theta_t)$, *as defined in* (17.6), *is minimal with respect to the backward kernel* M_{t-1}^{back} *when its density equals:*

$$\overleftarrow{m}_{t-1}(\theta_{t-1}|\theta_t) = \frac{\gamma_{t-1}(\theta_{t-1})m_t(\theta_t|\theta_{t-1})}{\int_\Theta \gamma_{t-1}(\theta'_{t-1})m_t(\theta_t|\theta'_{t-1})\mathrm{d}\theta'_{t-1}}. \tag{17.7}$$

For this optimal backward kernel, the expression of the potential function reduces to:

$$G_t(\theta_{t-1}, \theta_t) = \frac{\gamma_t(\theta_t)}{\int_\Theta \gamma_{t-1}(\theta'_{t-1})m_t(\theta_t|\theta'_{t-1})\mathrm{d}\theta'_{t-1}}. \tag{17.8}$$

Proof We rewrite the denominator of (17.6) as:

$$\gamma_{t-1}(\theta_{t-1})m_t(\theta_t|\theta_{t-1}) = \widetilde{\gamma}_t(\theta_t)\overleftarrow{m}_{t-1}(\theta_{t-1}|\theta_t)$$

where $\widetilde{\gamma}_t(\theta_t) = \int_\Theta \gamma_{t-1}(\theta_{t-1})m_t(\theta_t|\theta_{t-1})\mathrm{d}\theta_{t-1}$ is the marginal density of θ_t (up to constant L_t), and $\overleftarrow{m}_{t-1}$ is the conditional density of θ_{t-1} given θ_t, which equals (17.7). Thus

$$G_t(\theta_{t-1}, \theta_t) = \frac{\gamma_t(\theta_t)}{\widetilde{\gamma}_t(\theta_t)} \times \frac{m_{t-1}^{\mathrm{back}}(\theta_{t-1}|\theta_t)}{\overleftarrow{m}_{t-1}(\theta_{t-1}|\theta_t)}$$

and

$$\mathbb{E}\left[G_t(\Theta_{t-1}, \Theta_t) \mid \Theta_t = \theta_t\right] = \frac{\gamma_t(\theta_t)}{\widetilde{\gamma}_t(\theta_t)}$$

which is also the expression of $G_t(\theta_{t-1}, \theta_t)$ when $m_{t-1}^{\mathrm{back}}(\theta_{t-1}|\theta_t) = \overleftarrow{m}_{t-1}(\theta_{t-1}|\theta_t)$. We conclude by remarking that $\mathrm{Var}\left[G_t(\Theta_{t-1}, \Theta_t)\right] \geq \mathrm{Var}\left\{\mathbb{E}\left[G_t(\Theta_{t-1}, \Theta_t) \mid \Theta_t\right]\right\}$. □

Once again, the interest of this type of optimality result is mostly qualitative. For instance, the random distribution considered above for $G_t(\Theta_{t-1}, \Theta_t)$ is only an asymptotic (as $N \to +\infty$) approximation of the distribution of the weights in the corresponding SMC sampler. More importantly, it is not trivial to elicit practical recipes from this result. To see this, let's consider the following scenarios:

- If we choose M_t so that it leaves invariant $\mathbb{P}_{t-1}(\mathrm{d}\theta)$, (17.8) reduces to $\gamma_t(\theta_t)/\gamma_{t-1}(\theta_t)$, the potential function we used in the Feynman-Kac model based on invariant kernels. In other words, an SMC sampler based on invariant kernels is already using, implicitly, optimal backward kernels.

- If we could choose M_t so that

$$\int_{\theta_{t-1}\in\Theta} \mathbb{P}_{t-1}(\mathrm{d}\theta_{t-1})M_t(\theta_{t-1}, \mathrm{d}\theta_t) = \mathbb{P}_t(\mathrm{d}\theta_t) \tag{17.9}$$

  then (17.8) would equal one, and we would be in the ideal scenario where, at each time t, the N particles are i.i.d. simulations from the target $\mathbb{P}_t(\mathrm{d}\theta_t)$. Unfortunately, in most cases it is difficult to construct a Markov kernel such that (17.9) holds. (One exception is $M_t(\theta_{t-1}, \mathrm{d}\theta_t) = \mathbb{P}_t(\mathrm{d}\theta_t)$, but then if we are able to simulate from each $\mathbb{P}_t(\mathrm{d}\theta)$ efficiently, there is no point deriving an SMC sampler for this sequence of targets.)
- We may try in practice to construct M_t so that (17.9) holds approximately. On the other hand, in various practical scenarios the next target $\mathbb{P}_t(\mathrm{d}\theta)$ is chosen adaptively; for instance, in tempering SMC, the exponent λ_t is typically determined numerically at the next step (see next section). In those cases, this recommendation is not practical, and we may fall back on choosing M_t so that it leaves invariant $\mathbb{P}_{t-1}(\mathrm{d}\theta)$.

By and large, it is often simpler to stick with SMC samplers based on invariant kernels. That said, they are specific cases where the more general formalism may lead to more efficient algorithms. For instance, imagine a situation where one is able to construct kernels M_t that leave $\mathbb{P}_{t-1}(\mathrm{d}\theta_{t-1})$ invariant only approximately, while being must faster to simulate from than exactly invariant kernels. In that case, the more general formalism allows us to use those kernels in an SMC sampler. The difficulty is then to construct a backward kernel which approximates as closely as possible (17.7).

Exercises

17.1 *Prove Propositions 17.1 and 17.2. Generalise Proposition 17.2 to the case where* $V(\theta) = (c/2)\|\theta\|^2$.

17.2 *Explain how you may use the thermodynamic identity,* (3.2), *in order to construct an alternative estimate of the log of the normalising constant within tempering SMC. Explain why, when* $\delta = \lambda_t - \lambda_{t-1} \to 0$, *the respective contributions of iteration t to both estimates (i.e. something like* $\delta N^{-1}\sum_{n=1}^N V(\Theta_t^n)$ *and the log of* $N^{-1}\sum_{n=1}^N \exp\{-\delta V(\Theta_t^n)\}$*) should become identical.*

17.3 *Show that the acceptance probability of the MCMC-ABC kernel (Algorithm 17.4) is bounded by the probability (for a fixed input* Θ*) that* $d(Y, y^\star) \le \epsilon$. *Discuss the practical implications of this result.*

17.4 *Consider the one-hit ABC kernel (Algorithm 17.5). Show that the probability (for fixed* Θ, $\widetilde{\Theta}$*) that* $\widetilde{K} \le K$ *is* $\widetilde{q}/(q\widetilde{q} - q - \widetilde{q})$, *where* q *(resp.* $\widetilde{q}$*) is the probability* *that $d(Y, y^\star) \leq \epsilon$, conditional on Θ (resp. $\widetilde{\Theta}$) Use this result to establish that this kernel fulfils the detailed balance condition.*

17.5 *Go through the proof of the unbiasedness of L_t^N in Sect. 16.4.1, and explain which parts do not hold any more in situations where either Markov kernel M_t, or potential function G_t, depends on particles $X_{t-1}^{1:N}$. Explain why this is relevant to adaptive SMC samplers.*

17.6 *In the general Feynman-Kac construction based on backward kernels (Sect. 17.4) we assumed that kernel $M_t(\theta_{t-1}, \mathrm{d}\theta_t)$ admits a density (with respect to a measure that does not depend on θ_{t-1}; as a counter-example, consider a Metropolis-Hastings kernel). Repeat the derivations of the end of Sect. 17.1 in case M_t does not admit a density: propose a new expression for $G_t(\theta_{t-1}, \theta_t)$, and derive the optimal backward kernel (Proposition 17.3).*

Python Corner

To run a SMC sampler, we must define first the sequence of distributions we wish to approximate. For IBIS, this means defining a particular static model. This may be done in `particles` by sub-classing `StaticModel` (from module `smc_samplers`) and defining method `logpyt` which computes the log-likelihood of datapoint y_t (given $y_{0:t-1}$ and θ):

```
from particles import smc_samplers as ssp

class ToyModel(ssp.StaticModel):
 def logpyt(self, theta, t):  # density of Y_t given theta and Y_{0:t-1}
   return stats.norm.logpdf(self.data[t], loc=theta['mu'],
                                scale = theta['sigma'])
```

Argument `theta` should be a structured array (as discussed in the previous Python corner): both `theta['mu']` and `theta['sigma']` behaves like $(N,)$−shaped `NumPy` float arrays.

We may now run IBIS as follows:

```
T = 50
my_data = stats.norm.rvs(loc=3.14, size=T)
my_prior = dists.StructDist({'mu': dists.Normal(scale=10.),
                             'sigma': dists.Gamma()})
my_static_model = ToyModel(data=my_data, prior=my_prior)
fk_ibis = ssp.IBIS(my_static_model)
alg = particles.SMC(fk=fk_ibis, N=1000)
alg.run()
```

As usual, IBIS is simply defined as particular Feynman-Kac model, the definition of which we give below:

```
class IBIS(FKSMCsampler):
    """FeynmanKac class for IBIS algorithm.
    """
    mutate_only_after_resampling = True

    def logG(self, t, xp, x):
        lpyt = self.model.logpyt(x.theta, t)
        x.lpost += lpyt
        return lpyt

    def compute_post(self, x, t):
        x.lpost = self.model.logpost(x.theta, t=t)

    def M0(self, N):
        x0 = MetroParticles(theta=self.model.prior.rvs(size=N))
        self.compute_post(x0, 0)
        return x0

    def M(self, t, Xp):
        # in IBIS, M_t leaves invariant p(theta|y_{0:t-1})
        comp_target = lambda x: self.compute_post(x, t-1)
        return Xp.Metropolis(comp_target, mh_options=self.mh_options)
```

This is short and hopefully easy to read. However, a few details are hidden in `MetroParticles`, a class that defines collections of particles for SMC samplers.

In `particles`, a SMC sampler is an instance of class `SMC`, which stores the N current particles X_t^n in attribute `X`. Up to now, we considered examples where this `X` was a `NumPy` float array, of shape $(N,)$ or (N, d). However, if you look at the source of `particles`, you will see that class `SMC` does not apply any `NumPy`-specific operation on `X`, apart from "fancy indexing", i.e. something along the lines of:

```
resampled_X = X[A]   # A is an array of N indices (ancestors)
```

where array `A` has been generated first by one of the resampling algorithms. The new array, `resampled_X` contains the resampled particles. This "fancy indexing" operation works on `NumPy` arrays, but nothing prevents us from defining our own classes that implement the same functionality. (Specifically, one may define the meaning of the indexing operation, i.e. `X[A]`, for any class by defining special method `__getitem__`).

Going back to SMC samplers, there are several reasons why it is convenient to define a custom class to store the particles of such algorithms. SMC samplers simulate parameter values, whose components typically have names, as discussed in the Python corner of the previous chapter. In addition, on top of the components of the parameters, we need to store extra fields for e.g. the log-target density. However, we cannot treat all the fields in the same way; for instance, when implementing a random walk step, it does not make sense to apply that step to these extra fields. Hence the custom class must "know" which field is a component parameter, and which field is an extra field. Finally, we may implement Metropolis steps as a method of such a custom class. This is essentially what `MetroParticles` does; see the documentation for more details (or better yet, the notebook tutorial on SMC samplers, in the on-line documentation.)

The fact that we could extend the functionality of `SMC` by simply defining a custom class for object `X` is an illustration of "duck typing": in Python, an object is suitable for a certain purpose as soon as it implements certain methods and properties (rather than because it has some rigid type). Or, as the saying goes: "If it walks like a duck and it quacks like a duck, then it must be a duck".

When we implemented IBIS and SMC tempering for the logistic regression example of Sect. 17.3, we had the pleasant surprise to observe that our program kept busy *all* the cores of the computer; or in other words that we were getting a $\times k$ speed-up (where k is the number of cores) for free.

Pure Python code is supposed to run on a single core, because of a feature called GIL (Global Interpreter Lock). However, since `NumPy` relies on libraries written in compiled languages, it is able to "release" the GIL, and implement `NumPy` operations on multiple cores. In our case, the bulk of the computation occurred in method `logpyt`, which computed the log-likelihood of a logistic regression for a single observation, using basic `NumPy` operations, on large arrays (of size $N \times d$, where d is the number of predictors).

Bibliographical Notes

An early version of SMC tempering (without resampling) appeared in Neal (2001). The IBIS algorithm was introduced by Chopin (2002). The general framework (including the Feynman-Kac formalism based on backward kernels) for SMC samplers was developed in the seminal paper of Del Moral et al. (2006); see also Del Moral et al. (2007). Adaptive tempering was introduced in Jasra et al. (2011). The idea of using the current set of particles in order to calibrate MCMC kernels is explored in particular in Fearnhead and Taylor (2013) and Buchholz et al. (2020); the latter reference considers specifically Langevin and HMC kernels. Heng et al. (2020) use optimal control theory in order to design good proposals for SMC samplers.

We also mention in passing PMC (population Monte Carlo) algorithms, developed by Cappé et al. (2004), Douc et al. (2007a) and Douc et al. (2007b), which may be viewed as a type of SMC samplers where the sequence of target distributions $\mathbb{P}_t(\mathrm{d}\theta_t)$ is constant, and equal to the distribution of interest.

Over the years, SMC samplers have found useful applications in the following areas: experimental designs (Amzal et al. 2006); prior sensitivity analysis and cross-validation (Bornn et al. 2010); Bayesian inference for long-memory processes (Chopin et al. 2013); variable selection, where $\Theta = \{0, 1\}^d$ (Schäfer and Chopin 2013); optimisation in binary spaces (Schäfer 2013); probabilistic graphical models (Naesseth et al. 2014; Olsson et al. 2019); classification and scoring through the PAC-Bayesian approach (Ridgway et al. 2014; see also Guedj (2019) for an introduction to the PAC-Bayesian approach); estimation of DSGE models in macroeconomics (Herbst and Schorfheide 2014); Bayesian model choice for hidden Markov models (Zhou et al. 2016), and un-normalised models (Everitt et al. 2017); computation of Gaussian orthant probabilities (Ridgway 2016); confidence sets in econometrics models (Chen et al. 2018); computation of expectations with respect to stochastic differential equations (Beskos et al. 2017), adapting to SMC the multi-level Monte Carlo method of Giles (2008). Many of these papers feature numerical experiments in which the considered SMC sampler outperforms MCMC, even in very high dimensions (e.g., 4000 in the Cox example of Buchholz et al. 2020).

Regarding rare-event estimation through SMC, see Cérou et al. (2012) and references therein, in particular for similar algorithms such as the cross-entropy method that were developed independently in operations research, e.g., Botev and Kroese (2008, 2012); see also the books of Rubinstein and Kroese (2004) and Kroese et al. (2011).

An area of application where SMC samplers gained a lot of traction among practitioners is likelihood-free inference, as discussed in Sects. 3.5 and 17.2.5. The first particle algorithm for likelihood-free inference was proposed by Sisson et al. (2007). However this algorithm is not, strictly speaking, an SMC sampler, but rather a sequential importance sampling algorithm, where the proposal at iteration t is a mixture of Gaussians centred at the N particles of the previous iteration; as a result, it has a $\mathcal{O}(N^2)$ cost per iteration. The class of ABC-SMC algorithm we described in this Chapter is based on Del Moral et al. (2012), who derived a proper SMC sampler for likelihood-free inference (with cost $\mathcal{O}(N)$), and Lee (2012) who introduced the r−hits kernels we discuss in Sect. 17.2.5. See also Didelot et al. (2011), Prangle et al. (2018), and Buchholz and Chopin (2019).

Theoretical papers that consider specifically SMC samplers include Beskos et al. (2014) (on the scaling of SMC tempering in high dimension) and Beskos et al. (2016) (on the convergence of adaptive SMC samplers).

Bibliography

Amzal, B., Bois, F. Y., Parent, E., & Robert, C. P. (2006). Bayesian-optimal design via interacting particle systems. *Journal of the American Statistical Association, 101*(474), 773–785.

Beskos, A., Crisan, D., & Jasra, A. (2014). On the stability of sequential Monte Carlo methods in high dimensions. *Annals of Applied Probability, 24*(4), 1396–1445.

Beskos, A., Jasra, A., Kantas, N., & Thiery, A. (2016). On the convergence of adaptive sequential Monte Carlo methods. *Annals of Applied Probability, 26*(2), 1111–1146.

Beskos, A., Jasra, A., Law, K., Tempone, R., & Zhou, Y. (2017). Multilevel sequential Monte Carlo samplers. *Stochastic Processes and Their Applications, 127*(5), 1417–1440.

Bornn, L., Doucet, A., & Gottardo, R. (2010). An efficient computational approach for prior sensitivity analysis and cross-validation. *Canadian Journal of Statistics, 38*(1), 47–64.

Botev, Z. I., & Kroese, D. P. (2008). An efficient algorithm for rare-event probability estimation, combinatorial optimization, and counting. *Methodology and Computing in Applied Probability, 10*(4), 471–505.

Botev, Z. I., & Kroese, D. P. (2012). Efficient Monte Carlo simulation via the generalized splitting method. *Statistics and Computing, 22*(1), 1–16.

Buchholz, A., & Chopin, N. (2019). Improving approximate Bayesian computation via Quasi-Monte Carlo. *Journal of Computational and Graphical Statistics, 28*(1), 205–219.

Buchholz, A., Chopin, N., & Jacob, P. E. (2020). Adaptive tuning of Hamiltonian Monte Carlo within Sequential Monte Carlo. *Bayesian Analysis (to appear)*.

Cappé, O., Guillin, A., Marin, J. M., & Robert, C. P. (2004). Population Monte Carlo. *Journal of Computational and Graphical Statistics, 13*(4), 907–929.

Cérou, F., Del Moral, P., Furon, T., & Guyader, A. (2012). Sequential Monte Carlo for rare event estimation. *Statistics and Computing, 22*(3), 795–808.

Chen, X., Christensen, T. M., & Tamer, E. (2018). Monte Carlo confidence sets for identified sets. *Econometrica, 86*(6), 1965–2018.

Chopin, N. (2002). A sequential particle filter method for static models. *Biometrika, 89*(3), 539–551.

Chopin, N., & Ridgway, J. (2017). Leave Pima Indians alone: Binary regression as a benchmark for Bayesian computation. *Statistical Science, 32*(1), 64–87.

Chopin, N., Rousseau, J., & Liseo, B. (2013). Computational aspects of Bayesian spectral density estimation. *Journal of Computational and Graphical Statistics, 22*(3), 533–557.

Del Moral, P., Doucet, A., & Jasra, A. (2006). Sequential Monte Carlo samplers. *Journal of the Royal Statistical Society: Series B (Statistical Methodology), 68*(3), 411–436.

Del Moral, P., Doucet, A., & Jasra, A. (2007). Sequential Monte Carlo for Bayesian computation. In J. M. Bernardo, M. J. Bayarri, J. O. Berger, A. P. Dawid, D. Heckerman, A. F. M. Smith, & M. West (Eds.), *Bayesian statistics 8. Oxford science publications* (Vol. 8, pp. 115–148). Oxford: Oxford University Press.

Del Moral, P., Doucet, A., & Jasra, A. (2012). An adaptive sequential Monte Carlo method for approximate Bayesian computation. *Statistics and Computing, 22*(5), 1009–1020.

Didelot, X., Everitt, R. G., Johansen, A. M., & Lawson, D. J. (2011). Likelihood-free estimation of model evidence. *Bayesian Analysis, 6*(1), 49–76.

Douc, R., Guillin, A., Marin, J.-M., & Robert, C. P. (2007a). Convergence of adaptive mixtures of importance sampling schemes. *Annals of Statistics, 35*(1), 420–448.

Douc, R., Guillin, A., Marin, J.-M., & Robert, C. P. (2007b). Minimum variance importance sampling via population Monte Carlo. *ESAIM: Probability and Statistics, 11*, 427–447.

Everitt, R. G., Johansen, A. M., Rowing, E., & Evdemon-Hogan, M. (2017). Bayesian model comparison with un-normalised likelihoods. *Statistics and Computing, 27*(2), 403–422.

Fearnhead, P., & Taylor, B. M. (2013). An adaptive sequential Monte Carlo sampler. *Bayesian Analysis, 8*(2), 411–438.

Giles, M. B. (2008). Multilevel Monte Carlo path simulation. *Operational Research, 56*(3), 607–617.

Guedj, B., (2019). A Primer on PAC-Bayesian Learning. arxiv preprint 1901.05353

Heng, J., Bishop, A. N., Deligiannidis, G., & Doucet, A. (2020). Controlled Sequential Monte Carlo. *Annals of Statistics (to appear)*.

Herbst, E., & Schorfheide, F. (2014). Sequential Monte Carlo sampling for DSGE models. *Journal of Applied Econometrics, 29*(7), 1073–1098.

Jasra, A., Stephens, D., Doucet, & Tsagaris, T. (2011). Inference for Lévy driven stochastic volatility models via Sequential Monte Carlo. *Scandinavian Journal of Statistics, 38*(1), 1–22.

Kroese, D. P., Taimre, T., & Botev, Z. I. (2011). *Handbook of Monte Carlo methods*. Hoboken, NJ: Wiley.

Lee, A. (2012). On the choice of MCMC kernels for approximate Bayesian computation with SMC samplers. In *Proceedings of the 2012 Winter Simulation Conference (WSC)* (pp. 304–315). Berlin: IEEE.

Naesseth, C. A., Lindsten, F., & Schön, T. B. (2014). Sequential Monte Carlo for graphical models. In Z. Ghahramani, M. Welling, C. Cortes, N. D. Lawrence, & K. Q. Weinberger (Eds.), *Advances in neural information processing systems 27* (pp. 1862–1870). Red Hook, NY: Curran Associates, Inc.

Neal, R. M. (2001). Annealed importance sampling. *Statistics and Computing, 11*(2), 125–139.
Olsson, J., Pavlenko, T., & Rios, F. L. (2019). Bayesian learning of weakly structural Markov graph laws using sequential Monte Carlo methods. *Electronic Journal of Statistics, 13*(2), 2865–2897.
Prangle, D., Everitt, R. G., & Kypraios, T. (2018). A rare event approach to high-dimensional approximate Bayesian computation. *Statistics and Computing, 28*(4), 819–834.
Ridgway, J. (2016). Computation of Gaussian orthant probabilities in high dimension. *Statistics and Computing, 26*(4), 899–916.
Ridgway, J., Alquier, P., Chopin, N., & Liang, F. (2014). PAC-Bayesian AUC classification and scoring. In Z. Ghahramani, M. Welling, C. Cortes, N. D. Lawrence, & K. Q. Weinberger (Eds.), *Advances in neural information processing systems 27* (pp. 658–666). Red Hook, NY: Curran Associates, Inc.
Rubinstein, R. Y., & Kroese, D. P. (2004). *The cross-entropy method: A unified approach to combinatorial optimization, Monte-Carlo simulation, and machine learning*. Berlin/Heidelberg: Springer.
Schäfer, C. (2013). Particle algorithms for optimization on binary spaces. *ACM Transactions on Modeling and Computer Simulation, 23*(1), Art. 8, 25.
Schäfer, C., & Chopin, N. (2013). Sequential Monte Carlo on large binary sampling spaces. *Statistics and Computing, 23*(2), 163–184.
Sisson, S. A., Fan, Y., & Tanaka, M. M. (2007). Sequential Monte Carlo without likelihoods. *Proceedings of the National Academy of Sciences of the United States of America, 104*(6), 1760–1765.
Zhou, Y., Johansen, A. M., & Aston, J. A. D. (2016). Toward automatic model comparison: An adaptive sequential Monte Carlo approach. *Journal of Computational and Graphical Statistics, 25*(3), 701–726.

Chapter 18
SMC^2, Sequential Inference in State-Space Models

Summary In Chap. 16, we discussed PMCMC algorithms, i.e., MCMC samplers that (a) rely on particle filters to approximate the intractable likelihood; yet (b) leave invariant the exact posterior distribution of the considered (state-space) model. Potentially, PMCMC algorithms suffer from the same limitations as all MCMC samplers: they do not offer an easy way to estimate marginal likelihoods; they are too expensive for sequential scenarios; and calibrating their tuning parameters may be cumbersome (recall the numerical experiments of Sect. 16.5.2).

In Chap. 17, we discussed SMC samplers, which, among other benefits, address these shortcomings in situations where the likelihood may be computed exactly. The next logical step is to develop SMC samplers for models with intractable likelihoods.

Such SMC samplers are often called SMC^2, and are the subject of this chapter. The first part presents a basic, generic version of SMC^2 algorithms. The second part develops a more elaborate variant, which makes it possible to perform simultaneously and sequentially parameter inference, model choice, and state prediction for state-space models.

18.1 Pseudo-Marginal SMC Samplers

The generic SMC sampler presented in Sect. 17.1 of Chap. 17 approximates a sequence of distributions

$$\mathbb{P}_t(\mathrm{d}\theta) = \frac{\gamma_t(\theta)}{L_t}\nu(\mathrm{d}\theta)$$

such that (a) $\nu(\mathrm{d}\theta)$ is easy to sample from; (b) function $\theta \to \gamma_t(\theta)/\gamma_{t-1}(\theta)$ (or $\gamma_0(\theta)$ for $t = 0$) may be computed pointwise; and (c) one is able to construct kernels M_t that leave invariant $\mathbb{P}_{t-1}(\mathrm{d}\theta)$.

We are now interested in situations where requirement (b) is not met, because the ratio $\gamma_t(\theta)/\gamma_{t-1}(\theta)$ is intractable for some reason. For instance, in the context of

© Springer Nature Switzerland AG 2020

N. Chopin, O. Papaspiliopoulos, *An Introduction to Sequential Monte Carlo*, Springer Series in Statistics, https://doi.org/10.1007/978-3-030-47845-2_18

sequential inference for state-space models, $\gamma_t(\theta)/\gamma_{t-1}(\theta) = p_t^\theta(y_t|y_{0:t-1})$, which is typically not computable.

To deal with this, we adapt the pseudo-marginal approach of Sect. 16.2 as follows. We assume we are able to construct an auxiliary variable Z, $Z \sim \mathbb{M}^\theta(\mathrm{d}z)$ (given $\Theta = \theta$), and joint (unnormalised, tractable) densities $\gamma_t(\theta, z)$, such that

$$\int_{\mathcal{Z}} \gamma_t(\theta, z)\mathbb{M}^\theta(\mathrm{d}z) = \gamma_t(\theta)$$

for all $t \geq 0$. In other words, $\gamma_t(\theta, Z)$ is an unbiased estimate of $\gamma_t(\theta)$, for any $\theta \in \Theta$.

We then define $\mathbb{P}_t(\mathrm{d}\theta, \mathrm{d}z)$ as:

$$\mathbb{P}_t(\mathrm{d}\theta, \mathrm{d}z) = \frac{\gamma_t(\theta, z)}{L_t}\nu(\mathrm{d}\theta)\mathbb{M}^\theta(\mathrm{d}z) ;$$

note that this distribution admits $\mathbb{P}_t(\mathrm{d}\theta)$ as a marginal distribution for variable Θ. Then we implement an SMC sampler that targets the sequence $\{\mathbb{P}_t(\mathrm{d}\theta, \mathrm{d}z)\}_{t\geq 0}$; and as a result, we are able to approximate the distributions $\mathbb{P}_t(\mathrm{d}\theta)$, since they are simply marginal distributions of the targets.

To check that such a pseudo-marginal SMC sampler is indeed implementable, we return to the requirements above. Requirements (a) and (b) are automatically met as soon as the auxiliary variable Z is chosen so that $\mathbb{M}^\theta(\mathrm{d}z)$ is easy to sample from, and the joint densities $\gamma_t(\theta, z)$ (or least the ratios $\gamma_t(\theta, z)/\gamma_{t-1}(\theta, z)$) are tractable. Requirement (c) implies that we are able to construct a Markov kernel that leaves invariant distribution $\mathbb{P}_t(\mathrm{d}\theta, \mathrm{d}z)$. For this, we simply use a pseudo-marginal MCMC kernel, e.g. a variant of Algorithm 16.1.

The basic framework we have just described is sufficient to build practical algorithms, such as e.g. a pseudo-marginal version of SMC tempering; see 18.1 for more details. Its main limitation is that it assumes that the auxiliary variable Z stays the same across iterations. The algorithm we develop in the following section, although based on the same ideas, will also feature auxiliary variables whose nature and distribution must evolve over time.

18.2 SMC^2 for Sequential Inference in State-Space Models

18.2.1 General Structure

As announced in the introduction, we now focus on the following problem: sequential inference for a given state-space model. Ideally, we would like to implement the IBIS algorithm, where $\gamma_t(\theta) = p_t^\theta(y_{0:t})$. However, we already know that the latter quantity is typically intractable.

Instead, we consider a pseudo-marginal version of IBIS, where the likelihood $p_t^\theta(y_{0:t})$ is estimated unbiasedly by a particle filter run until time t. That is, the algorithm carries forward N_θ parameter values, $\Theta_t^1, \ldots, \Theta_t^{N_\theta}$, and N_θ associated particle filters, which provide unbiased estimates of $p_t^\theta(y_{0:t})$, for $\theta = \Theta_t^m$, $m = 1, \ldots, N_\theta$. We call this algorithm SMC^2, for obvious reasons.

To fix ideas, we consider the same settings as in Sect. 16.4.2: the particle filters are guided filters, of size N_x, which generate up to time t variables $Z_t^m = (X_{0:t}^{m,1:N_x}, A_{1:t}^{m,1:N_x})$, and the corresponding likelihood estimate, of the form (adapted from (16.11)):

$$L_t^{N_x}(\Theta_t^m, X_{0:t}^{m,1:N_x}, A_{1:t}^{m,1:N_x}) = \left(\frac{1}{N}\sum_{n=1}^{N_x} G_0^{\Theta_t^m}(X_0^{m,n})\right)\prod_{s=1}^{t}\left(\frac{1}{N}\sum_{n=1}^{N_x} G_s^{\Theta_t^m}(X_{t-1}^{m,A_t^n}, X_t^{m,n})\right).$$

We have seen in Sect. 16.4 that PMCMC algorithms leave invariant the following extended distribution (adapting slightly the notations):

$$\pi_t(\mathrm{d}\theta, \mathrm{d}x_{0:t}^{1:N_x}, a_{1:t}^{1:N_x}) = \frac{1}{p_t(y_{0:t})}\nu(\mathrm{d}\theta)\psi_t^\theta(\mathrm{d}x_{0:t}^{1:N_x}, a_{1:t}^{1:N_x})L_t^{N_x}(\theta, x_{0:t}^{1:N_x}, a_{1:t}^{1:N_x}) \tag{18.1}$$

where $\psi_t^\theta(\cdot)$ is the distribution of the random variables generated by a particle filter associated with parameter θ. This distribution admits the true posterior $\mathbb{P}_t(\mathrm{d}\theta \mid y_{0:t})$ as a marginal.

Notation/Terminology We follow the same convention as in Chap. 16: when writing joint distributions, we simply write e.g. $a_{1:t}^{1:N_x}$, instead of $\mathrm{d}a_{1:t}^{1:N_x}$ for discrete random variables. This avoids writing explicitly the dominating measure for these variables, which would simply be the counting measure.

It seems natural to define our SMC^2 algorithm as an SMC sampler that targets the sequence of π_t's. One technical difficulty is that the successive distributions π_t do not have the same support. In order to perform an importance sampling step from π_{t-1} to π_t, we must first extend the space, by simulating $(X_t^{1:N}, A_t^{1:N})$. This boils down to moving one step forward the associated particle filter. Formally, let $\psi_t^\theta(\mathrm{d}x_t^{1:N_x}, a_t^{1:N_x}|x_{0:t-1}^{1:N_x}, a_{1:t-1}^{1:N_x})$ denote the distribution of the variables generated at time t conditional on the previous steps; i.e.

$$\psi_t^\theta\left(\mathrm{d}x_t^{1:N_x}, a_t^{1:N_x} \,\middle|\, x_{0:t-1}^{1:N_x}, a_{1:t-1}^{1:N_x}\right) = \prod_{n=1}^{N_x} W_{t-1}^{a_t^n} M_t^\theta(x_{t-1}^{a_t^n}, \mathrm{d}x_t^n).$$

Then define the importance sampling weight function as follows:

$$\frac{\pi_t(\mathrm{d}\theta, \mathrm{d}x_{0:t}^{1:N_x}, a_{1:t}^{1:N_x})}{\pi_{t-1}(\mathrm{d}\theta, \mathrm{d}x_{0:t-1}^{1:N_x}, a_{1:t-1}^{1:N_x})\psi_t^\theta(\mathrm{d}x_t^{1:N_x}, a_t^{1:N_x}|x_{0:t-1}^{1:N_x}, a_{1:t-1}^{1:N_x})}$$

$$\propto \left\{\frac{1}{N}\sum_{n=1}^{N_x} G_t^\theta(x_{t-1}^{a_t^n}, x_t^n)\right\} \tag{18.2}$$

(where the normalising constant is $p_t(y_t|y_{0:t-1})$.)

All the ingredients are now in place to properly define our SMC² sampler:

- We initialise the algorithm by sampling N_θ particles from the prior $\nu(\mathrm{d}\theta)$. For each particle Θ_0^m, we perform iteration 0 of a particle filter (generating variables $X_0^{m,1:N_x}$) in order to obtain an unbiased estimate of $p_0^\theta(y_0)$ for $\theta = \Theta_0^1, \ldots, \Theta_0^{N_\theta}$.
- Before reweighting the particles at time t, we move forward one step (i.e. perform iteration t of) the N_θ particle filters. Then we reweight the particles according to (18.2).
- At the beginning of any iteration t where the ESS gets too low, we resample the particles, and move them according to a kernel K_t that leaves invariant the current target distribution, π_t; this kernel must be a PMCMC kernel (e.g. PMMH, or Particle Gibbs).

Algorithm 18.1 summarises this generic algorithm.

Input of SMC²

- Number of particles N_x and N_θ.
- A parametric family of state-space models, with parameter $\theta \in \Theta$.
- A prior distribution $\nu(\mathrm{d}\theta)$ for parameter Θ, from which one may sample from.
- A (θ-indexed) class of particle filters, which, for a given θ and when run until time t, provide an unbiased estimate of the likelihood $p_t^\theta(y_{0:t})$ of the considered model. These particle filters are associated with Feynman-Kac models such that particles X_t^n are simulated according to kernel M_t^θ (or distribution $\mathbb{M}_0^\theta$ at time 0), and reweighed according to function G_t^θ. Therefore, the variables generated at time t, i.e. $(X_t^{1:N_x}, A_t^{1:N_x})$, are sampled from:

$$\psi_t^\theta\left(\mathrm{d}x_t^{1:N_x}, a_t^{1:N_x}\,\middle|\, x_{0:t-1}^{1:N_x}, a_{1:t-1}^{1:N_x}\right) = \prod_{n=1}^{N_x} W_{t-1}^{a_t^n} M_t^\theta(x_{t-1}^{a_t^n}, \mathrm{d}x_t^n)$$

where $W_{t-1}^n = G_{t-1}^\theta(x_{t-2}^{a_{t-1}^n}, x_{t-1}^n)/\sum_{m=1}^N G_{t-1}^\theta(x_{t-2}^{a_{t-1}^m}, x_{t-1}^m)$.

(continued)

- A sequence of PMCMC kernels K_t that leaves invariant distribution (18.1).
- The usual inputs of an SMC algorithm: choice of a resampling scheme, threshold $\mathrm{ESS}_{\min}$.

Algorithm 18.1: SMC^2 (sequential inference for state-space models)

```
Operations referring to m must be performed
  for m = 1, ..., N_θ.
Θ_0^m ~ ν(dθ)
▷ Perform Step 0 for each of the N_θ particle
  filters:
X_0^{m,1:N_x} ~ ψ_0^{Θ_0^m}(dx_0^{1:N_x})
w_0^m ← N^{-1} Σ_{n=1}^{N_x} G_0^{Θ_0^m}(X_0^{m,n})
W_0^m ← w_0^m / Σ_{m=1}^{N_θ} w_0^m
for t = 1 to T do
    if ESS(W_{t-1}^{1:N_θ}) < ESS_min then
        ▷ Move the particles through PMCMC kernel
          K_t:
        (Θ_t^m, X̃_{0:t-1}^{m,1:N_x}, Ã_{1:t-1}^{m,1:N_x}) ~
          K_t(((Θ_{t-1}^m, X_{0:t-1}^{m,1:N_x}, A_{1:t-1}^{m,1:N_x}), d(θ, x_{0:t-1}^{1:N_x}, a_{1:t-1}^{1:N_x})))
        (X_{0:t-1}^{1:N_x}, A_{1:t-1}^{1:N_x}) ← (X̃_{0:t-1}^{1:N_x}, Ã_{1:t-1}^{1:N_x})
        w_{t-1}^m ← 1
    else
        Θ_t^m ← Θ_{t-1}^m
    ▷ Perform step t for each of the N_θ particle
      filters:
    (X_t^{m,1:N_x}, A_t^{m,1:N}) ~ ψ_t^{Θ_t^m}(dx_t^{1:N_x}, a_t^{1:N_x} | x_{0:t-1}^{1:N_x}, a_{1:t-1}^{1:N_x})
    w_t^m ← w_{t-1}^m {N^{-1} Σ_{n=1}^{N_x} G_t^{Θ_t^m}(X_{t-1}^{m,A_t^{m,n}}, X_t^{m,n})}
    W_t^m ← w_t^m / Σ_{l=1}^{N_θ} w_t^l
```

18.2.2 Choice of the PMCMC Kernel

We have seen that we may use any type of PMCMC kernel (that is any kernel that leaves invariant (18.1)) to rejuvenate the particles in Algorithm 18.1. In practice, we recommend the following approach by default: a PMMH kernel (Algorithm 16.4) based on a random walk proposal. This has the following advantages. First, we may use the same tricks as for basic SMC samplers to adapt the random walk proposal to the current set of particles (Sect. 17.2.1). Second, it makes it possible to reduce the overall memory cost of SMC^2.

To see this, consider the general description of SMC^2 in the previous section: particles are of the form $(\Theta_t^m, X_{0:t}^{m,1:N_x}, A_{1:t}^{m,1:N_x})$. The memory size required to store those N_θ particles at time t is $\mathcal{O}(tN_x N_\theta)$, and thus may get prohibitively large at some point.

However, with a PMMH kernel, there is no need to store the complete history of the N_θ particle filters: a quick inspection of the algorithm reveals that it is sufficient to store (a) the variables generated at the latest iteration (variables $(X_t^{m,1:N_x}, A_t^{m,1:N_x})$ at time t) and (b) the current likelihood estimates. This reduces the memory cost to $\mathcal{O}(N_\theta N_x)$, which stays constant over time, and is typically reasonable.

Of course, our default recommendation is not necessarily optimal. In certain cases, one may obtain better performance by moving the particles through a Particle Gibbs kernel for instance, notwithstanding the aforementioned memory issue. (However, see Remark 12.1 and the end of the Python corner of Chap. 12 for storing more efficiently the whole genealogy of a particle filter.)

18.2.3 Choice of N_x

Regarding N_x, we expect the same type of trade-off as for PMMH (Sect. 16.5): increasing N_x increases the CPU cost, but it may also improve performance (i.e., reduce the variance of the importance sampling steps, and improve the mixing of the mutation steps). A difficulty specific to SMC^2 is that we have to consider performance at different time steps.

For simplicity, we assume from now on that PMMH kernels are used to move the particles. Recall that, for PMMH, we recommended in Sect. 16.5 to set N_x so that the variance of likelihood estimates is $\ll 1$. And we saw in Sect. 14.2.3 that this variance behaves like $(1 + C/N_x)^t$, for estimates of the likelihood up to time t. This suggests to take $N_x = \mathcal{O}(t)$ to ensure that, if implemented at time t, the chosen PMMH kernel mixes sufficiently well. This leads us to consider increasing N_x in the course of an SMC^2 algorithm. The questions are then: when and how?

Regarding when, a basic recipe is to monitor the acceptance rate of PMMH kernels. When it gets too low (say, below 10%), then we double N_x. This recipe was proposed in the original SMC^2 paper (Chopin et al. 2013). It works well in simple cases, but it is not entirely satisfactory, in particular because it increases N_x "too late"; i.e., after we applied several steps of PMMH, rather than before. Further research is required to elicit a more robust criterion to decide when to increase N_x; see Chopin et al. (2015) for some work in that direction.

Regarding how, two options exist. To fix ideas, we denote by N_x the current number of x-particles, and $\tilde{N}_x$ the value we wish to change to; e.g. $\tilde{N}_x = 2N_x$ in the above recipe. A first option is to generate, for each $m = 1, \ldots, N_\theta$, a new particle filter, of size $\tilde{N}_x$. Then, we may exchange the 'old' particle filters (of size N_x) with the new ones, by doing an importance sampling step, with weights equal to the new likelihood estimates divided by the old ones. The corresponding weight function is:

$$\frac{L_t^{\tilde{N}_x}(\theta, \tilde{x}_{0:t}^{1:\tilde{N}_x}, \tilde{a}_{1:t}^{1:\tilde{N}_x})}{L_t^{N_x}(\theta, x_{0:t}^{1:N_x}, a_{1:t}^{1:N_x})}$$

where tildes are used to distinguished the variables generated by the new particle filter. (See Exercise 18.2.)

A second option is to recognise that the CSMC kernel (Algorithm 16.5), which rejuvenates the particle system conditional on a selected trajectory, may do so for any value of N_x. Hence we may change N_x whenever we apply this kernel. In fact, we could use a CSMC kernel for this sole purpose while still moving the particles through PMMH kernels. To apply the CSMC kernel, we must first sample a star trajectory, using Algorithm 16.6. Thus, we can no longer apply the simple trick for reducing memory usage mentioned in the previous section, i.e., keeping only variables $X_t^{m,1:N_x}$, for each m, rather than the complete system $\left(X_{0:t}^{m,1:N_x}, A_{1:t}^{m,1:N_x}\right)$. This is the main drawback of this approach (although, again, see Remark 12.1 on this point). The main advantage of the second approach is that it does not affect the importance weights.

18.3 Numerical Experiments

The main aim of the numerical experiment we carry out here is to illustrate the long-term behaviour of SMC^2. We consider a univariate stochastic volatility model with leverage effect; that is

$$X_t = \mu + \rho(X_{t-1} - \mu) + \sigma U_t, \qquad U_t \sim \mathcal{N}(0,1) \tag{18.3}$$

$$Y_t = \exp(X_t/2) V_t, \qquad V_t \sim \mathcal{N}(0,1) \tag{18.4}$$

and $\mathrm{Corr}(U_t, V_t) = \phi$, $\theta = (\mu, \rho, \sigma, \phi)$; parameter ϕ captures the leverage effect. To put this model into state-space form, note that $V_t | U_t = u_t \sim \mathcal{N}(\rho u_t, 1 - \rho^2)$, hence the distribution of Y_t given $X_t = x_t$ and $X_{t-1} = x_{t-1}$

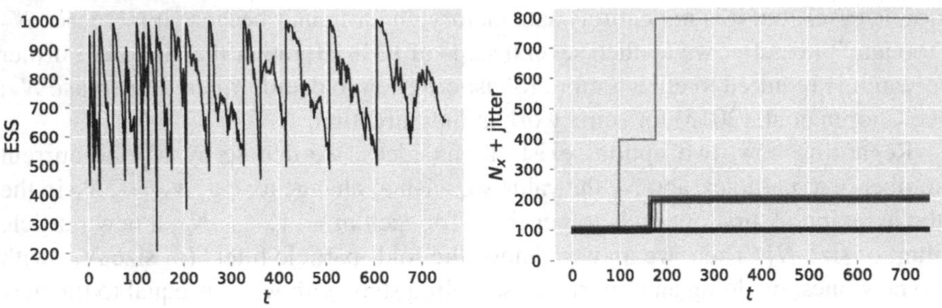

Fig. 18.1 Behaviour of SMC^2 (non-QMC version) when applied to the stochastic volatility model with leverage described in the text. Left: ESS versus time (single run); Right: N_x vs time for 25 runs. Jitter is added to aid visualisation

is $\mathcal{N}\left(e^{x_t/2}\rho u_t, e^{x_t}(1-\rho^2)\right)$, with $u_t = \{x_t - \mu - \rho(x_{t-1} - \mu)\}/\sigma$; see also Exercise 2.4.

We consider the same data as in Sect. 10.5.2, and use the following prior: $\mu \sim \mathcal{N}(0, 2^2)$, $\sigma^2 \sim \mathcal{G}(2, 2)$, $\rho \sim \mathcal{B}eta(9, 1)$, and $\phi \sim \mathcal{U}[-1, 1]$ (independently).

We consider two versions of SMC^2. In the first version, a standard bootstrap filter is used to estimate the likelihood for a given θ. In the second version, the SQMC version of a bootstrap filter is used. In both cases, we take $N_\theta = 10^3$, and initialise N_x to 100. The implementation follows the guidelines of Sect. 18.2.2. Resampling occurs at times t such that the ESS drops below $N_\theta/2$. When this happens, particles are moved through five random walk PMCMC steps, calibrated using the recipe discussed in Sect. 17.2.1 (i.e., the covariance of the random walk proposal is set to a fraction of the empirical covariance matrix of the weighted particle sample). The value of N_x is doubled using the exchange step each time the acceptance rate of the random walk steps falls below 10%. Each version of the algorithm is run 25 times.

The left panel of Fig. 18.1 plots the ESS versus t in a typical run of the non-QMC version. We observe the same behaviour as for the IBIS algorithm: resampling steps occur at a decreasing frequency. The right panel shows how N_x evolved in the course of the algorithm for the 25 runs of the non-QMC version: N_x was increased to 200 before time 200 for about half of the runs, and there is one run where N_x reaches 800. In contrast, all but one of the runs of the SQMC version kept N_x fixed to 100 (results not shown). This is of course due to the fact that SQMC offers lower-variance estimate for the likelihood.

Figure 18.2 compares the variability of the estimates of the marginal likelihood, as a function of time, for both versions of the algorithm. The SQMC version produces lower-variance estimates, although the variance reduction is not as significant as for basic filtering. In effect, the QMC variant of SMC^2 is a hybrid algorithm, that mixes Monte Carlo (for the θ-particles) and QMC (for the x-particles). Hence the Monte Carlo effect tends to dominate. We now concentrate on the results obtained from the SQMC version.

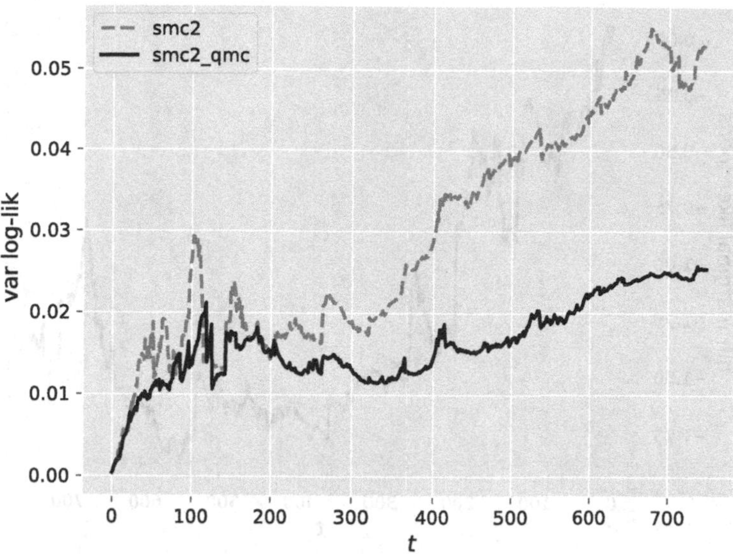

Fig. 18.2 Empirical variance (across 25 runs) of the estimate of the log of the marginal likelihood, for the two versions of SMC^2 discussed in the text (i.e., based on a standard or QMC version of the bootstrap filter)

The variances reported in Fig. 18.2 are remarkably low, which means we are able to compute marginal likelihoods $p_t(y_{0:t})$ at all times quite accurately. This makes it possible to perform sequential model choice. For instance, Fig. 18.3 plots the log-ratio of marginal likelihoods for comparing the considered model against a more basic model, with no leverage: $\phi = 0$. (To compute the marginal likelihoods of the basic model, we use SMC^2 with exactly the same settings as for the considered model.) We see that, the more data we have, the more evidence we have against a leverage effect.

This absence of a leverage effect is corroborated by Fig. 18.4, which represents the time evolution of the posterior distribution of each parameter: at all times value 0 seems well within the support of the marginal posterior distribution of ϕ. We also observe that parameter ρ is hardly identifiable (despite the informative prior); this is (in our experience) a reasonably common occurrence in stochastic volatility modelling.

Finally, to assess the variability over the 25 runs, we plot in Fig. 18.5 the 25 empirical CDFs that may be computed from the weighted particle sample at the final time; these CDFs approximate the true posterior CDF of each parameter. Except for parameter ρ, the variability is rather small, and not much larger than what one would obtain from an i.i.d. sample of the same size (see Exercise 18.3 for more comments on this point).

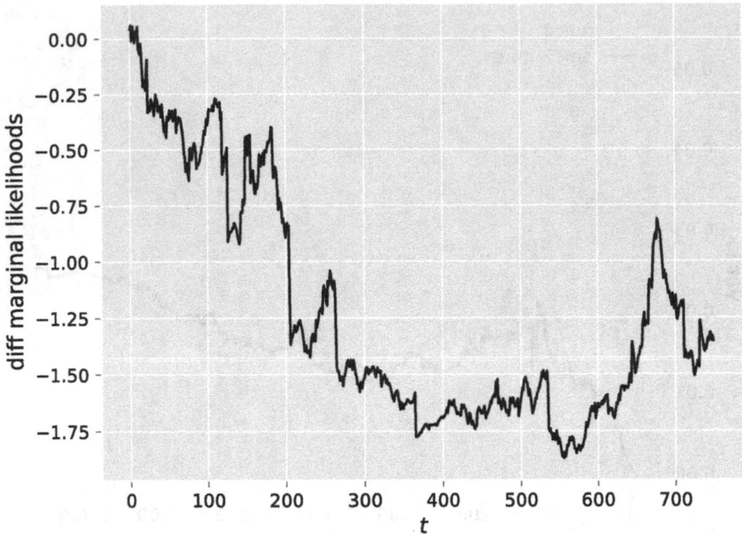

Fig. 18.3 Sequential model choice: log-ratio of marginal likelihoods versus time, for comparing the considered model (stochastic volatility with leverage) against the basic stochastic volatility model, in which $\phi = 0$. The latter seems preferred at all times

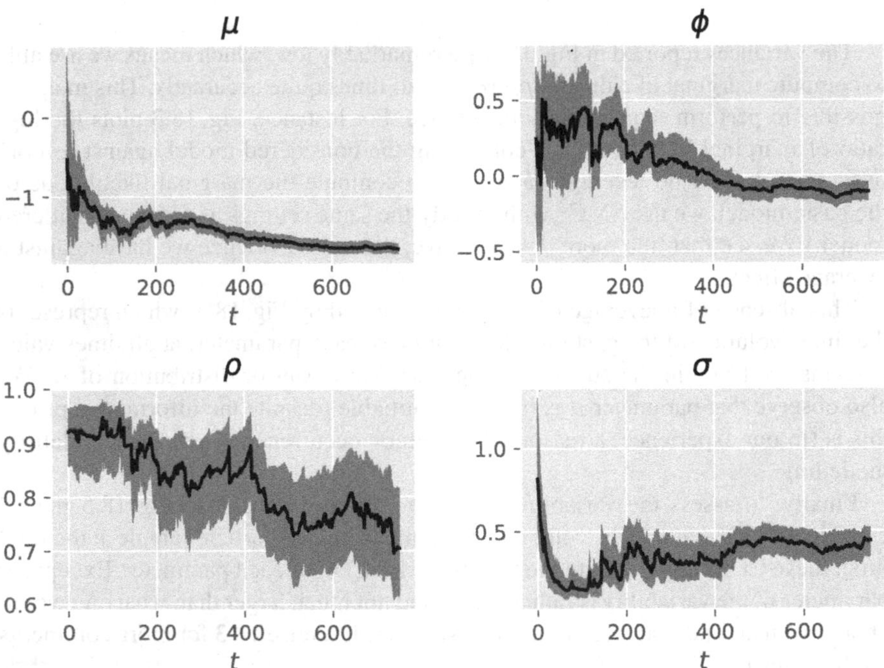

Fig. 18.4 Sequential inference: median (black line) and inter-quartile range (grey area) of the posterior distribution of each parameter at each time t. Estimated from a single run of SMC^2

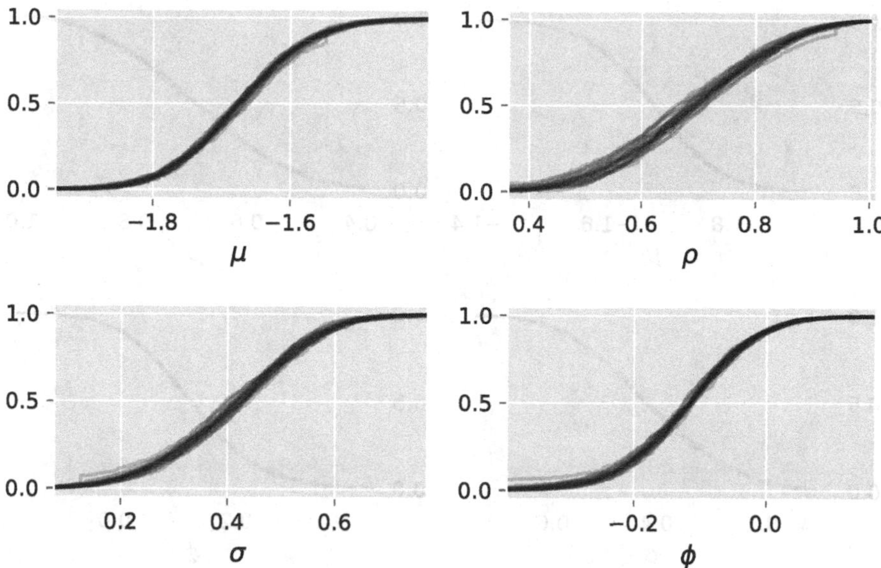

Fig. 18.5 Posterior CDFs for each parameter, as estimated by 25 runs of the (SQMC version of) SMC^2 for $N = 10^3$. To aid visualisation, CDFs are plotted using alpha blending (so as to make them appear slightly transparent)

To reduce this variability, one may of course increase N_θ; see Fig. 18.6, which corresponds to 25 runs of the algorithm for $N_\theta = 10^4$ (while all the other settings are left unchanged). This time the empirical CDFs are indistinguishable.

However running a single instance of SMC^2 for $N_\theta = 10^4$ took about 20 hours (compared to 2 hours for $N_\theta = 10^3$), which is of course rather expensive. We obtain the same level of variance reduction by simply averaging out the 25 runs of SMC^2 for $N_\theta = 10^3$, at a much lower cost in case one has access to a multi-core machine (as discussed in the Python corner of Chap. 11).

Exercises

18.1 *Show that the basic framework for pseudo-marginal SMC samplers laid out in Sect. 18.1 is sufficient to construct an SMC^2 tempering algorithm for parameter estimation in state-space models, which would work along the following lines: sample θ-particles from the prior $\nu(d\theta)$; for each particle generate a particle filter until the final time T and reweight according to a fraction of the likelihood estimate; that is, at time t, the target distribution should be of the form (in relation to* (18.1)*):*

$$\pi_t(d\theta, dx_{0:T}^{1:N_x}, a_{1:T}^{1:N_x}) \propto \nu(d\theta)\psi_T(dx_{0:T}^{1:N_x}, a_{1:T}^{1:N_x})\{L_T^{N_x}(\theta, x_{0:T}^{1:N_x}, a_{1:T}^{1:N_x})\}^{\lambda_t}$$

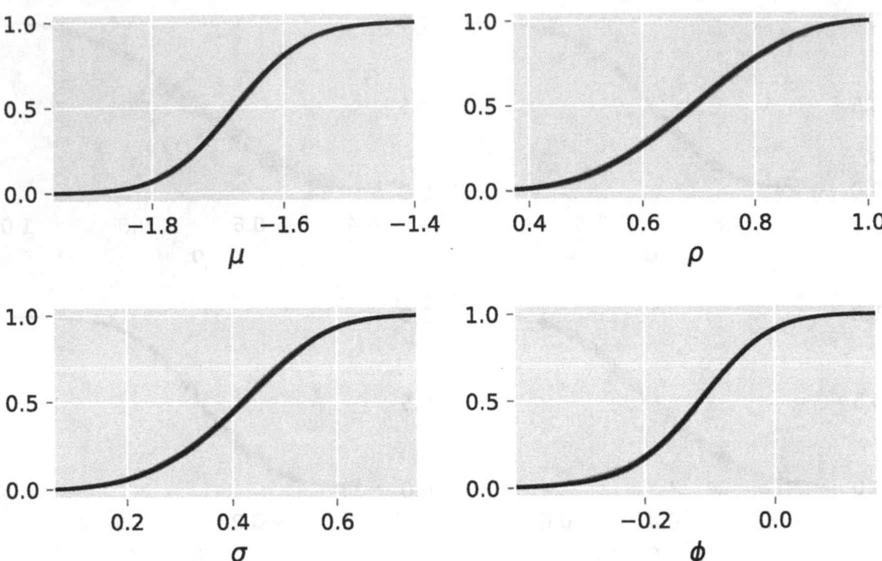

Fig. 18.6 Same plot as Fig. 18.5, for $N = 10^4$. The 25 runs are now nearly indistinguishable

for some sequence $0 = \lambda_0 < \ldots < \lambda_T = 1$. *Compute the marginal distribution of variable* θ *(with respect to distribution* π_t*). Comment; in particular discuss whether this algorithm is a pseudo-marginal variant of an ideal tempering algorithm which would target at time t the tempered distribution:*

$$\propto \nu(\mathrm{d}\theta) \left\{ p_T^\theta(y_{0:T}) \right\}^{\lambda_t} .$$

Discuss the validity of the algorithm (in particular does the final output remain valid in some sense, despite the previous remark). What type of PMCMC kernels one may or may not use in this kind of SMC^2 algorithm?

18.2 *Give the proposal distribution and the target distribution of the importance sampling step described as the "exchange" step in Sect. 18.2.3.*

18.3 *In relation to the discussion around Fig. 18.5, this exercise discusses how to assess the variability of empirical CDFs obtained from multiple runs. A first approach is to plot empirical CDFs obtained from k i.i.d. samples, of size N and simulated from an arbitrary distribution (say a Gaussian fitted to the particle sample). Implement this first approach for* $N = 10^3$ *and compare the result to Fig. 18.5. Discuss. A second approach is to remark that the average of N i.i.d. Bernoulli variables with mean p have variance* $p(1-p)/N$. *How can we use this remark to assess visually the variability of the empirical CDFs of Fig. 18.5, relative to that of i.i.d. samples?*

Python Corner

It should hardly be a surprise at this stage: SMC^2 is implemented in `particles` as a particular Feynman-Kac model. Going back to our recurring stochastic volatility example (see the previous Python corners):

```
from particles import smc_samplers as ssp
from particles import state_space_models as ssm

_, data = ssm.StochVol().simulate(100)  # simulate data (default parameters)
fk_smc2 = ssp.SMC2(ssm_cls=ssm.StochVol, data=data, prior=my_prior,
                   init_Nx=50, ar_to_increase_Nx=0.1)
alg_smc2 = particles.SMC(fk=fk_smc2, N=500)
alg_smc2.run()
```

The argument `N` when instantiating SMC is actually N_θ, the number of θ-particles. The number of x-particles, N_x, is initialized to 50 (argument `init_Nx`); then N_x is doubled every time that the average acceptance rate of the PMMH kernel used to rejuvenate the particles goes below 10% (argument `ar_to_increase_Nx`). By default, a random walk PMMH kernel is used.

By construction, SMC^2 generates a large number of particle filters (hence a large number of `SMC` objects). The funny thing about creating many instances of a given class is that any small inefficiency tends to add up and blow in your face. In particular, before deciding to implement probability distributions as custom objects (in module `distributions`), we used the "frozen" distributions implemented in `SciPy` (module `stats`) to specify state-space models. The corresponding implementation of SMC^2 was particularly slow. When profiling the code, we discovered that a large part of the CPU time was devoted ... to text operations. It turns out that, each time one creates a frozen distribution in `SciPy`, some text operation is performed to generate the documentation. And since a single run of SMC^2 required to create $\mathcal{O}(T N_\theta)$ frozen distributions, these text operations created a bottleneck. This is one of our motivations for creating our own random probability objects.

In the same spirit, `SMC` initially stored a few basic summaries across time, such as the ESS, the estimate of the normalising constant, and so on. The total memory cost of these summaries is negligible for a single instance, but when running SMC^2 for $N_\theta = 10^4$, it adds up to several Gigabytes. Now class `SMC` has an option to avoid saving any kind of summaries.

Bibliographical Notes

This chapter is based partly on Chopin et al. (2013), who introduced the SMC^2 algorithm discussed in Sect. 18.2; see also Fulop and Li (2013) who derived a similar algorithm, Duan and Fulop (2015) for an SMC^2 tempering algorithm (discussed in

Exercise 18.1), and Chopin et al. (2015) for how to make SMC^2 adaptive in a more robust way.

Since its introduction, SMC^2 has been applied to various problems, such as inference on low-count time-series models (Drovandi and McCutchan 2016), inference on stochastic kinetic models (Golightly and Kypraios 2018), high-dimensional filtering (Naesseth et al. 2019), historical climate reconstruction (Carson et al. 2018), and likelihood-free estimation (Kerama et al. 2020).

Bibliography

Carson, J., Crucifix, M., Preston, S., & Wilkinson, R. D. (2018). Bayesian model selection for the glacial-interglacial cycle. *The Journal of the Royal Statistical Society, Series C (Applied Statistics), 67*(1), 25–54.

Chopin, N., Jacob, P. E., & Papaspiliopoulos, O. (2013). SMC^2: An efficient algorithm for sequential analysis of state space models. *Journal of the Royal Statistical Society: Series B (Statistical Methodology), 75*(3), 397–426.

Chopin, N., Ridgway, J., Gerber, M., & Papaspiliopoulos, O. (2015). Towards automatic calibration of the number of state particles within the SMC^2 algorithm. *ArXiv e-print 1506.00570.*

Drovandi, C. C., & McCutchan, R. A. (2016). Alive SMC^2: Bayesian model selection for low-count time series models with intractable likelihoods. *Biometrics, 72*(2), 344–353.

Duan, J.-C., & Fulop, A. (2015). Density-tempered marginalized sequential Monte Carlo samplers. *Journal of Business & Economic Statistics, 33*(2), 192–202.

Fulop, A., & Li, J. (2013). Efficient learning via simulation: A marginalized resample-move approach. *Journal of Econometrics, 176*(2), 146–161.

Golightly, A., & Kypraios, T. (2018). Efficient SMC^2 schemes for stochastic kinetic models. *Statistics and Computing, 28*(6), 1215–1230.

Kerama, I., Thorne, T., & Everitt, R. G. (2020). Rare event ABC-SMC^2. *arxiv.*

Naesseth, C. A., Lindsten, F., & Schön, T. B. (2019). High-dimensional filtering using nested sequential Monte Carlo. *IEEE Transactions on Signal Processing, 67*(16), 4177–4188.

Chapter 19
Advanced Topics and Open Problems

Summary In this final chapter, we discuss a few advanced topics that were not covered in previous chapters, and some open problems that will hopefully be addressed in the near future.

19.1 SMC in High Dimension

We have already mentioned (Chap. 10) that the bootstrap filter performs poorly when the data-points are very informative. In that case, the observation density $f_t(y_t|x_t)$ becomes very peaky, and most particles get a negligible weight. This issue arises particularly when the dimension of the observations Y_t is large. One extreme example is weather forecasting, where the dimension may exceed 10^7. Particle filtering is obviously not a viable option for this type of application. In fact, even Kalman filtering is too expensive for such problems, and approximations (such as the ensemble Kalman filter) must be used instead; see Reich and Cotter (2015) for more background on this topic, which we also touched upon in Chap. 7. A less extreme example is multivariate stochastic volatility, for which the bootstrap filter becomes unreliable pretty quickly when the dimension increases (say beyond 10).

Thus, for high-dimensional state-space models, it is critical to design guided particle filters, or more precisely, to construct proposal distributions that alleviate weight degeneracy. Unfortunately, there is currently no general recipe to construct such good proposals. We discussed in Sect. 10.3.2 recipes based on Taylor expansions, but, they go only so far, as evidenced by the numerical experiments of Sects. 10.5.2 and 10.5.3.

A few recent papers made interesting contributions to this problem. Guarniero et al. (2017) and Heng et al. (2020) proposed iterative schemes that refine progressively the proposal distributions (within a parametric family) so as to reduce the

© Springer Nature Switzerland AG 2020
N. Chopin, O. Papaspiliopoulos, *An Introduction to Sequential Monte Carlo*, Springer Series in Statistics, https://doi.org/10.1007/978-3-030-47845-2_19

variance of the particle estimates. These approaches work off-line (for a fixed dataset). They may be used for instance within PMCMC (Chap. 16); that is, at each MCMC iteration, perform several steps of these iterative schemes so as to construct a particle filter that generates a likelihood estimate with a low variance.

Another interesting approach is the nested SMC algorithm of Naesseth et al. (2015); this is an SMC^2-like approach, which uses an inner SMC algorithm to build an approximation to the optimal proposal at time t. This type of approach is of course expensive, but for certain complicated problems, it may still represent a much better CPU vs variance trade-off than the bootstrap filter.

Still, we believe that more research on building good proposal distributions is warranted, given how crucial is this particular problem for the performance (and thus adoption) of particles in high-dimensional or otherwise complicated models.

19.2 Parallel Computation

We have already discussed parallel computation in the Python corner of Chap. 10: CPU cores do not get faster nowadays, but their number in modern computers keeps increasing. Thus there is a growing interest in "parallelising" algorithms, that is, to find ways to implement algorithms on k cores in such a way that the total running time is divided by k.

At first sight, SMC algorithms are easy to parallelise (especially relative to, say, MCMC). The only operation in an SMC algorithm that requires simultaneous access to the N particles is the computation of the sum of the weights in the resampling step. All the other operations treat the N particles independently and are thus "embarrassingly parallel". Murray et al. (2016) proposed resampling schemes that avoid computing the overall sum of the weights; one scheme amounts to run a Metropolis sampler in order to sample the ancestor variables. Another approach is to try to reduce as much as possible the interaction between the particle "islands" (the collections of particles run on independent cores); see Vergé et al. (2015) and Guldas et al. (2017).

One challenging scenario for parallel SMC is when the CPU time of the operations performed for each particle is random and varies greatly. A standard SMC algorithm has to wait until the N particles are available at time t, before progressing to time $t + 1$. Paige et al. (2014) proposed an asynchronous SMC algorithm where particles evolve "at their own pace", without waiting for the other particles.

At any rate, we believe that the advent of parallel computation will help greatly the adoption of SMC samplers, as an alternative to MCMC. As we explained in Chap. 17, the bulk of the computation for an SMC sampler is typically the computation of the N weights, which is particularly simple to parallelise (and may even be done automatically by certain libraries).

19.3 Variance Estimation, Particle Genealogy

It is common practice to run an SMC algorithm many times to assess the variability of its output. This is precisely what we did so in most of our numerical experiments. This basic approach is obviously a bit cumbersome.

It is actually possible to estimate the variance of particle estimates from a *single* run of an SMC algorithm, as proposed by Chan and Lai (2013), Lee and Whiteley (2018) and Du and Guyader (2019). Consider the un-weighted (for simplicity) particle estimate $\hat{\varphi}_N := N^{-1} \sum_{n=1}^{N} \varphi(X_t^n)$ at time t; its asymptotic variance may be estimated by:

$$\frac{1}{N} \sum_{n=1}^{N} \left[\sum_{m:B_0^m = n} \{\varphi(X_t^m) - \hat{\varphi}_N\} \right]^2 \tag{19.1}$$

where the inner sum equals zero if the indexing set is empty, and B_0^n stands for the index of the ancestor of X_t^n at time 0, along the notation we introduced in Chap. 12; you might find beneficial to recall Fig. 12.2.

The intuition behind this his expression is that distinct trees (set of lineages with a common ancestor) behave essentially like i.i.d. variables. This suggests that this approach suffers from the path coalescence problem discussed in Sect. 12.1.2; that is, the fact that the number of distinct roots drops quickly as times progresses. In fact, the variance estimate above is zero as soon as the N trajectories have coalesced to a single ancestor; in this case the quantity above under-estimates the true variance. Olsson and Douc (2019) proposed truncating the genealogy: replace B_0^m in (19.1) by B_s^n for some $s < t$. This introduces a bias, but it does make the variance estimate much more applicable. Choosing the truncation time s however seems delicate.

This kind of method motivates the theoretical study of particle genealogies. A recent paper by Koskela et al. (2020) makes an elegant connection between these genealogies and Kingman's coalescent; in particular they show that the time to coalescence of a particle system of size N is $\mathcal{O}(N)$. This improves on the previous result of Jacob et al. (2015), who showed that the time to coalescence is $\mathcal{O}(N \log N)$.

19.4 Continuous-Time State-Space Models

We have only considered discrete time state-space models in this book. However, models involving a continuous-time process $\{X_t\}$ are common. In fact, a few of the models we have discussed in this book were originally conceived as continuous-time processes and the versions that are commonly used in applications are discretisations of these processes.

Often observations Y_t are collected at discrete times, which leads to the so-called continuous-discrete filtering problem, see Papaspiliopoulos (2011) for an overview. One straightforward way is to discretise the state process, and then apply a standard particle filter to the discretised model. The issue is then how to choose the discretisation step δ; a large δ implies a large discretisation bias, while a small δ makes the particle filter expensive, since the number of time steps is $\mathcal{O}(\delta^{-1})$. A series of interesting articles in the late 1990s and early 2000s developed a theoretical understanding of this interplay and the overall errors in the particle filter as a function of δ, see for example Del Moral and Guionnet (2001).

Fearnhead et al. (2008), building on the exact discretisation paradigm for continuous-time processes developed in Beskos et al. (2006), showed how to design *exact* particle filters that do not suffer from any discretisation bias. For the simplest illustration, consider a bootstrap filter for a model where we observe $Y_t = f(X_t) + U_t$ at integer times. Then, the weights have a simple, tractable expression, and the only practical difficulty is to sample exactly the marginal distribution of X_t given X_{t-1}, again at integer times. For a guided filter, one may have to deal with an extra difficulty: the importance weights are then intractable. In that case, it may be possible to derive unbiased estimates of such weights, and then design an exact algorithm based on random weights; see Sect. 8.8, and Fearnhead et al. (2008), Fearnhead et al. (2010).

An interesting framework that connects the exact computations for the finite state-space of Chap. 6 and those for linear-Gaussian state-space models of Chap. 7 is developed in Papaspiliopoulos and Ruggiero (2014). They show that a certain class of continuous-discrete filtering problems can be solved by a mix of these type of exact computations, yielding so-called computable filters in which the computational effort does not grow linearly but polynomially with the time index.

Particle filters can be defined for continuous-time data, and this is the so-called continuous-continuous filtering problem. Fearnhead et al. (2010) using random weight particle filtering have shown how to build particle filters for a sub-class of continuous-continuous filtering problems without even resorting to time-discretisations. For a recent work on the continuous-continuous filtering problem see Arnaudon and del Moral (2018).

For rare event simulation (such as stochastic process hitting times) and filtering problems formulated in continuous time a collection of methods that have a significant overlap with the SMC framework have been developed in the Applied Mathematics and Physics literature, known collectively as diffusion Monte Carlo. They are also based on (continuous-time formulations of) Feynman-Kac models, they involve particle mutations (typically according to the state dynamics) and resampling. For a unified presentation of this methodology and a historical overview, we refer to Hairer and Weare (2014).

Bibliography

Arnaudon, M., & del Moral, P. (2018). A duality formula and a particle Gibbs sampler for continuous time Feynman-Kac measures on path spaces. *arXiv e-prints 1805.05044*.

Beskos, A., Papaspiliopoulos, O., Roberts, G. O., & Fearnhead, P. (2006). Exact and computationally efficient likelihood-based estimation for discretely observed diffusion processes. *Journal of the Royal Statistical Society: Series B (Statistical Methodology), 68*(3), 333–382. With discussions and a reply by the authors.

Chan, H. P., & Lai, T. L. (2013). A general theory of particle filters in hidden Markov models and some applications. *Annals of Statistics, 41*(6), 2877–2904.

Del Moral, P., & Guionnet, A. (2001). On the stability of interacting processes with applications to filtering and genetic algorithms. *Annales de l'Institut Henri Poincaré, Probabilités et Statistiques, 37*(2), 155–194.

Du, Q., & Guyader, A. (2019). Variance estimation in adaptive sequential Monte Carlo. *arXiv e-prints 1909.13602*.

Fearnhead, P., Papaspiliopoulos, O., & Roberts, G. O. (2008). Particle filters for partially observed diffusions. *Journal of the Royal Statistical Society: Series B (Statistical Methodology), 70*(4), 755–777.

Fearnhead, P., Papaspiliopoulos, O., Roberts, G. O., & Stuart, A. (2010). Random-weight particle filtering of continuous time processes. *Journal of the Royal Statistical Society: Series B (Statistical Methodology), 72*(4), 497–512.

Guarniero, P., Johansen, A. M., & Lee, A. (2017). The iterated auxiliary particle filter. *Journal of the American Statistical Association, 112*(520), 1636–1647.

Guldas, H., Cemgil, T., Whiteley, N., & Heine, K. (2017). A practical introduction to butterfly and adaptive resampling in sequential monte carlo. In Y. Zhao (Ed.), *17th IFAC Symposium on System Identification SYSID 2015 – Beijing, China, 19–21 October 2015. IFAC-PapersOnLine* (Vol. 28, pp. 787–792). Amsterdam: Elsevier.

Hairer, M., & Weare, J. (2014). Improved diffusion Monte Carlo. *Communications on Pure and Applied Mathematics, 67*(12), 1995–2021.

Heng, J., Bishop, A. N., Deligiannidis, G., & Doucet, A. (2020). Controlled sequential Monte Carlo. *Annals of Statistics (to appear)*.

Jacob, P. E., Murray, L. M., & Rubenthaler, S. (2015). Path storage in the particle filter. *Statistics and Computing, 25*(2), 487–496.

Koskela, J., Jenkins, P. A., Johansen, A. M., & Spanò, D. (2020). Asymptotic genealogies of interacting particle systems with an application to sequential Monte Carlo. *Annals of Statistics, 48*(1), 560–583.

Lee, A., & Whiteley, N. (2018). Variance estimation in the particle filter. *Biometrika, 105*(3), 609–625.

Murray, L. M., Lee, A., & Jacob, P. E. (2016). Parallel resampling in the particle filter. *Journal of Computational and Graphical Statistics, 25*(3), 789–805.

Naesseth, C., Lindsten, F., & Schon, T. (2015). Nested sequential Monte Carlo methods. In F. Bach & D. Blei (Eds.), *Proceedings of the 32nd International Conference on Machine Learning. Proceedings of Machine Learning Research* (Vol. 37, pp. 1292–1301). Lille: PMLR.

Olsson, J., & Douc, R. (2019). Numerically stable online estimation of variance in particle filters. *Bernoulli, 25*(2), 1504–1535.

Paige, B., Wood, F., Doucet, A., & Teh, Y. W. (2014). Asynchronous anytime sequential Monte Carlo. In Z. Ghahramani, M. Welling, C. Cortes, N. D. Lawrence, & K. Q. Weinberger (Eds.), *Advances in neural information processing systems 27* (pp. 3410–3418). Red Hook, NY: Curran Associates, Inc.

Papaspiliopoulos, O. (2011). Monte Carlo probabilistic inference for diffusion processes: A methodological framework. In *Bayesian time series models* (pp. 82–103). Cambridge: Cambridge University Press.

Papaspiliopoulos, O., & Ruggiero, M. (2014). Optimal filtering and the dual process. *Bernoulli, 20*(4), 1999–2019.

Reich, S., & Cotter, C. (2015). *Probabilistic forecasting and Bayesian data assimilation*. New York: Cambridge University Press.

Vergé, C., Dubarry, C., Del Moral, P., & Moulines, E. (2015). On parallel implementation of sequential Monte Carlo methods: The island particle model. *Statistics and Computing, 25*(2), 243–260.

Index

© Springer Nature Switzerland AG 2020

N. Chopin, O. Papaspiliopoulos, *An Introduction to Sequential Monte Carlo*, Springer Series in Statistics, https://doi.org/10.1007/978-3-030-47845-2

Printed in the United States
by Baker & Taylor Publisher Services

Printed in the United States
by Baker & Taylor Publisher Services